Iterative Methods for Approximate Solution
of Inverse Problems

Mathematics and Its Applications

Managing Editor:

M. HAZEWINKEL

Centre for Mathematics and Computer Science, Amsterdam, The Netherlands

Iterative Methods for Approximate Solution of Inverse Problems

by

A.B. Bakushinsky

Russian Academy of Science, Moscow,
Russia

and

M.Yu. Kokurin

Mary State University, Yoshkar-Ola,
Russia

A C.I.P. Catalogue record for this book is available from the Library of Congress.

ISBN 978-90-481-6798-2
ISBN 978-1-4020-3122-9

Published by Springer,
P.O. Box 17, 3300 AA Dordrecht, The Netherlands.

Sold and distributed in North, Central and South America
by Springer,
101 Philip Drive, Norwell, MA 02061, U.S.A.

In all other countries, sold and distributed
by Springer,
P.O. Box 322, 3300 AH Dordrecht, The Netherlands.

Printed on acid-free paper

To our teachers

Contents

Acknowledgments

We express sincere gratitude to our pupils and colleagues for numerous interesting and helpful discussions of the presented results.

Introduction

The concept of an inverse problem (IP) have gained widespread acceptance in modern applied mathematics, although it is unlikely that any rigorous formal definition of this concept exists. Most commonly, by IP is meant a problem of determining various quantitative characteristics of a medium such as density, sound speed, electric permittivity and conductivity, shape of a solid body by observations over physical (e.g., sound, electromagnetic, heat, gravitational) fields in the medium [3, 4, 26, 27, 30, 33, 42, 67, 72, 81, 101, 108, 116, 120, 124, 125]. The fields may be of natural appearance or specially induced, stationary or depending on time. An important and frequently addressed class of IPs arises when it is required to determine some physical characteristics of a bounded part of the medium while out of this part the characteristics are known. To be more specific, let a homogeneous medium contain a bounded inhomogeneous inclusion with unknown characteristics that are beyond the reach of immediate measurements; properties of the homogeneous background therewith are given. It is required to determine characteristics of the inclusion using results of measurements of physical fields in the composite medium formed by the background and the inclusion. As this takes place, the measurements are usually made in a region of observation located far from the inhomogeneity to be recovered. As a rule, a formal mathematical model of IP under consideration can be derived with relative ease. This is most often done with the use of a suitable model of the field propagation phenomenon for inhomogeneous media. In this context, the problem of determination of field values in a region of observation by given characteristics of the medium is called the direct problem. In many instances, direct problems can be reduced to classical partial differential equations of second order supplemented with necessary initial and boundary conditions. Elliptic equations usually arise in mathematical models of the gravitational field, stationary temperature fields and time–harmonic sound fields, equations of parabolic type describe phenomena of heat conduction, hyperbolic equations govern propagation of nonstationary scalar wave fields, the system of Maxwell's equations describes electromagnetic fields, etc. Since properties of the homogeneous medium outside the inclusion are fixed, any solution method for the direct problem naturally defines a mapping that takes each set of admissible physical or geometric characteristics of the inclusion to the field values on the region of observation. Therefore the process of solving IP formally is equivalent to exact or approximate determination of a preimage of given field values under this mapping. A significant feature of applied IPs is that fields are typically measured approximately and hence true field values are not in our disposal. We can hope only to get some estimates for the measured physical values together with corresponding bounds of the measurement errors. A strict formulation of an operator equation connected with IP involves as an obligatory part a detailed description of functional spaces of feasible images and admissible preimages of the problem's operator. These spaces are often called the observation space and the solution space respectively; they typically arise to be infinite–dimensional Hilbert or Banach spaces. A fundamental fact to regularization theory is that observation and solution spaces can't be taken quite arbitrarily. In practice, their choice is usually strictly determined by an available technique of registering field characteristics and by qualitative requirements on interpretations of the measurements. A solution to such operator equation generally can't be expressed in an explicit form so that the use of approximate methods of numerical analysis is the only way to get the solution or its approximation. Unfortunately, equations associated with IPs usually don't satisfy the regularity condition, which is typical for construction and analysis of classical solution methods for smooth operator equations. The regularity of an equation with a differentiable operator means that the derivative of this operator must have a continuous inverse or pseudo–inverse mapping. If this condition is violated, then we say that the equation is irregular.

Systematic studies of irregular scalar equations and finite–dimensional systems of nonlinear equations were originated by Schröder's work [129]. By now, theory of numerical methods for nonlinear finite–dimensional systems has accumulated a wide variety of results on solving irregular problems (see, e.g., [47, 64, 69] and references therein). Certain of approaches developed within the context of this theory were extended to abstract operator equations in infinite–dimensional spaces. Most of current papers on solving finite irregular systems involve special conditions on a nonlinearity of an equation; we point to the very popular condition of 2–regularity ([69]) and its modifications and to various conditions on the Jacobian at the solution. Let us emphasize that analogs of such conditions in the infinite–dimensional case require that ranges of operators' derivatives be closed subspaces of an observation space. Unfortunately, attempts to apply the mentioned approaches to equations of typical IPs have not met with success so far. The matter is likely that the conditions on derivatives' degeneration cited above are not characteristic for equations associated with IPs because operators of typical IPs have compact derivatives and hence ranges of the derivatives generally are not closed. This means that such equations are ill–posed problems. Therefore one would expect that the use of ideas and technique of regularization theory will have a favorable effect on the problem of constructing numerical methods for irregular equations and IPs.

It seems likely that an immediate application of regularization theory to constructing effective iterative methods for nonlinear irregular equations was successful so far only when applied to equations with operators monotone in the sense of duality. Specifically, in [17, 82] the authors developed the iterative regularization principle for constructing and studying iterative methods applicable to irregular monotone operator equations and variational inequalities. The technique proposed here allows to modify classical iterative schemes and transform them in a uniform style into stable methods for solving irregular monotone equations and inequalities. In application to ill–posed convex optimization problems, the iterative regularization principle has been developed in [145].

Further development of regularization theory inspires a permanent interest to studies of regularizing properties of classical iterative processes such as gradient iteration, conjugate gradient methods, Newton–type methods, etc. It is well known that in application to regular operator equations, these methods possess stability with respect to errors of input data. Moreover, some of the methods demonstrate very rapid convergence in the regular case. When applied to linear ill–posed equations, regularizing properties of gradient type methods and conjugate gradient processes were extensively studied in [42, 45, 110]. In the event that the equation is nonlinear with a differentiable operator, it has been possible to obtain similar results only by imposing very restrictive structural conditions on the operator's nonlinearity near a solution [36, 53, 54, 73, 74, 76, 118, 126–128]. The current stage of researches in this direction is presented in detail in [42]. It should be pointed out that the mentioned conditions can be effectively verified only in several special cases, while their fulfillment for wide classes of practical IPs is highly conjectual. The authors of [56, 57, 63] underline the conjectual character of related conditions in application to irregular equations associated with inverse problems of potential theory and inverse scattering problems.

An approach to the iterative solution of irregular operator equations, which we follow in this book, differs essentially from the lines of inquiry mentioned above. We abandon of any a priori structural conditions on an equation except for standard smoothness assumptions; instead, we enter into the structure of iterative methods an additional parameter (an element of the solution space) in order to enhance capabilities of controlling over iterations. Formally speaking, the controlling element can be taken arbitrarily from the solution space but convergence of the methods is guaranteed only if this element lies in an appropriate admissible subset of the solution space. The subset is given in an explicit form in terms of a derivative of the operator at the solution. If the controlling element is chosen from the admissible subset, then for the proposed iterative methods, along with the convergence, we obtain rate of convergence estimates. In the general case where this element not necessarily lies in its admissible subset, we establish stability of the methods with respect to small deviations of the controlling element from the subset. These results resemble well–known theorems on local convergence of classical iterative processes for solving smooth regular equations. Taking, for instance, the standard gradient method in application to a nonlinear equation with a continuously invertible operator, we observe that the convergence is guaranteed if an initial approximation lies in a small neighborhood of a solution. Here the starting point can be considered as an analog of the controlling parameter, and a small ball centered at the solution serves as an analog of the admissible subset of controlling parameters in our methods. The same is true for the Gauss–Newton and Newton–Kantorovich methods.

Let us give a short introduction to the contents of this book.

In Chapter 1 we collect for easy reference several basic results concerning linear and nonlinear operators in Hilbert and Banach spaces. We also give an elementary introduction into theory of ill–posed problems and discuss interrelations between this theory and the approach to irregular problems, which we develop in the book.

Chapter 2 contains necessary material on parametric families of approximating operators and related regularization algorithms for linear ill–posed equations in Hilbert and Banach spaces. Much attention is given to varied kinds of rate of convergence estimates for approximating operator families and to necessary and sufficient conditions for such estimates.

In Chapter 3 we address to the problem of a linearization of nonlinear irregular equations with smooth operators. The irregular character of equations implies that solutions of formally linearized problems if exist, can't approximate solutions of the original equations with acceptable accuracy. Using parametric approximations of solutions to linear ill–posed equations, we propose several parametric approximation schemes applicable to nonlinear irregular equations without any restrictions on their structure. Then we study the obtained parametric schemes and derive error estimates for approximations generated by them. When applied to nonlinear IPs, the technique of parametric approximations allows to construct in a uniform style well–defined approximations to solutions of these problems and to get corresponding error estimates.

Chapters 4 and 5 hold a central position in the book. In Chapter 4, the parametric approximations technique of Chapter 3 is used for constructing modifications of classical Gauss–Newton and Newton–Kantorovich methods in Hilbert and Banach spaces. Generalized Gauss–Newton type and Newton–Kantorovich type methods derived in this chapter generate stable approximations of solutions to nonlinear irregular equations with arbitrary differentiable operators. We establish rate of convergence estimates of the proposed iterative processes and analyze in detail conditions necessary and sufficient for these estimates.

In Chapter 5 we construct and study iterative processes possessing stability with respect to a choice of starting points in a neighborhood of a solution and to small variations of input data. In particular, for the gradient process we propose a series of modifications applicable to smooth irregular equations without any structural conditions on the operator. Here, the stability of the processes means that iterative points generated by them are attracted to a small neighborhood of the solution, as an iteration number increases. The processes of Chapter 4 don't possess this property, so to get an acceptable approximation to a solution in the case of noisy input data, we have to endow those processes with suitable stopping rules. Stable iterative processes can also be derived on the basis of parametric approximations; necessary technique is developed in Chapter 5, as well. Along with iterative methods, in Chapters 4 and 5 we analyze several continuous approximation procedures as their formal limiting analogs. Within the framework of these procedures, a solution of an irregular equation is approximated by a trajectory of an operator differential equation or a dynamical system naturally connected with the original equation. Two types of continuous methods are considered; methods of the first type are based on stopping of the trajectories at an appropriate point, the others possess the property of Lyapounov's stability and generate trajectories attracting to a small neighborhood of the solution.

Chapter 6 is devoted to numerical implementation of selected iterative processes from those derived in the previous chapters. We present results of computational experiments with inverse problems of gravimetry and inverse acoustic problems in various settings. The presented results allow the reader to make own impression concerning practical effectiveness of computational schemes developed in the book.

We included into the main text only references to papers closely connected with the discussed material. For additional information on the topics presented here the reader should consult the literature cited in Notes (Chapter 7). The notes contain some references to further developments and to alternative approaches in contemporary studies of IPs and irregular operator equations.

Each chapter of the book is divided into sections. Numbers of theorems, lemmas, conditions, remarks, examples, definitions and figures are consecutive in each chapter and indicate, in addition, the chapter number. References to sections include the chapter number. Numbering of formulae is consecutive and independent in each section. References to a formula from outside a section or a chapter where the formula is situated consist of two or three numbers and indicate the section number or the chapter and the section numbers. Constants are numerated consecutively and independently in each section.

Chapter 1

IRREGULAR OPERATOR EQUATIONS
AS ILL–POSED PROBLEMS

The aim of this chapter is to introduce necessary mathematical background for the succeeding exposition. For the reader's convenience, in Section 1.1 we collect frequently used facts concerning theory of linear operators and non-linear analysis. In Section 1.2 we sketch the basics of contemporary theory of ill–posed problems. We recall fundamental notions of this theory such as regularization algorithm and regularization parameter. We also discuss interrelations between these notions and main ideas of the book.

1.1   Preliminaries

In this section we give a brief overview of notions and results frequently used in the book. We only cite definitions and main statements that we shall need in next chapters. Most of the statements are standard and their proofs as well as deeper analysis are available, e.g., in [38–40, 58, 65, 123]. We take for granted that the reader is familiar with definitions and basic properties of Hilbert and Banach spaces, and operators in these spaces.

By $\|\cdot\|_X$ we denote a norm of a Hilbert or Banach space X. If X is a Hilbert space , then we have

$$\|x\|_X^2 = (x, x)_X, \quad x \in X,$$

where $(\cdot, \cdot)_X$ stands for the inner product of X. As a rule, throughout the book we deal with real Hilbert and Banach spaces but addressing to results of spectral theory of linear operators, without special indication we suppose these spaces to be complexified in the usual way.

Suppose X_1 and X_2 are Banach spaces; then by $L(X_1, X_2)$ we denote the space of all linear bounded operators $A\colon X_1 \to X_2$ with the norm

$$\|A\|_{L(X_1, X_2)} = \sup\{\|Ax\|_{X_2}\colon \|x\|_{X_1} \le 1\}.$$

It is well known that $L(X_1, X_2)$ is a Banach space. For short sake, we set $L(X) \equiv L(X, X)$. The subsets

$$N(A) = \{x \in X_1 : Ax = 0\}, \quad R(A) = \{y \in X_2 : y = Ax, x \in X_1\}$$

are called the null space and the range of an operator $A \in L(X_1, X_2)$ respectively. Each operator with a finite–dimensional range is called finite–dimensional. Below, $\sigma(A)$ and $\rho(A) = \mathbf{C} \backslash \sigma(A)$ are the spectrum and the resolvent subset of an operator $A \in L(X)$;

$$R(\lambda, A) = (\lambda E - A)^{-1}, \quad \lambda \in \rho(A)$$

is the resolvent operator for A; E stands for the identity mapping; $\mathbf{C}, \mathbf{R}, \mathbf{Z}, \mathbf{N}$ are the sets of complex, real, integer, and natural numbers. Throughout, $[x]$ and $\{x\} = x - [x]$ are the integer and fractional part of a real number x. By $\text{int} G$ we denote the interior of a subset $G \subset X$, $\overline{G}$ is the closure of G in the sense of the norm $\| \cdot \|_X$. Let $\Omega_r(x)$ denote the ball $\{y \in X : \|y - x\|_X \leq r\}$ centered at $x \in X$ and of radius r. Also, let $\Omega_r^0(x) = \{y \in X : \|y - x\|_X < r\}$. The image of a ball $\Omega_r(0)$ under an affine mapping $F(x) = Ax - b$ $(A \in L(X), b \in X)$ is called an ellipsoid.

An open connected subset of $\mathbf{R}^n$ is said to be a domain. A domain D is said to be star–shaped with respect to a point $r^0 \in D$ if the intersection of any ray originated at r^0 with D is an interval.

Let X, X_1, X_2 be Hilbert spaces, $A \in L(X_1, X_2)$, and $A^* \in L(X_2, X_1)$ the operator adjoint of A. An operator $A \in L(X)$ is called selfadjoint if $A^* = A$. The spectrum of a selfadjoint operator lies on the real axis of $\mathbf{C}$. If $A \in L(X)$ is a selfadjoint operator and $(Ax, x)_X \geq 0$ $\forall x \in X$, then we say that A is nonnegative and write $A \geq 0$. An important class of selfadjoint nonnegative operators in a Hilbert space is formed by orthoprojectors. An operator $P \in L(X)$ is called an orthoprojector if $P^* = P$ and $P^2 = P$. For each orthoprojector P, the range $R(P) = \mathcal{M}$ is a closed subspace of X, and we have $\|P\|_{L(X)} = 1$ whenever $P \neq O$, the null operator. Moreover,

$$(x - Px, y)_X = 0 \quad \forall y \in \mathcal{M}.$$

We say that P is an orthoprojector onto $\mathcal{M}$ and write $P = P_{\mathcal{M}}$.

Now let X be a Banach space. An operator $P \in L(X)$ is called a projector (in general, nonorthogonal) onto the subspace $\mathcal{M} = R(P)$ if $P^2 = P$. The subspace $\mathcal{M}$ is closed, and for each projector P, $\|P\|_{L(X)} \geq 1$. Let Q be a subset of X; then, by definition, $x + Q = \{x + y : y \in Q\}, x \in X$.

We shall use some definitions and facts connected with calculi of linear bounded operators in Hilbert and Banach spaces. Let X be a Hilbert space and $A \in L(X)$ a selfadjoint operator with $\sigma(A) \subset [M_0, M_1]$, then we have the spectral decomposition:

$$A = \int\limits_{M_0-0}^{M_1} \lambda dE_\lambda = \int\limits_{-\infty}^{\infty} \lambda dE_\lambda.$$

Here $\{E_\lambda\}_{\lambda \in (-\infty, \infty)}$ is the family of spectral projectors of A. The projector–valued function $\lambda \to E_\lambda$ is continuous in λ from the right so that in the sense of strong convergence, $E_{\lambda+0} = E_\lambda$. Also, recall that $E_\lambda E_\mu = E_\lambda$ when $\lambda \leq \mu$, and

$$E_\lambda = \begin{cases} O, & \lambda < \inf\{(Ax, x)_X : \|x\|_X = 1\}, \\ E, & \lambda \geq \sup\{(Ax, x)_X : \|x\|_X = 1\}. \end{cases}$$

Gap points of the function $\lambda \to E_\lambda$ correspond to eigenvalues of A. Namely, if $E_\lambda \neq E_{\lambda-0}$ then λ is an eigenvalue of A, and $E_\lambda - E_{\lambda-0}$ is the orthoprojector onto a proper subspace of λ.

Let a function $\varphi : (-\infty, \infty) \to \mathbf{C}$ be measurable, bounded, and finite almost everywhere (a.e.) with respect to the spectral family $\{E_\lambda\}$, that is, with respect to all Lebesgue–Stieltjes measures associated with the functions $\lambda \to \|E_\lambda x\|_X^2$, where x ranges over X. Then the function $\varphi(A)$ of a selfadjoint operator $A \in L(X)$ is defined as follows:

$$\varphi(A) = \int\limits_{M_0-0}^{M_1} \varphi(\lambda) dE_\lambda, \quad \varphi(A)x = \int\limits_{M_0-0}^{M_1} \varphi(\lambda) dE_\lambda x, \tag{1}$$

$$x \in D(\varphi(A)).$$

Moreover,

$$\|\varphi(A)x\|_X^2 = \int\limits_{M_0-0}^{M_1} |\varphi(\lambda)|^2 d\|E_\lambda x\|_X^2, \quad x \in D(\varphi(A)). \tag{2}$$

Each Borel measurable function satisfies all the above–listed conditions on $\varphi(\lambda)$. The spectral projectors $E_\mu, \mu \in (-\infty, \infty)$ are also functions of A: $E_\mu = e_\mu(A)$, where

$$e_\mu(\lambda) = \begin{cases} 1, & \lambda \leq \mu, \\ 0, & \lambda > \mu. \end{cases}$$

The operator $\varphi(A)$ is bounded iff

$$\operatorname*{ess\,sup}_{\{E_\lambda\}} |\varphi(\lambda)| < \infty.$$

Here ess sup is taken over the family of all Lebesgue–Stieltjes measures associated with the functions $\lambda \to \|E_\lambda x\|_X^2, x \in X$. In more detail,

$$\operatorname*{ess\,sup}_{\{E_\lambda\}} |\varphi(\lambda)| = \inf c$$

over all constants c such that $|\varphi(\lambda)| \leq c$ is fulfilled a.e. on $\mathbf{R}$ with respect to measures generated by all the functions $\lambda \to \|E_\lambda x\|_X^2, x \in X$. Therefore,

$$\operatorname*{ess\,sup}_{\{E_\lambda\}} |\varphi(\lambda)| \leq \sup_{\lambda \in \sigma(A)} |\varphi(\lambda)|.$$

According to the spectral radius formula,

$$\|\varphi(A)\|_{L(X)} = \operatorname*{ess\,sup}_{\{E_\lambda\}} |\varphi(\lambda)|. \tag{3}$$

In particular, for the resolvent $R(\lambda, A)$ of a selfadjoint operator $A \in L(X)$ in a Hilbert space X we have the estimate

$$\|R(\lambda, A)\|_{L(X)} \leq \frac{1}{\operatorname{dist}(\lambda, \sigma(A))} \quad \forall \lambda \in \rho(A).$$

In general, the operator $\varphi(A)$ is unbounded, but if the right part of (3) is finite and $\varphi(\lambda)$ is a real–valued function, then $\varphi(A)$ again is a bounded selfadjoint operator. If ess sup in (3) is infinite, then $\varphi(A)$ is unbounded with the domain of definition

$$D(\varphi(A)) = \left\{ x \in X : \int_{M_0-0}^{M_1} |\varphi(\lambda)|^2 d\|E_\lambda x\|_X^2 < \infty \right\}, \tag{4}$$

where $\overline{D(\varphi(A))} = X$. Representations (2) and (4) remain valid if $\varphi(\lambda)$ is not a.e. finite with respect to the family $\{E_\lambda\}$; in this case the lineal $D(\varphi(A))$ is not dense in X. Further, let functions $\varphi(\lambda), \psi(\lambda)$ be measurable, bounded, and defined a.e. with respect to the spectral family $\{E_\lambda\}$. Suppose that

$$\operatorname*{ess\,sup}_{\{E_\lambda\}} (|\varphi(\lambda)| + |\psi(\lambda)|) < \infty.$$

Then operators $\varphi(A), \psi(A) \in L(X)$ commute, and

$$\varphi(A)\psi(A) = \psi(A)\varphi(A) = (\varphi \cdot \psi)(A). \tag{5}$$

Let a real–valued function $\varphi(\lambda)$ be measurable, bounded, and finite a.e. with respect to the spectral family $\{E_\lambda\}$ of a selfadjoint operator $A \in L(X)$, and a real or complex–valued function $\psi(\lambda)$ be measurable and finite a.e. with respect to the spectral family $\{F_\mu\}$ of the operator $B = \psi(A)$. Then the composite function $\omega(\lambda) = \psi(\varphi(\lambda))$ is also measurable and finite a.e. with respect to $\{E_\lambda\}$, and $\omega(A) = \psi(\varphi(A))$. The above assertions can be naturally extended to unbounded selfadjoint operators [123].

We now address to the operator calculus in a Banach space X. Functions of an operator $A \in L(X)$ in a Banach space are usually defined by the Riesz–Dunford integral formula. Suppose a function $\varphi(\lambda)$ is analytic on a domain containing the spectrum $\sigma(A)$, and a contour Γ lies in this domain and surrounds $\sigma(A)$. Then $\varphi(A)$ is defined by the Riesz–Dunford formula

$$\varphi(A) = \frac{1}{2\pi i} \int_\Gamma \varphi(\lambda) R(\lambda, A) d\lambda, \tag{6}$$

where the integral is meant in the sense of Bochner. We obviously have $\varphi(A) \in L(X)$. Here and in the subsequent exposition, all contours in $\mathbf{C}$ are supposed to be positively oriented, that is, oriented counterclockwise. If $A \in L(X)$ and the functions $\varphi(\lambda)$ and $\psi(\lambda)$ are analytic on a domain containing $\sigma(A)$, then the operators $\varphi(A)$ and $\psi(A)$ satisfy equality (5). Besides, the spectral mapping theorem asserts that

$$\sigma(\varphi(A)) = \varphi(\sigma(A)) \equiv \{z \in \mathbf{C} : z = \varphi(\lambda), \lambda \in \sigma(A)\}.$$

For a selfadjoint operator $A \in L(X)$ in a Hilbert space X and an analytic function $\varphi(\lambda)$, the definitions (1) and (6) are equivalent.

Let us turn to theory of fractional powers of linear operators in a Banach space X. Since the function $\varphi(\lambda) = \lambda^p, p \notin \mathbf{N} \cup \{0\}$ is not analytic on a neighborhood of $\lambda = 0$, we can't define the function $\varphi(A) = A^p, p \notin \mathbf{N} \cup \{0\}$ of an operator $A \in L(X)$ with $0 \in \sigma(A)$ immediately by formula (6). Recall that for each $p \in \mathbf{N}$, the power A^p is defined as $A^p = A \cdots A$ (p times). The following condition is needed if we are to define A^p for an arbitrary exponent $p > 0$.

Condition 1.1. *There exist $\varphi_0 \in (0, \pi)$ and $r_0 > 0$ such that*

$$\sigma(A) \subset K(\varphi_0),$$
$$K(\varphi_0) \equiv \{\lambda \in \mathbf{C} \backslash \{0\} : |\arg\lambda| < \varphi_0\} \cup \{0\} \tag{7}$$

and the estimate is valid:

$$\|R(\lambda, A)\|_{L(X)} \leq \frac{r_0}{|\lambda|} \quad \forall \lambda \in \mathbf{C} \backslash K(\varphi_0). \tag{8}$$

Condition 1.1 and its variants are often called sectorial conditions. Let us now highlight several classes of operators $A \in L(X)$ satisfying Condition 1.1.

Example 1.1. 1) Selfadjoint nonnegative operators $A^* = A \geq 0$ in a Hilbert space X. Condition 1.1 is satisfied with an arbitrary $\varphi_0 \in (0, \pi)$.

2) Accretive operators in a complex Hilbert space, i.e., operators $A \in L(X)$ satisfying

$$\mathrm{Re}(Ax, x)_X \geq 0 \quad \forall x \in X.$$

Here φ_0 can be taken arbitrarily from $(\pi/2, \pi)$.

3) Spectral operators of scalar type such that $\sigma(A) \subset K(\psi_0)$, $\psi_0 \in (0, \pi)$ ([40]). The value φ_0 can be taken arbitrarily from (ψ_0, π).

4) Operators satisfying the condition

$$\|R(-t, A)\|_{L(X)} \leq \frac{c_1}{t} \quad \forall t > 0.$$

In 3) and 4), X is an arbitrary Banach space.

Suppose $p \in (0, 1)$, and operator A satisfies Condition 1.1; then A^p is defined as follows ([24, 97]):

$$A^p = \frac{\sin \pi p}{\pi} \int\limits_0^\infty t^{p-1}(tE + A)^{-1} A \, dt. \tag{9}$$

It follows from (8) that the integral in (9) exists in the sense of Bochner, and (9) defines an operator $A^p \in L(X)$. For arbitrary $p > 1$, $p \notin \mathbf{N}$, with $m = [p]$ we set

$$A^p = A^m \cdot A^{p-m} \equiv A^{p-m} \cdot A^m.$$

Thus the linear bounded operator A^p is well defined for each $p > 0$.

Given an operator $A \in L(X)$ satisfying Condition 1.1, consider the regularized operator $A_\varepsilon = A + \varepsilon E$. It is readily seen that for each $\varepsilon > 0$, the power A_ε^p can be defined by the integral formula (6) for all $p > 0$. It is sufficient to put $\Gamma = \Gamma_\varepsilon$, where the contour Γ_ε surrounds $\sigma(A_\varepsilon) = \{\lambda + \varepsilon : \lambda \in \sigma(A)\}$ but doesn't contain the point $\lambda = 0$ inside. Let us recall the following useful results (see [28, 97]).

Lemma 1.1. *Suppose X is a Banach space and an operator $A \in L(X)$ satisfies Condition 1.1. Then for each $p \in (0, 1)$,*

$$\|A_\varepsilon^p - A^p\|_{L(X)} \leq c_2 \varepsilon^p \quad \forall \varepsilon > 0, \tag{10}$$

where $c_2 = c_2(p, r_0)$.

Lemma 1.2. *Let X be a Banach space. Suppose $\lambda \in \rho(A)$ and $A, B \in L(X)$.*

1) Assume that $\|BR(\lambda, A)\|_{L(X)} < 1$. *Then* $\lambda \in \rho(A + B)$ *and*

$$R(\lambda, A + B) = R(\lambda, A) \sum_{k=0}^{\infty} (BR(\lambda, A))^k. \tag{11}$$

2) Let $\|R(\lambda, A)B\|_{L(X)} < 1$. *Then* $\lambda \in \rho(A + B)$ *and*

$$R(\lambda, A + B) = \sum_{k=0}^{\infty} (R(\lambda, A)B)^k R(\lambda, A). \tag{12}$$

The series in (11) and (12) converge in the norm of $L(X)$.

Let us now turn to necessary facts of nonlinear analysis. Let X, Y be Banach spaces, $F : D(F) \subset X \to Y$ a nonlinear operator, and $D(F)$ an open subset of X. The operator F is said to be Fréchet differentiable at the point $x \in D(F)$ if there exists a linear operator $F'(x) \in L(X, Y)$ such that

$$F(x + h) - F(x) = F'(x)h + \omega(x, h), \quad x + h \in D(F),$$

where

$$\lim_{\|h\|_X \to 0} \frac{\|\omega(x, h)\|_Y}{\|h\|_X} = 0.$$

The operator $F'(x)$ is called the Fréchet derivative of F at x.

Suppose $x \in D(F)$ and $h \in X$. If the Gâteaux differential

$$\mathcal{D}F(x, h) = \lim_{t \to 0} \frac{1}{t}(F(x + th) - F(x))$$

is linear in h, that is, there exists $F_g'(x) \in L(X, Y)$ such that

$$\mathcal{D}F(x, h) = F_g'(x)h \quad \forall h \in X,$$

then F is said to be Gâteaux differentiable at x and $F_g'(x)$ is called the Gâteaux derivative of the operator F at the point x. Each Fréchet differentiable operator $F : X \to Y$ is also Gâteaux differentiable, and $F_g'(x) = F'(x)$. Further, if the Gâteaux derivative $F_g'(x)$ exists for all x from a neighborhood of x_0, and the mapping $x \to F_g'(x)$ is continuous from X into $L(X, Y)$ at $x = x_0$, then F is Fréchet differentiable at x_0.

Fréchet and Gâteaux derivatives $F^{(m)}$ and $F_g^{(m)}$ of order $m \geq 2$ can be defined inductively as corresponding derivatives of the mappings $F^{(m-1)}$ and $F_g^{(m-1)}$ respectively.

If the mapping $z \to F'(z)h$ $(h \in X)$ is continuous from the segment $\{x + \theta h : \theta \in [0, 1]\} \subset X$ into Y, then the integral mean value theorem gives

$$F(x + h) - F(x) = \int_0^1 F'(x + th)h\, dt. \tag{13}$$

Hereafter, we shall frequently use the following statement.

Lemma 1.3. *Let operator* $F : D(F) \subset X \to Y$ *be Fréchet differentiable on a convex subset* $Q \subset D(F)$, *and the derivative* $F'(x)$ *be Lipschitz continuous on* Q, *i.e.,*

$$\|F'(x) - F'(y)\|_{L(X,Y)} \le L\|x - y\|_X \quad \forall x, y \in Q.$$

Then

$$\|F(x) - F(y) - F'(y)(x - y)\|_Y \le \frac{1}{2}L\|x - y\|_X^2 \quad \forall x, y \in Q. \qquad (14)$$

In the case where $Y = \mathbf{R}$, an operator $F = J : D(J) \subset X \to Y$ is called a functional on X. The Fréchet or Gâteaux derivative $J'(x)$ of the functional J at $x \in D(J)$ is an element of the dual space $X^* = L(X, \mathbf{R})$. If X is a Hilbert space, it is convenient to identify $J'(x) \in X^*$ with the element $j(x) \in X$ such that

$$J'(x)h = (j(x), h)_X \quad \forall h \in X.$$

By the Riesz representation theorem, the element $j(x)$ is uniquely determined. In particular, $J'(x) \in X$ is called the Fréchet derivative of the functional J at the point x if

$$J(x + h) = J(x) + (J'(x), h)_X + \omega(x, h); \quad x + h \in D(J),$$

where

$$\lim_{\|h\|_X \to 0} \frac{|\omega(x, h)|}{\|h\|_X} = 0.$$

We shall also use properties of the metric projection of a point onto a subset of a Hilbert space. Let D be a closed convex subset of a Hilbert space X. Then there exists a unique point $z \in D$ such that

$$\|x - z\|_X = \min\{\|x - y\|_X : y \in D\}.$$

We call z the projection of x onto D and write $z = P_D(x)$. For a subset $Q \subset X$ and a point $x \in X$, we denote

$$\mathrm{dist}(x, Q) = \inf\{\|x - y\|_X : y \in Q\}.$$

If D is a closed convex subset of X, then

$$\mathrm{dist}(x, D) = \|x - P_D(x)\|_X.$$

In addition, we have the following inequalities:

$$(x - P_D(x), y - P_D(x))_X \leq 0 \quad \forall x \in X \quad \forall y \in D, \tag{15}$$

$$\|P_D(x_1) - P_D(x_2)\|_X \leq \|x_1 - x_2\|_X \quad \forall x_1, x_2 \in X. \tag{16}$$

The notions of convex and strongly convex functionals play an important role in theory of nonlinear optimization problems. Let X be a Banach space. A functional $J : D(J) \subset X \to \mathbf{R}$ is called convex on a convex subset $Q \subset D(J)$ if

$$J(\alpha x + (1 - \alpha)y) \leq \alpha J(x) + (1 - \alpha)J(y) \quad \forall x, y \in Q \quad \forall \alpha \in [0, 1].$$

If there exists a constant $\kappa > 0$ such that

$$J(\alpha x + (1 - \alpha)y) \leq \alpha J(x) + (1 - \alpha)J(y) - \alpha(1 - \alpha)\kappa\|x - y\|_X^2$$

$$\forall x, y \in Q \quad \forall \alpha \in [0, 1],$$

then the functional J is said to be strongly convex on Q with the constant of strong convexity κ. Suppose a functional $J(x)$ is differentiable on a convex subset Q of a Hilbert space X, and the derivative $J'(x) \in X$ is continuous in $x \in Q$. For the functional $J(x)$ to be strongly convex on Q with a constant $\kappa > 0$ it is necessary and sufficient to have

$$(J'(x) - J'(y), x - y)_X \geq 2\kappa\|x - y\|_X^2 \quad \forall x, y \in Q. \tag{17}$$

Furthermore, (17) is equivalent to the inequality

$$J(x) \geq J(y) + (J'(y), x - y)_X + \kappa\|x - y\|_X^2 \quad \forall x, y \in Q. \tag{18}$$

Conditions (17) and (18) with $\kappa = 0$ are necessary and sufficient for the functional $J(x)$ to be convex on Q.

Let us consider an optimization problem

$$\min_{x \in Q} J(x), \tag{19}$$

where Q is a closed convex subset of a Hilbert space X. Let Q_* be the solution set of (19):

$$Q_* = \left\{ x^* \in Q : J(x^*) = \inf_{x \in Q} J(x) \right\}.$$

Suppose the functional $J(x)$ is continuous and strongly convex on Q; then $Q_* = \{x^*\}$ is a singleton and

$$\kappa\|x - x^*\|_X^2 \leq J(x) - J(x^*) \quad \forall x \in Q.$$

If $J(x)$ is a continuous convex functional, then Q_* is a convex closed (possibly empty) subset of Q.

1.2   Irregular Equations and Contemporary Theory of Ill–Posed Problems

From the formal point of view, our approach to constructing iterative methods for irregular equations need not be connected with regularization theory. Nevertheless this approach undoubtedly owes its origin to the progress in contemporary theory of ill–posed problems. In the present section we recall some basic concepts of this theory and discuss their interrelations with the consequent material. We are not aimed at giving a full and rigorous description of basic results of the theory and refer for a detailed presentation to textbooks and monographs on this subject ([17, 42, 48, 49, 52, 59, 68, 78, 106, 107, 133, 137, 139, 145, 146]).

Consider an equation

$$F(x) = 0, \quad x \in X_1, \tag{1}$$

where $F : X_1 \to X_2$ is a nonlinear operator acting in a pair of Hilbert or Banach spaces (X_1, X_2). It is sometimes useful to study along with (1) the minimization problem

$$\min_{x \in X_1} \frac{1}{2} \|F(x)\|_{X_2}^2. \tag{2}$$

The problems (1) and (2) are not equivalent; each solution of (1) is evidently a global minimizer of the problem (2) while a solution to (2) do not necessarily satisfy (1). Throughout the book we usually suppose a solution to (2) also satisfies equation (1) but in several cases it is more convenient to deal with the optimization problem (2) rather than with (1).

In short, the ill–posedness of the problems (1), (2) means that analyzing problems close in some sense to (1) and (2), we can't guarantee that solutions to these perturbed problems are close to corresponding solutions of the original ones. Moreover, solutions to the perturbations of ill–posed problems do not necessarily exist at least in the original space X_1. The aforesaid assumes clear definitions of feasible perturbations of the original problems as well as detailed descriptions of the spaces X_1 and X_2. As an example, let us consider (1) and (2) with the affine operator $F(x) = Ax - b$, where $A \in L(X_1, X_2), b \in X_2$. If operators A and A^*A are not continuously invertible, then the problems (1) and (2) are ill–posed relative to the natural class of perturbations $\widetilde{F}(x) = \widetilde{A}x - \widetilde{b}$. On the other hand, in computational mathematics it is the practice to call irregular each linear equation $Ax = b$ for which A^{-1} or $(A^*A)^{-1}$ is not a continuous operator. Therefore in the case under consideration, the irregularity of equation (1) or variational problem (2) and their ill–posedness are in effect the same things. In the nonlinear case, at least for smooth operators $F(x)$, the situation is similar. Let us clarify this in a few words. Nonlinear problems (1) and (2) are typically called irregular if the derivative $F'(x)$ or the product

$F'^*(x)F'(x)$ is not continuously invertible for x from a neighborhood of the solution x^*. Throughout this book we shall understand the (ir)regularity of (1) and (2) according to the following refined definition.

Definition 1.1. *An equation (1) or a variational problem (2) with a Fréchet differentiable operator $F(x)$ is called regular if there exists an open neighborhood Ω of the solution x^* such that the derivative $F'(x) : X_1 \rightarrow X_2, x \in \Omega$ has a bounded inverse operator $F'(x)^{-1}$ defined on the all of X_2, or the product $F'^*(x)F'(x) : X_1 \rightarrow X_1, x \in \Omega$ has a bounded inverse $(F'^*(x)F'(x))^{-1}$ defined on X_1. Otherwise the problems (1) and (2) are called irregular.*

Suppose (1) and (2) are irregular in this sense, and a point $x_0 \in X_1$ is chosen sufficiently close to the solution x^*. Let us consider for the original problem a linearization, which in case of (1) has the form

$$F(x_0) + F'(x_0)(x - x_0) = 0, \quad x \in X_1, \qquad (3)$$

and for (2) looks as

$$\min_{x \in X_1} \frac{1}{2} \|F(x_0) + F'(x_0)(x - x_0)\|_{X_2}^2. \qquad (4)$$

Observe that the affine operator $F(x_0) + F'(x_0)(x - x_0)$ in (3) and (4) is close to $F(x)$ uniformly in x from a small neighborhood of x^*. Since in the irregular case operators $F'(x_0)$ and $F'^*(x_0)F'(x_0)$ need not be continuously invertible, the linearized problems (3) or (4) may have no solutions. Further, even if these solutions exist they can differ significantly from the solution x^* of the original problem (1) or (2). Therefore the problems (1) and (2) are ill–posed on the pare of spaces (X_1, X_2) with respect to pointwise perturbations of $F(x)$. So we may hope that the ideology and technique of regularization theory will be useful for constructing solution methods for these problems. The concept of a regularization algorithm holds a central position in regularization theory. Below we shall provide a schematic description of this notion; for detailed discussions we refer to [17, 42, 68, 106, 133, 137, 139, 146].

Let us immerse an individual operator $F(x)$ into a space $\mathfrak{F}$ supplied with a metric ρ. For instance, in the case of an affine operator $F(x) = Ax - b$ we can define $\mathfrak{F}$ as the space of all operators $\widetilde{F}(x) = A\hat{x} - \widetilde{b}$ with $\widetilde{b} \in X_2$ having the same $A \in L(X_1, X_2)$. It is natural to define the metric on $\mathfrak{F}$ as follows: $\rho(\widetilde{F}, F) = \|\widetilde{b} - b\|_{X_2}$. Further, given a nonlinear problem (1) or (2) with an operator $F \in \mathfrak{F}$, by G we denote the mapping that takes each $\widetilde{F} \in \mathfrak{F}$ to a solution of the perturbed problem

$$\widetilde{F}(x) = 0, \quad x \in X_1$$

or the problem

$$\min_{x \in X_1} \frac{1}{2} \|\widetilde{F}(x)\|_{X_2}^2.$$

Since these problems are not necessarily uniquely solvable, G is in general a set–valued mapping defined on a subset of $\mathfrak{F}$. To simplify considerations, it is convenient to fix some single–valued selector of G and to denote by G just the selector chosen. For this reason, in the sequel we can restrict ourselves with single–valued operators G and formally identify the problems (1) and (2) with the problem of finding $G(F)$.

Definition 1.2. *The problem (1) or (2) or, equivalently, the problem of evaluating an element $x^* = G(F)$ is called ill–posed on the class $\mathfrak{F}$ with respect to the pare of spaces (X_1, X_2), if at least one of the following conditions is violated:*

1) the domain of definition $D(G)$ of the mapping G is the all of $\mathfrak{F}$;

2) the mapping $G : \mathfrak{F} \to X_1$ is continuous in the sense of the metric ρ and the norm $\| \cdot \|_{X_1}$.

The original problem is called well–posed on $\mathfrak{F}$ with respect to (X_1, X_2) if both conditions 1) and 2) are satisfied.

The fact that the definition involves a reference to the pare (X_1, X_2) is of fundamental importance for regularization theory because a suitable change of this pare can transform an ill–posed problem into a well–posed one, and vice versa. Let us recall in this connection that solution and observation spaces of applied inverse problems (in our notation, the spaces X_1 and X_2) are primarily determined by available measuring technique and by requirements on inter-pretation of experimental data. Hence in our context it is natural to consider (X_1, X_2) as a fixed pare of Hilbert or Banach spaces.

Definition 1.3. *A family of mappings*

$$\mathfrak{R}_{\alpha(\cdot, \delta)}(\cdot) : \mathfrak{F} \to X_1, \quad \delta \geq 0$$

is called a regularization algorithm for the original problem (1) or (2) if for each operator $F \in D(G)$ the following condition is satisfied:

$$\lim_{\delta \to 0} \sup_{\substack{\widetilde{F} \in \mathfrak{F} \\ \rho(\widetilde{F}, F) \leq \delta}} \|\mathfrak{R}_{\alpha(\widetilde{F}, \delta)}(\widetilde{F}) - G(F)\|_{X_1} = 0. \tag{5}$$

Neither the equality $D(G) = \mathfrak{F}$ nor continuity of the mapping $G : D(G) \to X_1$ is assumed here so that the problem of evaluating $G(F)$ is ill–posed. In (5), the value $\alpha = \alpha(\widetilde{F}, \delta) > 0$ is called the regularization parameter. Let $\widetilde{F}(x)$ be an available approximation to the true input operator $F(x)$. Also, suppose $\rho(\widetilde{F}, F) \leq \delta$, where $\delta > 0$ is an error level of input data. It follows from (5) that the element $\mathfrak{R}_{\alpha(\widetilde{F},\delta)}(\widetilde{F})$, obtained by the input data $\widetilde{F}(x)$ and the error level δ, can be taken as an adequate approximation to the solution $x^* = G(F)$.

In practice, regularization algorithms are frequently created according to the following general scheme. First, a family of continuous mappings

$$\mathfrak{R}_\alpha : \mathfrak{F} \to X_1, \quad \alpha \in (0, \alpha_0]$$

is constructed such that

$$\lim_{\alpha \to 0} \|\mathfrak{R}_\alpha(F) - G(F)\|_{X_1} = 0 \quad \forall F \in D(G). \tag{6}$$

Equality (6) means that the family of continuous operators $\{\mathfrak{R}_\alpha\}_{\alpha \in (0,\alpha_0]}$ defined of the all of $\mathfrak{F}$ approximates the discontinuous mapping G on $D(G)$ as $\alpha \to 0$. Next, a function $\alpha = \alpha(\widetilde{F}, \delta)$ is sought in order to ensure the equality (5). The fundamental Vinokourov's theorem ([17]) establishes existence of this sort of coordination between α, $\widetilde{F}$, and δ. Let us remark that in many instances the regularization parameter can be chosen as a function of the error level δ but not of the approximate operator $\widetilde{F}(x)$ so that $\alpha(\widetilde{F}, \delta) = \alpha(\delta)$ ([17]).

Practical implementation of the outlined abstract scheme meets a number of theoretical and computational difficulties. By time, satisfactory means of tackling these difficulties are known only in the case where X_1, X_2 are Hilbert spaces, operator $F(x)$ is affine, and the class $\mathfrak{F}$ consists of affine operators only.

In past several decades, it has been possible to proceed considerably towards constructing implementable regularization algorithms $\mathfrak{R}_{\alpha(\delta)}(\widetilde{F})$ in the case of monotone operators $F(x)$ [17, 82]. On the contrary, in the case where $F(x)$ is a smooth operator of general form, for approximation of $G(F)$ only abstract schemes converging in the weak sense were derived so far. By now, the classical Tikhonov's scheme [17, 42, 106, 137, 139] remains the sole universal approach to constructing such approximation methods when applied to equations with general smooth operators. In past years, theoretical aspects of Tikhonov's method have been studied in detail both in the case of linear equations and in general nonlinear situation. Nevertheless it is unlikely that any effectively implementable regularization algorithms for general nonlinear equations can be derived immediately from Tikhonov's scheme. Indeed, according to this scheme, the approximating element $\mathfrak{R}_{\alpha(\delta)}(\widetilde{F})$ for a solution of (1) or (2) is defined as a global minimizer or ε–minimizer to the Tikhonov functional of the form

$$\widetilde{\Psi}_{\alpha(\delta)}(x) = \frac{1}{2}\|\widetilde{F}(x)\|_{X_2}^2 + \frac{1}{2}\alpha(\delta)\|x - \xi\|_{X_1}^2, \quad x \in X_1 \quad (\xi \in X_1).$$

The functional $\widetilde{\Psi}_{\alpha(\delta)}(x)$ is to be minimized on the all of the solution space X_1 or, in numerically implementable versions of Tikhonov's method, on an appropriate finite–dimensional subspace of X_1. In essence, the last problem is equivalent to the problem of finding a global minimizer of an arbitrary smooth function of several variables. Therefore in its general form, the problem of minimization $\widetilde{\Psi}_{\alpha(\delta)}(x)$ is unlikely to be effectively solvable with the present–day computational technique.

In the book we propose the following program of tackling these difficulties.

1) For construction of approximating families $\{\mathfrak{R}_\alpha\}_{\alpha\in(0,\alpha_0]}$ in the case of nonlinear irregular problems (1) and (2) we intend to use iterative procedures. More precisely, let the regularization parameter α take discrete values $\alpha = 1, 1/2, \ldots 1/n, \ldots$. We set

$$\mathfrak{R}_{1/(n+1)}(F) = I_n\mathfrak{R}_{1/n}(F), \quad F \in \mathfrak{F}, \tag{7}$$

where I_n stands for an operator of the iteration. Apart from the iteration number, I_n may depend on its own inner parameters. If possible, I_n should be comparable in complexity with operators of classical iterative methods for regular equations.

2) So far, it has been possible to guarantee the equality (5) over sufficiently wide spaces $\mathfrak{F}$ only for special classes of operators $F(x)$, namely, for affine and monotone operators in a Hilbert space. Even when dealing with linear equations (1) in a Banach space we need to narrow down the space of feasible exact and approximate operators in one way or another [17]. We shall be able to treat maximally wide class of smooth nonlinear operators $F(x)$ if we relax condition (6), which requires that the elements $\mathfrak{R}_{1/n}(F)$ must approximate $G(F)$ for each operator $F \in D(G)$ as $n \to \infty$.

3) For the iterative approximating families $\{\mathfrak{R}_\alpha\}$, $\alpha = 1, 1/2, \ldots, 1/n, \ldots$ that we propose in the book, the equality (6) is valid not for all $F \in D(G)$ but merely for $F \in D(G)$ satisfying appropriate additional conditions in terms of inner parameters of I_n and the element $G(F)$. Let $x^* = G(F)$ be a solution to (1) or (2). The mentioned additional conditions lie in the fact that an element $\xi \in X_1$, which enters into the composition of $I_n = I_n(\xi)$ as a controlling parameter, must belong to an explicitly given admissible subset $\mathbb{M}(x^*)$. We shall see below that the subset $\mathbb{M}(x^*)$ of admissible elements ξ can be taken as an ellipsoid centered at the solution x^* with upper restrictions on the diameter of $\mathbb{M}(x^*)$, or as a finite–dimensional affine subspace containing x^*. If necessary, these restrictions are amplified by conditions on the derivative $F'(x^*)$ at the solution like Condition 1.1. Let us emphasize that the admissible subset $\mathbb{M}(x^*)$ is not a singleton whenever $F'(x^*) \neq O$. The last condition is satisfied for typical infinite–dimensional operator equations.

4) One may consider the condition $\xi \in \mathbb{M}(x^*)$ as an a priori assumption concerning the problem, however this assumption essentially differs from the

above–mentioned conditions on a nonlinearity of $F(x)$ near a solution ([42]). Our condition amounts to assumptions on available information about the unknown solution x^* and doesn't pose any restrictions on the structure of the original operator $F(x)$ near x^*. Turning to the classical special case of regular operators $F(x)$, it is not difficult to find out that our condition is analogous to the assumption that an initial approximation in traditional iterative methods must be chosen from a small ball centered at a solution ([34, 77, 94, 145]). Assumptions of this kind are typical for convergence theorems of classical iterative processes in the regular situation. Our convergence theorems differ from their classical analogs by the fact that in the irregular case, in addition to the standard condition that a starting point must lie in a small ball centered at x^*, we assume that $\xi \in \mathbb{M}(x^*)$, where $\mathbb{M}(x^*)$ typically has an empty interior unlike a ball. This drawback is partially compensated by stability of proposed iterative methods with respect to deviations of ξ from the admissible subset $\mathbb{M}(x^*)$ in the following sense. Suppose $I_n = I_n(\widetilde{\xi})$, where

$$\operatorname{dist}(\widetilde{\xi}, \mathbb{M}(x^*)) \leq \Delta;$$

then for the iterative process (7) there exists a stopping rule $N = N(\delta, \Delta)$ such that

$$\lim_{\delta, \Delta \to 0} \sup_{\substack{\widetilde{F} \in \mathfrak{F} \\ \rho(\widetilde{F}, F) \leq \delta}} \|\mathfrak{R}_{1/N(\delta, \Delta)}(\widetilde{F}) - G(F)\|_{X_1} = 0. \qquad (8)$$

Moreover, it is possible to derive qualified estimates for the approximation error

$$\|\mathfrak{R}_{1/N(\delta, \Delta)}(\widetilde{F}) - G(F)\|_{X_1}$$

in terms of the error levels δ and Δ.

Chapter 2

REGULARIZATION METHODS
FOR LINEAR EQUATIONS

Linear equations of the first kind, that is, equations $Ax = f$ involving compact operators $A \in L(X_1, X_2)$ in Hilbert or Banach spaces X_1, X_2 are typical linear ill–posed problems. In this chapter, our aim is to present a unified approach to construction of approximating families $\{\mathfrak{R}_\alpha\}$ and regularization algorithms for such problems in the irregular case when A is not necessarily a compact operator but continuous invertibility of A is not guaranteed. The irregular equation $Ax = f$ can be considered as an abstract model for a linear ill-posed inverse problem. In Section 2.1 we present a technique of constructing affine approximating families and regularization algorithms for linear ill–posed operator equations in a Hilbert space with approximate input data (A, f). Special emphasis is placed on sufficient and necessary conditions for convergence of the methods at a given rate. In Section 2.2 we generalize this technique to the case of linear equations in a Banach space. In particular, we prove that the power sourcewise representation of an initial discrepancy is sufficient for convergence of proposed methods at the power rate. In Section 2.3 it is shown that the power sourcewise representation is actually close to a necessary condition for the convergence at the power rate. The algorithms and investigation technique developed in the present chapter will be used in Chapter 3 for justification of the formalism of universal linear approximations for smooth nonlinear operator equations in Hilbert and Banach spaces. In Chapter 4, on the basis of this technique we propose and study a class of iterative methods for solving nonlinear irregular equations of general form.

2.1   General Scheme of Constructing
Affine Regularization Algorithms
for Linear Equations in Hilbert Space

In this section we consider linear operator equations

$$Ax = f, \quad x \in X_1 \quad (f \in X_2), \tag{1}$$

where X_1 and X_2 are Hilbert spaces and $A \in L(X_1, X_2)$. To start with, assume that $X_1 = X_2 = X$ and $A = A^* \geq 0$. Suppose $f \in R(A)$; then the set X_* of solutions to equation (1) is nonempty. It is readily seen that X_* is a closed affine subspace of X. Let us define the family $\mathfrak{F}$ of all feasible exact and approximate data (A, f) in equation (1) as $\mathfrak{F} = L_+(X) \times X$, where $L_+(X)$ is the cone of all selfadjoint nonnegative operators from $L(X)$. The norms of $L(X)$ and X induce a metric on $\mathfrak{F}$ and turn the family $\mathfrak{F}$ into a metric space. Throughout, we shall estimate the distance between exact input data (A, f) and their approximations (A_h, f_δ) separately in A and f. More exactly, we shall assume that

$$\|A_h - A\|_{L(X)} \leq h, \quad \|f_\delta - f\|_X \leq \delta,$$

where the pare (h, δ) characterizes error levels in the input data (A, f). It is a well–known fact that the family of regularizing mappings for equation (1) can be constructed in the form $\mathfrak{R}_{\alpha(h,\delta)} : \mathfrak{F} \to X$, where the regularization parameter $\alpha(h, \delta) = \alpha(A_h, f_\delta; h, \delta)$ depends on the error levels h, δ but not on the approximate elements (A_h, f_δ). Moreover, the approximating operator $\mathfrak{R}_\alpha$ can be chosen in such a way that $\mathfrak{R}_\alpha(A, \cdot) : X \to X$ is an affine continuous mapping for each $A \in L_+(X)$. The present chapter is devoted to studying precisely this class of approximating operators and related regularization algorithms $\mathfrak{R}_{\alpha(h,\delta)} : \mathfrak{F} \to X$ for equation (1). The interest to regularization procedures of this type arises out of the simplicity of their implementation when applied to equations (1) with exact operators A. In this event, evaluation of an approximate solution is reduced to finding an image of the element f_δ under the continuous affine map $\mathfrak{R}_{\alpha(0,\delta)}(A, \cdot)$. Clearly, the last operation is stable in computational sense.

We begin our detailed discussion by construction of approximating operators $\mathfrak{R}_\alpha$ for equation (1). Let us fix an initial estimate $\xi \in X$ of the solution to (1) and choose a real or complex–valued function $\Theta(\lambda, \alpha)$ such that for each $\alpha \in (0, \alpha_0]$, the function $\Theta(\cdot, \alpha)$ is Borel measurable on a segment $[0, M]$, where $[0, M] \supset \sigma(A)$. Also, suppose $\Theta(\lambda, \alpha)$ satisfies the following condition.

Condition 2.1. *There exist constants $p_0 > 0$ and $c_1 = c_1(\alpha_0, p_0, M) > 0$ such that for all $p \in [0, p_0]$,*

$$\sup_{\lambda \in [0,M]} |1 - \Theta(\lambda, \alpha)\lambda| \lambda^p \leq c_1 \alpha^p \quad \forall \alpha \in (0, \alpha_0]. \tag{2}$$

The family of approximating operators

$$\mathfrak{R}_\alpha : \mathfrak{F} \to X, \quad \alpha \in (0, \alpha_0];$$
$$\mathfrak{R}_\alpha(A, f) = (E - \Theta(A, \alpha)A)\xi + \Theta(A, \alpha)f \tag{3}$$

is a key tool in constructing regularization algorithms for equation (1). Along with α, the initial estimate $\xi \in X$ serves as a parameter of the family $\{\mathfrak{R}_\alpha\}$. Following (3), let us construct the parametric family of elements

$$x_\alpha = (E - \Theta(A, \alpha)A)\xi + \Theta(A, \alpha)f, \quad \alpha \in (0, \alpha_0] \tag{4}$$

for approximation of a solution to (1) as $\alpha \to 0$. In other words, (4) defines an abstract solution method for equation (1) in the case of exact input data (A, f). Here $\Theta(\lambda, \alpha)$ is called the generating function for the parametric mapping (3) and parametric approximating scheme (4); the operator function $\Theta(A, \alpha)$ is defined according to (1.1.1). Clearly, the mapping $\mathfrak{R}_\alpha(A, f)$ is affine in its second argument. Application of the Riesz–Dunford formula (1.1.6) implies the continuity of $\mathfrak{R}_\alpha(A, f)$ in its first argument on $L_+(X)$. The least upper bound p^* for values of p_0 in Condition 2.1 is called the qualification of the scheme (4). Notice that infinite values of the qualification are not excluded. The following theorem characterizes approximating properties of the mapping (3) and the family (4) with respect to a solution of equation (1).

Theorem 2.1. *([17, 144]) Suppose Condition 2.1 is satisfied. Then*

$$\lim_{\alpha \to 0} \|x_\alpha - x^*\|_X = 0, \tag{5}$$

where x_* is the solution of (1) nearest to ξ, i.e.,

$$x^* = P_{X_*}(\xi). \tag{6}$$

If, in addition, the initial discrepancy has the form

$$x^* - \xi = A^p v, \quad v \in X \tag{7}$$

with $p, q \geq 0, p + q \in (0, p_0]$, then

$$\|A^q(x_\alpha - x^*)\|_X \leq l\alpha^{p+q} \quad \forall \alpha \in (0, \alpha_0]. \tag{8}$$

From (5) and (6) it follows that if we define a single–valued selector $G : \mathfrak{F} \to X$ of the multivalued mapping $(A, f) \to A^{-1}f$ by the equalities

$$D(G) = \{(A, f) \in \mathfrak{F} : f \in R(A)\}, \quad G(A, f) = P_{A^{-1}f}(\xi),$$

then (5) takes the form (1.2.6). It should be pointed out that the selector G is in general a discontinuous mapping. Indeed, letting for instance

$$X = \mathbf{R}^2 = \{x = (x_1, x_2)\},$$

$$A_\varepsilon x = (\varepsilon x_1, x_2), \quad f = (0,0), \quad \xi = (1,0),$$

we get $G(A_0, f) = (1,0)$ and $G(A_\varepsilon, f) = (0,0)$ whatever $\varepsilon > 0$. In this situation, $\{\mathfrak{R}_\alpha\}$ is a family of continuous mappings, that approximate G on the all of $D(G)$.

Relations similar to (7) in theory of ill–posed problems are known as source-wise representations. They are of frequent use in contemporary investigations of regularization methods for equations (1). In view of their role in regularization theory, these conditions have been repeatedly analyzed both in general setting and in application to operators A of special forms ([96]). From (1.1.4) and (1.1.5) it follows that $R(A^{p_2}) \subset R(A^{p_1})$ whatever $0 < p_1 < p_2$. In many cases of practical interest, the representation (7) can be interpreted as saying that the element $x^* - \xi$ is of higher smoothness compared to the smoothness of a typical element of the original space X. For instance, let $p \in \mathbf{N}$ and $A : L_2(\Omega) \to L_2(\Omega)$ be the Green operator of a linear regularly elliptic differential equation of order $2m$ under homogeneous boundary conditions on a domain $\Omega \subset \mathbf{R}^N$ with $\partial\Omega \in C^\infty$. Then we have $R(A^p) = W_2^{2pm}(\Omega)$ [96]. Therefore in this case condition (7) is equivalent to the inclusion $x^* - \xi \in W_2^{2pm}(\Omega)$. These remarks make obvious a key role of the qualification as one of the main numerical characteristics of the approximating procedure (4). In particular, if the qualification has a finite value, then the process (4) is saturable in the sense that the increase of the exponent p in (7) over some threshold doesn't imply the same increase of the exponent in estimate (8). Such an effect occurs when the exponent p or, in other words, the smoothness characteristic of $x^* - \xi$, comes over the qualification value p^*. On the contrary, if $p^* = \infty$, then we have the saturation free approximating process (4). This means that Theorem 2.1 implies arbitrarily high rate of convergence of the approximations x_α to x^* as $\alpha \to 0$ whenever the initial discrepancy $x^* - \xi$ is sufficiently smooth.

Let us now turn to the general case where $A \in L(X_1, X_2)$ is a linear operator acting from a Hilbert space X_1 into another Hilbert space X_2. We put $\mathfrak{F} = L(X_1, X_2) \times X_2$. Since for selfadjoint operators a developed functional calculus is available, it is more convenient to deal with equations (1) involving selfadjoint operators rather than operators $A \in L(X_1, X_2)$ of general form. Acting on both parts of (1) by the adjoint operator A^*, we get the symmetrized equation

$$A^* A x = A^* f, \quad x \in X_1 \tag{9}$$

with a selfadjoint and nonnegative operator $A^* A$. The following well–known assertion establishes a relation between solution sets of equations (1) and (9).

Lemma 2.1. *The solution set of (9) coincides with the set of solutions to the equation*

$$Ax = P_{\overline{R(A)}}f, \quad x \in X_1.$$

Solutions of equation (9) are usually called quasisolutions to the original problem (1). In the special case where $f \notin R(A)$ and $P_{\overline{R(A)}}f \in R(A)$, the equation (1) has no solutions but quasisolutions of (1) exist. Notice that the set of quasisolutions coincides with the set of minimizers of the problem

$$\min_{x \in X_1} \frac{1}{2} \|Ax - f\|_{X_2}^2.$$

Also, it is clear that the solution set of (1), if nonempty, coincides with the set of quasisolutions to (1). Below, speaking of equation (1) with an operator $A \in L(X_1, X_2)$ of general form, we shall bear in mind precisely the problem of finding some quasisolution of (1). By analogy with (6), we shall look for the quasisolution nearest to some fixed element $\xi \in X_1$. By $G : \mathfrak{F} \to X_1$ we denote the mapping that takes each pare of input data $(A, f) \in D(G)$,

$$D(G) = \{(A, f) \in \mathfrak{F} : P_{\overline{R(A)}}f \in R(A)\},$$

to the quasisolution of (1) nearest to $\xi \in X_1$. The example cited above proves that the mapping G can be discontinuous, even if the spaces X_1 and X_2 are finite–dimensional. Let X_{1*} denote the set of all quasisolutions to (1). We assume throughout that $X_{1*} \neq \emptyset$. Applying operators (3) to the symmetrized equation (9), we obtain a parametric family of approximating operators $\mathfrak{R}_\alpha :$ $\mathfrak{F} \to X_1$ for the mapping G defined above. Suppose the generating function $\Theta(\lambda, \alpha)$ satisfies Condition 2.1. We set

$$\mathfrak{R}_\alpha(A, f) = (E - \Theta(A^*A, \alpha)A^*A)\xi + \Theta(A^*A, \alpha)A^*f, \qquad (10)$$
$$(A, f) \in \mathfrak{F}.$$

The corresponding parametric scheme for approximation of a quasisolution to equation (1) looks as follows:

$$x_\alpha = (E - \Theta(A^*A, \alpha)A^*A)\xi + \Theta(A^*A, \alpha)A^*f, \quad \alpha \in (0, \alpha_0]. \qquad (11)$$

Theorem 2.2. *([17, 144]) Suppose Condition 2.1 is satisfied. Then*

$$\lim_{\alpha \to 0} \|x_\alpha - x^*\|_{X_1} = 0,$$

where

$$x^* = P_{X_{1*}}(\xi). \tag{12}$$

If, in addition, the initial discrepancy has the form

$$x^* - \xi = (A^*A)^p w, \quad w \in X_1 \tag{13}$$

with $p, q \geq 0, p + q \in (0, p_0]$, then

$$\|(A^*A)^q(x_\alpha - x^*)\|_{X_1} \leq l\alpha^{p+q} \quad \forall \alpha \in (0, \alpha_0]. \tag{14}$$

According to Theorem 2.2, the approximating operator family (10) satisfies equality (1.2.6).

Remark 2.1. If $q = 0$, then in the left parts of inequalities (8) and (14) we have the norm of the pointwise discrepancy $x_\alpha - x^*$. In the case where $q = 1$, the expressions in (8), (14) take the form $A(x_\alpha - x^*) = Ax_\alpha - f$, $A^*A(x_\alpha - x^*) = A^*Ax_\alpha - A^*f$ and give the usual discrepancy for equations (1) and (9) respectively.

Let us now consider several generating functions $\Theta(\lambda, \alpha)$ most frequently used in computational practice and specify the schemes (4) and (11) for these functions.

Example 2.1. The function

$$\Theta(\lambda, \alpha) = \frac{1}{\lambda + \alpha} \tag{15}$$

satisfies Condition 2.1 for all $p_0 \in (0, 1]$. Numerical implementation of the scheme (4), (15) is reduced to solving the equation

$$(A + \alpha E)x_\alpha = \alpha\xi + f, \quad \alpha \in (0, \alpha_0] \tag{16}$$

with the continuously invertible operator $A + \alpha E$. The method (16) for solving (1) is known as Lavrent'ev's method. When applied to an operator $A \in L(X_1, X_2)$, the scheme (11) leads to well–known Tikhonov's method

$$(A^*A + \alpha E)x_\alpha = \alpha\xi + A^*f, \quad \alpha \in (0, \alpha_0] \tag{17}$$

for finding a quasisolution to (1).

Example 2.2. The function

$$\Theta(\lambda, \alpha) = \frac{1}{\lambda}\left[1 - \left(\frac{\alpha}{\lambda + \alpha}\right)^N\right], \quad N \in \mathbf{N} \tag{18}$$

satisfies Condition 2.1 for each $p_0 \in (0, N]$. Note that if $N = 1$, then the function (18) takes the form (15). Evaluation of the approximation $x_\alpha, \alpha \in (0, \alpha_0]$ defined by (4) and (18) can be reduced to the following finite iterative process:

$$x_\alpha = x_\alpha^{(N)}, \tag{19}$$

where

$$x_\alpha^{(0)} = \xi; \quad (A + \alpha E)x_\alpha^{(k+1)} = \alpha x_\alpha^{(k)} + f, \quad k = 0, 1, \dots, N - 1. \tag{20}$$

The method (19)–(20) in theory of ill–posed problems is known as the iterated Lavrent'ev's method. The process (11), (18) can be implemented by means of the iteration

$$x_\alpha = x_\alpha^{(N)}, \tag{21}$$

$$x_\alpha^{(0)} = \xi; \tag{22}$$
$$(A^* A + \alpha E)x_\alpha^{(k+1)} = \alpha x_\alpha^{(k)} + A^* f, \quad k = 0, 1, \dots, N - 1.$$

The method (21)–(22) is called the iterated Tikhonov's method. It should be noted that Condition 2.1 for the function (18) is violated when $p_0 > N$. Therefore methods (16) and (17) have the qualification $p^* = 1$, and the qualification of methods (19)–(20) and (21)–(22) is $p^* = N$. This means that all these processes are saturable. Let us give several examples of the processes (4) and (11) free of this drawback.

Example 2.3. The function

$$\Theta(\lambda, \alpha) = \begin{cases} \dfrac{1}{\lambda}\left(1 - \exp\left(-\dfrac{\lambda}{\alpha}\right)\right), & \lambda \neq 0, \\[2mm] \dfrac{1}{\alpha}, & \lambda = 0 \end{cases} \tag{23}$$

satisfies Condition 2.1 for each $p_0 > 0$. The process (4) with the generating function (23) can be implemented as follows. Suppose an X–valued function $u = u(t), t \geq 0$ solves the abstract Cauchy problem

$$\frac{du}{dt} + Au = f, \quad u(0) = \xi; \tag{24}$$

then we set

$$x_\alpha = u\left(\frac{1}{\alpha}\right), \quad \alpha \in (0, \alpha_0]. \tag{25}$$

The scheme (11), (23) can be implemented in the similar way:

$$x_\alpha = u\left(\frac{1}{\alpha}\right), \quad \alpha \in (0, \alpha_0]; \tag{26}$$

$$\frac{du}{dt} + A^*Au = A^*f, \quad u(0) = \xi. \tag{27}$$

The methods (24)–(25) and (26)–(27) are often referred to as stabilization methods.

Example 2.4. Let $g\colon [0, M] \to \mathbb{R}$ be a bounded and Borel measurable function, continuous at $\lambda = 0$. Suppose that

$$\sup_{\lambda \in [\varepsilon, M]} |1 - \lambda g(\lambda)| < 1 \quad \forall \varepsilon \in (0, M]. \tag{28}$$

Then the function

$$\Theta(\lambda, \alpha) = \begin{cases} \dfrac{1}{\lambda}[1 - (1 - \lambda g(\lambda))^{1/\alpha}], & \lambda \neq 0, \\ \dfrac{g(0)}{\alpha}, & \lambda = 0 \end{cases} \tag{29}$$

defined for the discrete set of values $\alpha = 1, 1/2, \ldots, 1/n, \ldots$ satisfies (2) for all $p_0 > 0$. The procedure (4), (29) is reduced to the iterative process: if $\alpha = 1/n$, then we set

$$x_\alpha = x^{(n)}, \tag{30}$$

where

$$x^{(0)} = \xi; \\ x^{(k+1)} = x^{(k)} - g(A)(Ax^{(k)} - f), \quad k = 0, 1, \ldots, n - 1. \tag{31}$$

Inequality (28) and other conditions on $g(\lambda)$ listed above are satisfied, e.g., for the functions $g(\lambda) \equiv \mu_0 \in (0, 2/M)$ and $g(\lambda) = (\lambda + \mu_0)^{-1}$ with $\mu_0 > 0$. The scheme (11), (29) can be implemented similarly to (30)–(31):

$$x_\alpha = x^{(n)}, \quad x^{(0)} = \xi; \tag{32}$$

$$x^{(k+1)} = x^{(k)} - g(A^*A)A^*(Ax^{(k)} - f), \quad k = 0, 1, \ldots, n - 1. \tag{33}$$

In the case where $g(\lambda) \equiv \mu_0 \in (0, 2/M)$, the schemes (30)–(31) and (32)–(33) generate the simplest explicit iterative processes for finding a solution (quasisolution) of equation (1). When $g(\lambda) = (\lambda + \mu_0)^{-1}, \mu_0 > 0$, these schemes lead to the simplest implicit processes for approximation of a solution or quasisolution of (1).

Since generating functions from Examples 2.3 and 2.4 satisfy Condition 2.1 without any upper restrictions on p_0, we conclude that the procedures (24)–(25), (26)–(27), (30)–(31), and (32)–(33) have the qualification $p^* = \infty$ and hence these procedures are saturation free.

The foregoing considerations treat the case where the input data (A, f) of equation (1) are known without errors. Now suppose these data are given approximately. Let $A \in L(X_1, X_2)$. Assume that instead of the exact data (A, f) in (1) their approximations $(A_h, f_\delta) \in \mathfrak{F} = L(X_1, X_2) \times X_2$ are available such that

$$\|A_h - A\|_{L(X_1, X_2)} \le h, \quad \|f_\delta - f\|_{X_2} \le \delta. \tag{34}$$

In accordance with (11), an approximation to the quasisolution $x^* = P_{X_{1*}}(\xi)$ can be defined as follows:

$$x_{\alpha(h,\delta)}^{(h,\delta)} = \mathfrak{R}_{\alpha(h,\delta)}(A_h, f_\delta), \tag{35}$$

where

$$\mathfrak{R}_{\alpha(h,\delta)}(A_h, f_\delta) =$$
$$= (E - \Theta(A_h^* A_h, \alpha(h, \delta)) A_h^* A_h) \xi + \Theta(A_h^* A_h, \alpha(h, \delta)) A_h^* f_\delta.$$

In (35), the regularization parameter $\alpha = \alpha(h, \delta)$ should be coordinated with the error levels h, δ in order to ensure the convergence

$$\lim_{h,\delta \to 0} \|x_{\alpha(h,\delta)}^{(h,\delta)} - x^*\|_{X_1} = 0. \tag{36}$$

The equality (36) must be fulfilled uniformly with respect to a choice of approximations (A_h, f_δ) subject to the conditions (34); (36) then implies that the mapping $\mathfrak{R}_{\alpha(h,\delta)} : \mathfrak{F} \to X_1$ defines a regularization algorithm for the problem (1). Without loss of generality it can be assumed that the spectra $\sigma(A_h^* A_h), \sigma(A^* A)$ lie in a segment $[0, M]$. The following theorem points up a possible way of coordination of the regularization parameter α with the error levels (h, δ).

Theorem 2.3. *([17, 144]). Let Condition 2.1 be satisfied, and let*

$$\sup_{\lambda \in [0,M]} |\Theta(\lambda, \alpha)\sqrt{\lambda}| \le \frac{c_2}{\sqrt{\alpha}} \quad \forall \alpha \in (0, \alpha_0]. \tag{37}$$

Suppose that

$$\lim_{h,\delta \to 0} \alpha(h, \delta) = 0, \quad \lim_{h,\delta \to 0} \frac{h + \delta}{\alpha(h, \delta)} = 0.$$

Then uniformly with respect to a choice of (A_h, f_δ) subject to (34), the following is true:

1) Equality (36) is valid with x^* specified in (12).

2) If, in addition, the initial discrepancy has the sourcewise representation (13), and

$$\alpha(h, \delta) = c_3(h + \delta)^{1/(2p+1)},$$

then
$$\|x_{\alpha(h,\delta)}^{(h,\delta)} - x^*\|_{X_1} \le c_4(h+\delta)^{2p/(2p+1)}.$$

This estimate is order–optimal over the class of all equations (1) with quasisolutions x^* allowing for representation (13).

Sufficiently complete overview of the variety of other possible approaches to
coordination of a regularization parameter with error levels can be found, e.g.,
in [17, 42, 68, 106, 137, 139].

Remark 2.2. Condition (37) is fulfilled for all generating functions presented
in Examples 2.1–2.4.

Let us turn to more detailed analysis of the convergence rate of approximations (4) with exact data. It will be shown below that the sourcewise representation (7) is actually very close to a necessary condition for estimate (8). The
same is true for the process (11) and related representation (13) and estimate
(14). We shall need the following condition.

Condition 2.2. *There exists a constant* $c_5 = c_5(\tau, \alpha_0, M) > 0$ *such that*

$$\int_0^{\alpha_0} \alpha^{-2\tau-1}|1 - \Theta(\lambda, \alpha)\lambda|^2 d\alpha \ge \frac{c_5}{\lambda^{2\tau}} \tag{38}$$

$$\forall \lambda \in (0, M] \quad \forall \tau \in (0, p_0).$$

Theorem 2.4. *Let Condition 2.2 be satisfied. Suppose*

$$p > 0, \quad q \ge 0, \quad p+q \in (0, p_0]$$

and the estimate is valid

$$\|A^q(x_\alpha - x^*)\|_X \le l\alpha^{p+q} \quad \forall \alpha \in (0, \alpha_0], \tag{39}$$

where x_α *is defined by (4) and* x^* *is specified in (6). Then*

$$x^* - \xi \in R(A^{p-\varepsilon}) \quad \forall \varepsilon \in (0, p). \tag{40}$$

Proof. From (4) and (39) using the equality $f = Ax^*$ we get

$$\|A^q(x_\alpha - x^*)\|_X^2 =$$
$$= \|A^q(E - \Theta(A, \alpha)A)(x^* - \xi)\|_X^2 \le l^2 \alpha^{2(p+q)}. \tag{41}$$

From (1.1.2) and (41) it follows that for each $\omega \in (0, 2p)$,

$$\int_{0-0}^{M} \alpha^{-2(p+q)-1+\omega} \lambda^{2q} |1 - \Theta(\lambda, \alpha)\lambda|^2 d\|E_\lambda(x^* - \xi)\|_X^2 \leq$$

$$\leq l^2 \alpha^{-1+\omega} \quad \forall \alpha \in (0, \alpha_0].$$

Integrating this inequality over $\alpha \in (0, \alpha_0]$, we obtain

$$\int_0^{\alpha_0} \int_{0-0}^{M} \alpha^{-2(p+q)-1+\omega} \lambda^{2q} |1 - \Theta(\lambda, \alpha)\lambda|^2 d\|E_\lambda(x^* - \xi)\|_X^2 d\alpha < \infty.$$

Since the integrand in the last expression is nonnegative, by Fubini's theorem [123] we can change the order of integrals to get

$$\int_{0-0}^{M} \lambda^{2q} \left(\int_0^{\alpha_0} \alpha^{-2(p+q)-1+\omega} |1 - \Theta(\lambda, \alpha)\lambda|^2 d\alpha \right) d\|E_\lambda(x^* - \xi)\|_X^2 < \infty. \quad (42)$$

Substituting $p + q - \frac{\omega}{2}$ for τ in (38) yields

$$\int_0^{\alpha_0} \alpha^{-2(p+q)-1+\omega} |1 - \Theta(\lambda, \alpha)\lambda|^2 d\alpha \geq \frac{c_5}{\lambda^{2(p+q)-\omega}} \quad \forall \lambda \in (0, M].$$

Let μ^* be the Lebesgue–Stieltjes measure associated with the function $\nu^*(\lambda) = \|E_\lambda(x^* - \xi)\|_X^2$. Observe that (6) implies $x^* - \xi \perp N(A)$, that is,

$$(x^* - \xi, y)_X = 0 \quad \forall y \in N(A).$$

Therefore the function $\nu^*(\lambda)$ is continuous at the point $\lambda = 0$ and hence $\mu^*(\{0\}) = 0$. Inequality (42) now gives

$$\int_{0-0}^{M} \lambda^{-2(p-\frac{\omega}{2})} d\|E_\lambda(x^* - \xi)\|_X^2 = \int_0^{M} \lambda^{-2(p-\frac{\omega}{2})} d\|E_\lambda(x^* - \xi)\|_X^2 < \infty.$$

In view of (1.1.4), we have

$$x^* - \xi \in D(A^{-(p-\frac{\omega}{2})}) = R(A^{p-\frac{\omega}{2}}).$$

Since $\omega \in (0, 2p)$ can be taken arbitrarily small, the last inclusion implies (40). This completes the proof. $\qquad\square$

Example 2.5. Direct calculations show that (38) is fulfilled for all generating functions from Examples 2.1–2.3. In fact, for the function (18) the integral in (38) can be estimated as follows:

$$\int\limits_0^{\alpha_0} \alpha^{-2\tau-1}|1 - \Theta(\lambda, \alpha)\lambda|^2 d\alpha = \int\limits_0^{\alpha_0} \alpha^{-2\tau-1}\left(\frac{\alpha}{\lambda + \alpha}\right)^{2N} d\alpha =$$

$$= \frac{1}{\lambda^{2\tau}} \int\limits_0^{\alpha_0/\lambda} \frac{t^{2N-2\tau-1}dt}{(1+t)^{2N}} \geq \frac{c_6}{\lambda^{2\tau}}, \quad c_6 = \int\limits_0^{\alpha_0/M} \frac{t^{2N-2\tau-1}dt}{(1+t)^{2N}}, \quad \tau \in (0, N).$$

Here the original integral was transformed by the substitution $\alpha = \lambda t$. Assuming that $\tau > 0$, for the function (23) with the same substitution we obtain

$$\int\limits_0^{\alpha_0} \alpha^{-2\tau-1}|1 - \Theta(\lambda, \alpha)\lambda|^2 d\alpha =$$

$$= \int\limits_0^{\alpha_0} \alpha^{-2\tau-1} \exp\left(-\frac{2\lambda}{\alpha}\right) d\alpha \geq \frac{c_7}{\lambda^{2\tau}}, \quad c_7 = \int\limits_0^{\alpha_0/M} t^{-2\tau-1} \exp\left(-\frac{2}{t}\right) dt.$$

Let us now analyze the necessity of representation (7) for rate of convergence estimates of the methods (30)–(31). For this purpose we introduce the following discrete analog of Condition 2.2.

Condition 2.3. *The function $g(\lambda)$ satisfies conditions listed in Example 2.4 and there exists a constant $c_8 = c_8(\tau, \alpha_0, M) > 0$ such that*

$$\sum_{n=1}^{\infty} n^{2\tau-1}|1 - \lambda g(\lambda)|^{2n} \geq \frac{c_8}{\lambda^{2\tau}} \quad \forall \lambda \in (0, M] \quad \forall \tau > 0. \tag{43}$$

Theorem 2.5. *Let Condition 2.3 be satisfied. Suppose $p > 0, q \geq 0$ and there exists $l \geq 0$ such that*

$$\|A^q(x^{(n)} - x^*)\|_X \leq \frac{l}{n^{p+q}} \quad \forall n \in \mathbf{N}, \tag{44}$$

where $x^{(n)}$ are defined by (30)–(31) and x^* is specified in (6). Then (40) is valid.

Proof. As in the proof of Theorem 2.4, with the use of (44) we get

$$
\|A^q(x^{(n)} - x^*)\|_X^2 = \|A^q(E - Ag(A))^n(x^* - \xi)\|_X^2 =
$$
$$
= \int_{0-0}^{M} \lambda^{2q}(1 - \lambda g(\lambda))^{2n} d\|E_\lambda(x^* - \xi)\|_X^2 \le \frac{l^2}{n^{2(p+q)}} \quad \forall n \in \mathbf{N}. \tag{45}
$$

From (45) it follows that for each $\omega \in (0, 2p)$,

$$
\int_{0-0}^{M} n^{2(p+q)-1-\omega} \lambda^{2q} |1 - \lambda g(\lambda)|^{2n} d\|E_\lambda(x^* - \xi)\|_X^2 \le \frac{l^2}{n^{1+\omega}}.
$$

Summing these inequalities, we obtain

$$
\sum_{n=1}^{\infty} \int_{0-0}^{M} n^{2(p+q)-1-\omega} \lambda^{2q} |1 - \lambda g(\lambda)|^{2n} d\|E_\lambda(x^* - \xi)\|_X^2 < \infty.
$$

Next, using Levy's theorem [123] and the equality $\mu^*(\{0\}) = 0$, after substituting into (43) $\tau = p + q - \frac{\omega}{2}$ we get

$$
\sum_{n=1}^{\infty} \int_{0-0}^{M} n^{2(p+q)-1-\omega} \lambda^{2q} |1 - \lambda g(\lambda)|^{2n} d\|E_\lambda(x^* - \xi)\|_X^2 =
$$
$$
= \int_{0-0}^{M} \lambda^{2q} \left(\sum_{n=1}^{\infty} n^{2(p+q)-1-\omega} |1 - \lambda g(\lambda)|^{2n} \right) d\|E_\lambda(x^* - \xi)\|_X^2 \ge
$$
$$
\ge c_8 \int_{0-0}^{M} \lambda^{-2(p-\frac{\omega}{2})} d\|E_\lambda(x^* - \xi)\|_X^2.
$$

The obtained inequality in view of (1.1.4) gives the required inclusion

$$
x^* - \xi \in R(A^{p-\varepsilon}) \quad \forall \varepsilon \in (0, p).
$$

$\square$

Remark 2.3. It is readily seen that the conclusion of Theorem 2.5 remains valid if inequality (43) is true for all $\lambda \in (0, M_1]$ with some $M_1 \in (0, M]$.

Example 2.6. Condition 2.3 is satisfied for the functions $g(\lambda) \equiv \mu_0, \mu_0 \in (0, 2/M)$ and $g(\lambda) = (\lambda + \mu_0)^{-1}, \mu_0 > 0$ introduced in Example 2.4. Let us show this for the case where $g(\lambda) \equiv \mu_0$. The sum in (43) then takes the form

$$\sum_{n=1}^{\infty} n^{2\tau-1}(1 - \mu_0\lambda)^{2n}.$$

Using the estimate ([140])

$$c_9 \frac{\Gamma(1-t)}{(1-x)^{1-t}} \leq \sum_{n=1}^{\infty} \frac{x^n}{n^t} \leq c_{10} \frac{\Gamma(1-t)}{(1-x)^{1-t}}; \quad c_9 = c_9(t), \quad c_{10} = c_{10}(t)$$

$$\forall x \in [1/2, 1) \quad \forall t < 1,$$

where $\Gamma(\cdot)$ stands for the Euler gamma–function, and letting

$$x = |1 - \mu_0\lambda|^2, \quad t = -2\tau + 1,$$

we get

$$\sum_{n=1}^{\infty} n^{2\tau-1}(1 - \mu_0\lambda)^{2n} \geq c_9 \frac{\Gamma(2\tau)}{\lambda^{2\tau}(\mu_0(2 - \mu_0\lambda))^{2\tau}} \geq \frac{c_8}{\lambda^{2\tau}} \quad \forall \lambda \in (0, M_1]$$

for all sufficiently small positive numbers $M_1 \leq M$. It suffices to apply Remark 2.3.

Remark 2.4. Examples given in [42, 82] prove that the inclusion $\varepsilon \in (0, p)$ in Theorems 2.4 and 2.5 generally can't be replaced by the equality $\varepsilon = 0$. At the same time, such a replacement is possible if $p = p^*$, where p^* is the qualification of the method under consideration ([42]).

Converse statements for Theorem 2.2 can be obtained directly from Theorems 2.4 and 2.5 by formal replacement of the operator A with A^*A.

Theorems 2.4 and 2.5 establish that estimate (8) actually is not improvable and show that the rate of convergence of the process (4) is completely determined by the power p in representation (7) or, in other words, by smoothness characteristics of the initial discrepancy $x^* - \xi$.

The technique used in the proofs of Theorems 2.4 and 2.5 improves the ideas of [82], where these results were obtained for special generating functions in the case of $q = 0$. Another approach to converse theorems for the processes (4) and (11) was earlier developed in [42], where assertions of Theorems 2.4 and 2.5 were established with the use of the condition

$$\sup_{\lambda \in [0, M]} |\Theta(\lambda, \alpha)| \leq \frac{c_{11}}{\alpha} \quad \forall \alpha \in (0, \alpha_0] \tag{46}$$

instead of (38) and (43).

Notice that Theorems 2.4 and 2.5 written for $q = 1$, point to a possibility of determining smoothness characteristics of the unknown solution x^* simultaneously with the finding of x^*. To do this, it is sufficient to estimate the exponent $r > 0$ in the formula

$$\|Ax_\alpha - f\|_X = O(\alpha^r), \quad \alpha \to 0.$$

We conclude this section with a discussion of conditions necessary and sufficient for more slow, as compared to (8), estimates

$$\|x_\alpha - x^*\|_X \le l(-\ln\alpha)^{-p} \quad \forall \alpha \in (0, \alpha_0] \quad (p > 0) \tag{47}$$

and their iterated generalizations. For brevity sake, we consider only the process (4) with the exact input data $A^* = A \ge 0$, $f \in R(A)$. An extension of results to operators $A \in L(X_1, X_2)$ creates no difficulties.

Below we assume that $\alpha_0 \in (0, 1)$ and $\sigma(A) \subset [0, M]$, where $M \in (0, 1)$. We need the following condition on the generating function $\Theta(\lambda, \alpha)$.

Condition 2.4. *There exists a constant* $c_{12} = c_{12}(\alpha_0, p, M) > 0$ *such that for all* $p \ge 0$,

$$\sup_{\lambda \in (0, M]} |1 - \Theta(\lambda, \alpha)\lambda|(-\ln\lambda)^{-p} \le c_{12}(-\ln\alpha)^{-p}$$

$$\forall \alpha \in (0, \alpha_0].$$

In [61] it was proved that Condition 2.4 is valid for each function $\Theta(\lambda, \alpha)$ satisfying Condition 2.1. As a corollary, we get the following example.

Example 2.7. All functions $\Theta(\lambda, \alpha)$ from Examples 2.1–2.4 satisfy Condition 2.4.

Denote

$$\varphi(\lambda) = \begin{cases} (-\ln\lambda)^{-p}, & \lambda \in (0, 1), \\ 0, & \lambda = 0, \end{cases} \quad p > 0.$$

Since $\sigma(A) \subset [0, M]$ and $\varphi(\lambda)$ is bounded on the segment $[0, M]$, formula (1.1.1) defines an operator $(-\ln A)^{-p} \in L(X)$.

Let us give a condition sufficient for the estimate (47).

Theorem 2.6. *([61]) Suppose Condition 2.4 and the sourcewise representation*

$$x^* - \xi = (-\ln A)^{-p}v, \quad v \in X \tag{48}$$

are satisfied, where $p > 0$ and $x^* \in X_*$ is specified in (6). Then (47) is valid.

Proof. From (4) with the use of formula (1.1.3) and Condition 2.4 we get

$$
\begin{aligned}
\|x_\alpha - x^*\|_X &= \|(E - \Theta(A, \alpha)A)(x^* - \xi)\|_X \leq \\
&\leq \|(E - \Theta(A, \alpha)A)(-\ln A)^{-p}\|_{L(X)}\|v\|_X \leq \\
&\leq \|v\|_X \sup_{\lambda \in (0,M]} |1 - \Theta(\lambda, \alpha)\lambda|(-\ln \lambda)^{-p} \leq \\
&\leq l(-\ln \alpha)^{-p} \quad \forall \alpha \in (0, \alpha_0].
\end{aligned}
\tag{49}
$$

□

The next theorems (see [61] for the proofs) show that representation (48) approaches a necessary condition for (47) and hence can't be essentially weakened.

Theorem 2.7. *Let (46) be satisfied. Assume that*

$$
\|x_\alpha - x^*\|_X \leq l(-\ln \alpha)^{-p} \quad \forall \alpha \in (0, \alpha_0] \quad (p > 0),
\tag{50}
$$

where x_α is defined by (4) and $x^* \in X_*$ is specified in (6). Then

$$
x^* - \xi \in R((-\ln A)^{-(p-\varepsilon)}) \quad \forall \varepsilon \in (0, p).
\tag{51}
$$

Theorem 2.8. *Let (46) and conditions of Example 2.4 be satisfied. Assume that*

$$
\|x^{(n)} - x^*\|_X \leq l(\ln n)^{-p}, \quad n = 2, 3, \ldots \quad (p > 0),
\tag{52}
$$

where $x^{(n)}$ are defined by (31) and $x^* \in X_*$ is specified in (6). Then (51) is valid.

It is easy to see that the linear manifold of all elements $x^* - \xi$ of the form (48) with $p = p_1$ contains the manifold defined by the right part of (7) with $p = p_2$ whatever $p_1, p_2 > 0$ be. At the same time, both the representation (48) and the equality (7) defines narrowing manifolds when the exponent p increases.

Let us explain the structure of manifolds defined by (7) and (48) in more detail. Denote $S^1 = \{z \in \mathbf{C} : |z| = 1\}$. Suppose $A^* = A \geq 0$ is an operator in $X = L_2(S^1)$ having a complete system of eigenvectors $\{\exp(int)\}_{n \in \mathbf{Z}}, t =$

$\arg z \in [0, 2\pi]$. Given $s \geq 0$ define the Sobolev–Slobodetskii space [141]

$$H^s(S^1) = \Big\{ x : [0, 2\pi] \to \mathbf{C} : \tag{53}$$

$$\sum_{n \in \mathbf{Z}} |a_n(x)|^2 (1+n^2)^s < \infty, \quad a_n(x) = \int_0^{2\pi} \exp(-int)x(t)dt \Big\}.$$

The space (53) is connected with the classical Sobolev spaces and spaces of smooth functions by standard embedding theorems [1, 141]. Suppose the eigenvectors $\exp(int)$, $\exp(-int)$ correspond to the eigenvalue $\lambda_n, n \geq 0$, and $\lambda_n, n \in \mathbf{N}$ satisfy the condition

$$c_{13}n^{-r} \leq \lambda_n \leq c_{14}n^{-r}, \quad n \in \mathbf{N}, \quad r > 0;$$

then from (53) it follows that

$$R(A^p) = H^{pr}(S^1).$$

Therefore (7) is equivalent to the inclusion $x^* - \xi \in H^{pr}(S^1)$. Under the condition

$$c_{15}\exp(-n^r) \leq \lambda_n \leq c_{16}\exp(-n^r), \quad n \in \mathbf{N},$$

in the similar way we obtain

$$R((-\ln A)^{-p}) = H^{pr}(S^1).$$

Hence (7) and (48) imply an increased smoothness of the element $x^* - \xi$ in the sense of the scale $\{H^s(S^1)\}_{s=0}^{\infty}$, as compared with the smoothness of elements from the original space $L_2(S^1) = H^0(S^1)$. The power asymptotic behavior of eigenvalues is typical for the Green operators of regularly elliptic differential equations [141]; integral operators with the Poisson kernel provide examples of operators with the exponential decrease of eigenvalues ([60]).

Example 2.8. Let $C : D(C) \to X$ ($\overline{D(C)} = X$) be a selfadjoint unbounded operator in a Hilbert space X. Suppose $\sigma(C) \subset [a_0, \infty), a_0 > 0$. It is well known [58] that a solution $u = u(t)$ to the abstract Cauchy problem

$$\frac{du}{dt} + Cu = 0, \quad u(0) = u_0 \tag{54}$$

is uniquely determined for all $u_0 \in X$. We have

$$u(t) = U(t)u_0, \quad t \geq 0,$$

where $\{U(t)\}_{t \geq 0}$ is the semigroup of linear bounded operators with the generator $-C$, that is,

$$U(t) = \exp(-tC), \quad t \geq 0.$$

Under the above conditions, the problem (54) is well–posed with respect to perturbations of the initial state u_0 [58, 97]. Now consider the inverse problem of reconstructing the initial state $u_0 = x$ by a solution $u(t)$ given at the moment $t = T : u(T) = U(T)u_0$. This problem is ill–posed in most nontrivial cases [97, 105]. Setting $A = U(T)$ and $f = u(T)$, we reduce the inverse Cauchy problem to the form (1). In this event, the logarithmic sourcewise representation

$$x^* - \xi \in R((-\ln A)^{-p})$$

is equivalent to the inclusion

$$x^* - \xi \in R(C^{-p}) = D(C^p).$$

This means that $x^* - \xi$ has an increased smoothness, as compared with the smoothness of elements from the original space $X = D(C^0)$.

Theorems 2.7 and 2.8 allow for natural generalizations to the case of arbitrary number of logarithms in estimates (50), (52) and representations (48), (51). Let us introduce the sequence $\{M_n^*\}$:

$$M_1^* = 1; \quad M_{n+1}^* = \exp\left(-\frac{1}{M_n^*}\right), \quad n \in \mathbf{N}.$$

Pick numbers $N \in \mathbf{N}$ and $\alpha_0 \in (0, M_N^*)$. We intend to give conditions on the discrepancy $x^* - \xi$ necessary and sufficient for the iterated logarithmic estimate

$$\|x_\alpha - x^*\|_X \le l(\underbrace{\ln \ldots \ln(-\ln \alpha)}_{N})^{-p} \quad \forall \alpha \in (0, \alpha_0] \quad (p > 0). \tag{55}$$

The approximations x_α are defined by (4). For all $N > 1$ and $p = p_1 > 0$, estimate (55) is slower than (47) with arbitrary $p = p_2 > 0$. Assume that $\sigma(A) \subset [0, M]$, where $M \in (0, M_N^*)$. By analogy with the case $N = 1$, we introduce the following condition.

Condition 2.5. *For each $p \ge 0$ there exists a constant*

$$c_{17} = c_{17}(\alpha_0, p, M, N) > 0$$

such that

$$\sup_{\lambda \in (0, M]} |1 - \Theta(\lambda, \alpha)\lambda| (\underbrace{\ln \ldots \ln(-\ln \lambda)}_{N})^{-p} \le$$

$$\le c_{17} (\underbrace{\ln \ldots \ln(-\ln \alpha)}_{N})^{-p} \quad \forall \alpha \in (0, \alpha_0]. \tag{56}$$

The proof of the next theorem follows the scheme (49) and uses the technique developed in [42].

Theorem 2.9. *1) Let Condition 2.5 and the sourcewise representation*

$$x^* - {}^\circ\xi = (\underbrace{\ln \ldots \ln(-\ln A)}_{N})^{-p}v, \quad v \in X \quad (p > 0) \tag{57}$$

be satisfied, where $x^* \in X_*$ is specified in (6). Then (55) is valid.
2) Suppose (46) is fulfilled and

$$\|x_\alpha - x^*\|_X \le l(\underbrace{\ln \ldots \ln(-\ln \alpha)}_{N})^{-p} \quad \forall \alpha \in (0, \alpha_0],$$

where $p > 0$. Then for each $\varepsilon \in (0, p)$,

$$x^* - \xi \in R((\underbrace{\ln \ldots \ln(-\ln A)}_{N})^{-(p-\varepsilon)}).$$

Let us denote

$$\eta(\lambda, \alpha) = |1 - \Theta(\lambda, \alpha)\lambda|; \quad \lambda \in [0, M], \quad \alpha \in (0, \alpha_0].$$

Verification of Condition 2.5 can be facilitated with the following observation.

Lemma 2.2. *Let $\Theta(\lambda, \alpha)$ satisfy Condition 2.5 for $N = 1$. Suppose for all $\mu_0, \beta_0 \in (0, 1)$ there exists a constant $c_{18} = c_{18}(\mu_0, \beta_0) > 0$ such that*

$$\eta((-\ln \mu)^{-1}, (-\ln \beta)^{-1}) \ge c_{18}\eta(\mu, \beta) \tag{58}$$
$$\forall \mu \in (0, \mu_0] \quad \forall \beta \in (0, \beta_0].$$

Then the function $\Theta(\lambda, \alpha)$ satisfies Condition 2.5 for all $N \in \mathbf{N}$.

The proof is by induction on N. It is sufficient to make in (56) the substitution $\lambda = (-\ln \mu)^{-1}, \alpha = (-\ln \beta)^{-1}$ and to take into account (58).

When applied to an operator $A : L_2(S^1) \to L_2(S^1), A^* = A \ge 0$ having a complete system of eigenvectors $\{\exp(int), \exp(-int)\}_{n \in \mathbf{N} \cup \{0\}}$ and eigenvalues λ_n decreasing at the iterated exponential rate,

$$c_{19} \underbrace{\exp(-\exp \ldots \exp n^r)}_{N} \le \lambda_n \le c_{20} \underbrace{\exp(-\exp \ldots \exp n^r)}_{N}, \quad n \in \mathbf{N},$$

we have

$$R((\underbrace{\ln \ldots \ln(-\ln A)}_{N})^{-p}) = H^{pr}(S^1).$$

Hence (57) is equivalent to the inclusion

$$x^* - \xi \in H^{pr}(S^1).$$

Example 2.9. Simple calculations prove that all the functions $\Theta(\lambda, \alpha)$ considered in Examples 2.1–2.4 satisfy (58). Let us show this, e.g., for the function (18). In this event, (58) has the form

$$\frac{\ln \beta}{\ln \mu} + 1 \le c_{18}\left(\frac{\mu}{\beta} + 1\right) \quad \forall \mu \in (0, \mu_0] \quad \forall \beta \in (0, \beta_0]. \tag{59}$$

The substitution $\beta = 1/\gamma, \mu = 1/\nu$ reduces (59) to the form

$$\frac{\ln \gamma}{\ln \nu} + 1 \le c_{18}\left(\frac{\gamma}{\nu} + 1\right) \quad \forall \gamma \in [\gamma_0, \infty) \quad \forall \nu \in [\nu_0, \infty),$$

where $\gamma_0 = 1/\beta_0, \nu_0 = 1/\mu_0$. To end the proof it remains to set $\gamma = t\nu$ and to note that the resulting inequality

$$\frac{\ln t}{\ln \nu} + 2 \le c_{18}(t + 1)$$

is true for all $\nu \ge \nu_0$ and $t > 0$ with a sufficiently large constant $c_{18} > 0$.

Following [17, 42], the reader will easily get estimates for the rate of convergence of approximations (35) as $h, \delta \to 0$ in the condition that (57) is fulfilled and the regularization parameter $\alpha = \alpha(h, \delta)$ is suitably coordinated with the error levels h and δ.

2.2 General Scheme of Constructing Regularization Algorithms in Banach Space

In this section we consider linear operator equations

$$Ax = f, \quad x \in X, \tag{1}$$

where $A : X \to X$ is a bounded operator acting in a Banach space X. Suppose $f \in R(A)$; then the set X_* of solutions to equation (1) is a nonempty affine subspace of X. A direct extension of results presented in Section 2.1 to the case of a Banach space X is partially hampered by the absence of an immediate analog of the spectral decomposition and corresponding functional calculus for a sufficiently wide class of operators $A \in L(X)$. Among existing operator calculi in a Banach space, the Riesz–Dunford calculus seems to be most convenient for our purposes. According to this calculus, a function $\varphi(A)$ of an operator

A is defined by the integral formula (1.1.6). With the use of this definition of $\Theta(A, \alpha)$, in [7, 17] the approximating scheme

$$x_\alpha = \Re_\alpha(A, f) = (E - \Theta(A, \alpha)A)\xi + \Theta(A, \alpha)f, \quad \alpha \in (0, \alpha_0] \qquad (2)$$

was introduced and studied in the case where $\xi = 0$. The procedure (2) naturally extends the parametric scheme (1.4) to a Banach space X. We suppose the function $\Theta(\lambda, \alpha)$ is analytic in λ on a domain containing the spectrum $\sigma(A)$. As above, the controlling parameter $\xi \in X$ plays the role of an initial estimate of a solution to (1). In [7, 17] it was proved that the approximations x_α defined by (2) converge to a solution $x^* \in X_*$ as $\alpha \to 0$ whenever Condition 1.1 is satisfied and the representation (1.7) is fulfilled with $p = 1$. The behavior of x_α with approximately given f has been also analyzed. Let us remark that the restriction of our considerations with analytic generating functions $\Theta(\lambda, \alpha)$ doesn't lead to significant lose of generality because most popular computational algorithms of the form (2) are generated precisely by analytic functions (see Examples 2.1–2.4). This section and Section 2.3 are devoted to continuation of above–mentioned investigations. Specifically, we intend to obtain for the scheme (2) analogs of main statements of the theory presented in Section 2.1. The outlined program primarily involves the study of convergence $x_\alpha \to x^*$ as $\alpha \to 0$ in case of arbitrary, not necessarily natural, exponents $p > 0$ in the sourcewise representation (1.7). The program also provides for justification of equality (1.2.5) when both the operator A and the right hand side f in (1) are available with errors.

Following [83, 84] and Section 1.1, we assume throughout this section that operator A satisfies the sectorial Condition 1.1.

Let us introduce some notation. Fix a number $R_0 > \|A\|_{L(X)}$ and set

$$K(R_0, \varphi_0) = K(\varphi_0) \cap S(R_0); \quad S(r) = \{\lambda \in \mathbf{C} : |\lambda| \le r\}, \quad r > 0,$$

$$K_\alpha(R_0, d_0, \varphi_0) = K(R_0, \varphi_0) \cup S(\min\{R_0, d_0\alpha\}). \qquad (3)$$

Also, by γ_α we denote the boundary of $K_\alpha(R_0, d_0, \varphi_0), \alpha > 0$. We recall that $K(\varphi)$ is defined by (1.1.7). According to (1.1.7), $\sigma(A) \subset K(R_0, \varphi_0)$. It is obvious that estimate (1.1.8) remains valid with the angle $K(\varphi_0)$ replaced by the sector $K(R_0, \varphi_0)$.

We now turn to study the convergence rate of the process (2) in the case of exact input data (A, f). First of all, we need to specify in more detail the class of generating functions $\Theta(\lambda, \alpha)$ in (2). From here on, we suppose the following condition is satisfied.

Condition 2.6. *For each $\alpha \in (0, \alpha_0]$, the function $\Theta(\lambda, \alpha)$ is analytic in λ on an open subset $D_\alpha \subset \mathbf{C}$, where*

$$K_\alpha(R_0, d_0, \varphi_0) \subset D_\alpha$$

and $d_0 \in (0, 1)$ is a fixed constant.

Assume that the contour $\Gamma_\alpha \subset D_\alpha$ is chosen such that γ_α lies within the domain surrounded by Γ_α. Additionally suppose that there exists a constant $d_1 > d_0$ such that for all $\alpha \in (0, \alpha_0]$ the contour Γ_α doesn't contain the point $\lambda = -d_1\alpha$ inside. From Condition 2.6 it follows that $\Theta(A, \alpha)$ has the integral representation (1.1.6) with $\varphi(\lambda) = \Theta(\lambda, \alpha)$ and $\Gamma = \Gamma_\alpha$. Thus if Γ_α satisfies the above condition, then we have

$$\Theta(A, \alpha) = \frac{1}{2\pi i} \int_{\Gamma_\alpha} \Theta(\lambda, \alpha) R(\lambda, A) d\lambda, \quad \alpha \in (0, \alpha_0]. \tag{4}$$

It is clear that $\{\Re_\alpha(A, f)\}$ is a family of continuous mappings affine in the second argument. Here the continuity is meant with respect to variations of $A \in L(X)$ within the framework of Condition 1.1 and to arbitrary variations of f in X.

Let the initial discrepancy $x^* - \xi$ have the sourcewise representation

$$x^* - \xi = A^p v, \quad v \in X, \quad p > 0. \tag{5}$$

According to (2) and (1.1.5), for each $q \geq 0$,

$$A^q(x_\alpha - x^*) = -(E - \Theta(A, \alpha)A)A^{p+q}v. \tag{6}$$

Denote $m = [p + q]$ and $\mu = \{p + q\}$. By the definition given in Section 1.1, $A^{p+q} = A^m A^\mu$. From (6), for each $\varepsilon > 0$ we obtain

$$\|A^q(x_\alpha - x^*)\|_X \leq \|(E - \Theta(A, \alpha)A)A^m(A + \varepsilon E)^\mu v\|_X + \\ + \|(E - \Theta(A, \alpha)A)A^m[(A + \varepsilon E)^\mu - A^\mu]v\|_X. \tag{7}$$

Let us choose $\varepsilon = d_1\alpha$ and estimate individual terms in the right part of (7). From (4) it follows that

$$\|(E - \Theta(A, \alpha)A)A^m(A + \varepsilon E)^\mu v\|_X \leq \\ \leq c_1\|v\|_X \int_{\Gamma_\alpha} |1 - \Theta(\lambda, \alpha)\lambda||\lambda|^m|\lambda + \varepsilon|^\mu \|R(\lambda, A)\|_{L(X)}|d\lambda| \leq \\ \leq c_2\|v\|_X \int_{\Gamma_\alpha} |1 - \Theta(\lambda, \alpha)\lambda|(|\lambda|^{p+q-1} + \alpha^\mu|\lambda|^{m-1})|d\lambda|. \tag{8}$$

Having this estimate in mind, we impose on $\Theta(\lambda, \alpha)$ the following additional restriction (compare with Condition 2.1).

Condition 2.7. *There exists $p_0 > 0$ such that for all $p \in [0, p_0]$,*

$$\int_{\Gamma_\alpha} |1 - \Theta(\lambda, \alpha)\lambda| |\lambda|^{p-1} |d\lambda| \leq c_3 \alpha^p \quad \forall \alpha \in (0, \alpha_0], \tag{9}$$

where the constant c_3 is independent of α.

Remark 2.5. The least upper bound of the values p_0 from Condition 2.7 has the meaning of the qualification parameter p^* for (2) in a Banach space X.

Suppose Condition 2.7 is satisfied and $p + q \in (0, p_0]$. Since

$$|\lambda| \geq d_0 \alpha \quad \forall \lambda \in \Gamma_\alpha,$$

from (8) and (9) it follows that

$$\|(E - \Theta(A, \alpha)A)A^m(A + \varepsilon E)^\mu v\|_X \leq c_4 \|v\|_X \alpha^{p+q} \quad \forall \alpha \in (0, \alpha_0]. \tag{10}$$

Using (1.1.8) and (1.1.10), for the second summand in (7) we get the estimate

$$\|(E - \Theta(A, \alpha)A)A^m[(A + \varepsilon E)^\mu - A^\mu]v\|_X \leq$$
$$\leq c_5 \|v\|_X \varepsilon^\mu \|(E - \Theta(A, \alpha)A)A^m\|_{L(X)} \leq$$
$$\leq \frac{c_5}{2\pi} \|v\|_X \varepsilon^\mu \int_{\Gamma_\alpha} |1 - \Theta(\lambda, \alpha)\lambda| |\lambda|^m \|R(\lambda, A)\|_{L(X)} |d\lambda| \leq \tag{11}$$
$$\leq c_6 \|v\|_X \alpha^{p+q} \quad \forall \alpha \in (0, \alpha_0].$$

Combining (11) with estimates (7) and (10), we obtain

$$\|A^q(x_\alpha - x^*)\|_X \leq l\alpha^{p+q} \quad \forall \alpha \in (0, \alpha_0]. \tag{12}$$

Thus, we proved the following result.

Theorem 2.10. *Let Conditions 1.1, 2.6, and 2.7 be satisfied. Suppose representation (5) is fulfilled with $q \geq 0, p + q \in (0, p_0]$, approximations x_α are defined by (2), and $x^* \in X_*$ is specified by (5). Then estimate (12) is valid.*

Remark 2.6. It follows from Theorem 2.10 that a solution x^* possessing representation (5) is uniquely determined.

Let us now turn to the case where the input data (A, f) are available with errors. As a class $\mathfrak{F}$ that contains all feasible exact and approximate data we take $\mathfrak{F} = L(X) \times X$. Assume that instead of the original operator A and the element f some approximations $(A_h, f_\delta) \in \mathfrak{F}$ are given such that

$$\|A_h - A\|_{L(X)} \leq h, \quad \|f_\delta - f\|_X \leq \delta. \tag{13}$$

The levels of errors h, δ in (13) are also assumed to be known. According to (2), we define an approximation to the solution of (1) as follows:

$$
\begin{aligned}
x_{\alpha(h,\delta)}^{(h,\delta)} &= \mathfrak{R}_{\alpha(h,\delta)}(A_h, f_\delta) = \\
&= (E - \Theta(A_h, \alpha(h,\delta))A_h)\xi + \Theta(A_h, \alpha(h,\delta))f_\delta.
\end{aligned}
\tag{14}
$$

The regularization parameter $\alpha = \alpha(h, \delta)$ should be coordinated with the error levels h, δ in order to ensure the equality

$$\lim_{h,\delta \to 0} \|x_{\alpha(h,\delta)}^{(h,\delta)} - x^*\|_X = 0 \quad (x^* \in X_*) \tag{15}$$

uniformly with respect to a choice of approximations $(A_h, f_\delta) \in \mathfrak{F}$ subject to conditions (13). This will imply that the mapping

$$\mathfrak{R}_{\alpha(h,\delta)} : \mathfrak{F} \to X, \quad \mathfrak{R}_{\alpha(h,\delta)}(A_h, f_\delta) = x_{\alpha(h,\delta)}^{(h,\delta)}$$

defines a regularization algorithm for equation (1).

First of all, we shall prove that the operator function $\Theta(A_h, \alpha)$ in (14) can be presented in the form (4), provided the error level $h > 0$ is sufficiently small. The following lemma establishes a condition, under which the contour Γ_α surrounds both $\sigma(A)$ and $\sigma(A_h)$.

Lemma 2.3. *Assume that*

$$\frac{r_0 h}{d_0 \alpha} \leq \omega_0, \tag{16}$$

where $\omega_0 \in (0, 1)$, and r_0 is specified in Condition 1.1. Then the contour Γ_α surrounds the spectrum $\sigma(A_h)$ so that

$$\Theta(A_h, \alpha) = \frac{1}{2\pi i} \int_{\Gamma_\alpha} \Theta(\lambda, \alpha) R(\lambda, A_h) d\lambda.$$

Proof. For each $\lambda \in \mathbf{C} \backslash \text{int} K_\alpha(R_0, d_0, \varphi_0)$, by construction of $K_\alpha(R_0, d_0, \varphi_0)$ (see (3)) we have $|\lambda| \geq d_0 \alpha$. Setting in Lemma 1.2 $B = A_h - A$, with the use of (13) and (16) we get

$$\|BR(\lambda, A)\|_{L(X)} \leq \frac{r_0 h}{|\lambda|} \leq \frac{r_0 h}{d_0 \alpha} \leq \omega_0 < 1.$$

Therefore $\lambda \in \rho(A_h)$. Thus we see that γ_α and hence Γ_α contains all points of the spectrum $\sigma(A_h)$ inside. This completes the proof. $\qquad\square$

From here on, in this section we suppose condition (16) is satisfied. Also, assume that the sourcewise representation (5) is fulfilled. Let us estimate the norm $\|x_\alpha^{(h,\delta)} - x^*\|_X, \alpha \in (0, \alpha_0]$. We have

$$
\begin{aligned}
x_\alpha^{(h,\delta)} - x^* = & (E - \Theta(A_h, \alpha)A_h)(\xi - x^*) + \\
& + \Theta(A_h, \alpha)[(f_\delta - f) + (A - A_h)x^*].
\end{aligned}
\tag{17}
$$

By (5), (13), and (17), it follows that

$$
\begin{aligned}
\|x_\alpha^{(h,\delta)} - x^*\|_X \leq & \|\Theta(A_h, \alpha)\|_{L(X)}(\delta + \|x^*\|_X h) + \\
& + \|(E - \Theta(A_h, \alpha)A_h)A^p v\|_X.
\end{aligned}
\tag{18}
$$

Consider the individual summands in (18).

Combining (13), (16), and (1.1.11), we obtain

$$
\|R(\lambda, A_h)\|_{L(X)} \leq
$$

$$
\leq \|R(\lambda, A)\|_{L(X)} \sum_{k=0}^{\infty} \|(A_h - A)R(\lambda, A)\|_{L(X)}^k \leq
$$

$$
\leq \frac{r_0}{(1 - \omega_0)|\lambda|} = \frac{c_7}{|\lambda|} \quad \forall \lambda \in \Gamma_\alpha \quad \forall \alpha \in (0, \alpha_0].
$$

Consequently,

$$
\begin{aligned}
\|\Theta(A_h, \alpha)\|_{L(X)} \leq & \frac{1}{2\pi} \int_{\Gamma_\alpha} |\Theta(\lambda, \alpha)| \|R(\lambda, A_h)\|_{L(X)} |d\lambda| \leq \\
& \leq c_8 \int_{\Gamma_\alpha} \frac{|\Theta(\lambda, \alpha)|}{|\lambda|} |d\lambda| \quad \forall \alpha \in (0, \alpha_0].
\end{aligned}
\tag{19}
$$

In addition to Conditions 1.1, 2.6, and 2.7, suppose $\Theta(\lambda, \alpha)$ satisfies the following condition.

Condition 2.8. *For each $\alpha \in (0, \alpha_0]$,*

$$
\int_{\Gamma_\alpha} \frac{|\Theta(\lambda, \alpha)|}{|\lambda|} |d\lambda| \leq \frac{c_9}{\alpha},
$$

where the constant c_9 is independent of α.

According to (19) and Condition 2.8,

$$\|\Theta(A_h, \alpha)\|_{L(X)} \leq \frac{c_{10}}{\alpha} \quad \forall \alpha \in (0, \alpha_0]. \tag{20}$$

For the second summand in (18) we have

$$\|(E - \Theta(A_h, \alpha)A_h)A^p v\|_X \leq \|(E - \Theta(A, \alpha)A)A^p v\|_X + \\ + \|(\Theta(A_h, \alpha)A_h - \Theta(A, \alpha)A)A^p v\|_X. \tag{21}$$

Following the proof of Theorem 2.10 and taking into account (9), we get

$$\|(E - \Theta(A, \alpha)A)A^p v\|_X \leq c_{11}\|v\|_X \alpha^p \quad \forall \alpha \in (0, \alpha_0]. \tag{22}$$

Let us set $m = [p]$, $\mu = \{p\}$, and $\varepsilon = c_{12}\alpha$. Observe that

$$\|(\Theta(A_h, \alpha)A_h - \Theta(A, \alpha)A)A^p v\|_X \leq \\ \leq \|(\Theta(A_h, \alpha)A_h - \Theta(A, \alpha)A)A^m(A + \varepsilon E)^\mu v\|_X + \\ + \|(\Theta(A_h, \alpha)A_h - \Theta(A, \alpha)A)A^m[(A + \varepsilon E)^\mu - A^\mu]v\|_X. \tag{23}$$

For the first term in the right part of (23) we have

$$\|(\Theta(A_h, \alpha)A_h - \Theta(A, \alpha)A)A^m(A + \varepsilon E)^\mu v\|_X \leq \\ \leq \frac{1}{2\pi}\|v\|_X \int_{\Gamma_\alpha} |1 - \Theta(\lambda, \alpha)\lambda| \cdot \\ \cdot \|(R(\lambda, A) - R(\lambda, A_h))A^m(A + \varepsilon E)^\mu\|_{L(X)}|d\lambda|.$$

From (1.1.12) and (16) it follows that

$$\|(R(\lambda, A) - R(\lambda, A_h))A^m(A + \varepsilon E)^\mu\|_{L(X)} \leq \\ \leq \sum_{k=1}^{\infty}(\|R(\lambda, A)\|_{L(X)}h)^k\|R(\lambda, A)A^m\|_{L(X)}\|(A + \varepsilon E)^\mu\|_{L(X)} \leq \\ \leq \frac{c_{13}h}{(1 - \omega_0)|\lambda|}\|R(\lambda, A)A^m\|_{L(X)} \quad \forall \lambda \in \Gamma_\alpha \quad \forall \alpha \in (0, \alpha_0].$$

Using the identity

$$R(\lambda, A)A = -E + \lambda R(\lambda, A),$$

we obtain

$$\|R(\lambda, A)A^m\|_{L(X)} \leq c_{14}\nu_p(|\lambda|),$$

where, by definition,

$$\nu_p(t) = \begin{cases} t^{-1}, & [p] = 0, \\ 1, & [p] > 0, \end{cases} \quad t > 0.$$

Therefore,

$$\|(\Theta(A_h,\alpha)A_h - \Theta(A,\alpha)A)A^m(A+\varepsilon E)^\mu v\|_X \le$$
$$\le c_{15}\|v\|_X \nu_p(\alpha)h \quad \forall \alpha \in (0,\alpha_0]. \tag{24}$$

In the similar way, from Condition 1.1 and Lemma 1.1 we deduce

$$\|(\Theta(A_h,\alpha)A_h - \Theta(A,\alpha)A)A^m[(A+\varepsilon E)^\mu - A^\mu]v\|_X \le$$
$$\le c_{16}\|v\|_X \alpha^\mu \int_{\Gamma_\alpha} |1 - \Theta(\lambda,\alpha)\lambda| \|(R(\lambda,A) - R(\lambda,A_h))A^m\|_{L(X)} |d\lambda| \le$$
$$\le c_{17}\|v\|_X \nu_p(\alpha)\alpha^\mu h. \tag{25}$$

At last, the combination of (18) with (20)–(25) gives

$$\|x_\alpha^{(h,\delta)} - x^*\|_X \le c_{18}\left(\frac{h+\delta}{\alpha} + (\nu_p(\alpha)h + \alpha^p)\|v\|_X\right). \tag{26}$$

We proved the following statement.

Theorem 2.11. *Let Conditions 1.1 and 2.6–2.8 be satisfied, and let the initial discrepancy $x^* - \xi$ have the sourcewise representation (5). Suppose the regularization parameter $\alpha = \alpha(h,\delta)$ is connected with the error levels h,δ such that (16) is fulfilled and*

$$\alpha(h,\delta) \in (0,\alpha_0], \quad \lim_{h,\delta\to 0} \alpha(h,\delta) = \lim_{h,\delta\to 0} \frac{h+\delta}{\alpha(h,\delta)} = 0. \tag{27}$$

Then equality (15) and estimate (26) with $\alpha = \alpha(h,\delta)$ are valid.

Theorem 2.11 establishes conditions, sufficient for the mapping

$$\mathfrak{R}_{\alpha(h,\delta)} : \mathfrak{F} \to X$$

to be a regularization algorithm for equation (1).

Let us now analyze a possibility to relax condition (5). Instead of (5) assume that for some $p > 0$ there exist elements $\tilde{v}, \tilde{w} \in X$ such that

$$x^* - \xi = A^p\tilde{v} + \tilde{w}, \quad \|\tilde{w}\|_X \le \Delta. \tag{28}$$

Here Δ has the meaning of an error measure of the true sourcewise representation (5). It follows from (14) that

$$\|x_\alpha^{(h,\delta)} - x^*\|_X \leq c_{19}\|\Theta(A_h, \alpha)\|_{L(X)}(h + \delta) +$$
$$+ \|(E - \Theta(A_h, \alpha)A_h)A^p\tilde{v}\|_X + \|E - \Theta(A_h, \alpha)A_h\|_{L(X)}\Delta.$$

Repeating with evident changes the proof of Theorem 2.11, we obtain

$$\|x_\alpha^{(h,\delta)} - x^*\|_X \leq c_{20}\left(\frac{h + \delta}{\alpha} + (\nu_p(\alpha)h + \alpha^p)\|\tilde{v}\|_X + \Delta\right). \tag{29}$$

Therefore the following theorem is true.

Theorem 2.12. *Let Conditions 1.1 and 2.6–2.8 be satisfied, and the discrepancy $x^* - \xi$ possess the approximate sourcewise representation (28). Suppose the regularization parameter $\alpha = \alpha(h, \delta)$ is coordinated with the error levels h, δ such that (16) and (27) are fulfilled. Then estimate (29) is valid with $\alpha = \alpha(h, \delta)$. Moreover,*

$$\limsup_{h,\delta\to 0} \|x_{\alpha(h,\delta)}^{(h,\delta)} - x^*\|_X \leq c_{20}\Delta. \tag{30}$$

Inequality (30) implies that the algorithm (14) is stable with respect to perturbations of the sourcewise representation (5).

Theorem 2.12 allows to establish the convergence of $x_{\alpha(h,\delta)}^{(h,\delta)}$ as $h, \delta \to 0$ without any assumptions concerning a sourcewise representation of $x^* - \xi$. Let X be a reflexive Banach space. According to [38], Condition 1.1 implies that X can be presented as the direct sum

$$X = \overline{R(A)} \oplus N(A).$$

Hence each element $x \in X$ is uniquely representable in the form $x = y + z$, where $y \in \overline{R(A)}$ and $z \in N(A)$. Therefore for each $\xi \in X$ there exists a unique element $x^* \in X_*$ with $x^* - \xi \in \overline{R(A)}$. Consequently for arbitrarily small $\varepsilon > 0$ we can find v_ε and w_ε such that

$$x^* - \xi = Av_\varepsilon + w_\varepsilon, \quad \|w_\varepsilon\|_X \leq \varepsilon.$$

Putting into (29) $p = 1$ and $\Delta = \varepsilon$, we conclude that

$$\|x_\alpha^{(h,\delta)} - x^*\|_X \leq c_{20}\left(\frac{h + \delta}{\alpha} + (h + \alpha)\|v_\varepsilon\|_X + \varepsilon\right).$$

Now, by (27),

$$\limsup_{h,\delta \to 0} \|x_{\alpha(h,\delta)}^{(h,\delta)} - x^*\|_X \le c_{20}\varepsilon.$$

Since $\varepsilon > 0$ can be chosen arbitrarily small, we get the following statement.

Theorem 2.13. *Let X be a reflexive Banach space and $x^* \in X_*$ satisfy $x^* - \xi \in \overline{R(A)}$. Suppose Conditions 1.1, 2.6, 2.8 and inequality (9) with $p = 1$ are satisfied. Then (15) is valid.*

In case where $h = \delta = 0$ and $\alpha(h, \delta) = \alpha \to 0$, the assertion of Theorem 2.13 takes the form akin to (1.2.6), if we define $G : \mathfrak{F} \to X$ by the equalities

$$D(G) = \{(A, f) \in \mathfrak{F} : f \in R(A)\}, \quad G(A, f) = A^{-1}f \cap \{\xi + \overline{R(A)}\}.$$

It is significant that (1.2.6) is true not on the all of $D(G)$ but on its proper subset defined by Condition 1.1 on operator A. The equality (1.2.5) is true just on the same subset of $D(G)$. At the end of Section 1.2 we have already pointed out the necessity and utility of narrowing the class of elements $F \in D(G)$, for which (1.2.5) and (1.2.6) must be fulfilled, from $D(G)$ to its appropriate subset.

The presented scheme in combination with theory of projection methods [77, 94] allows to construct in a uniform style various finite–dimensional regularization algorithms for equations (1) with compact operators $A \in L(X)$. By finite–dimensional we mean regularization algorithms involving finite–dimensional approximations of operators A_h and elements f_δ. Let us choose two families of finite–dimensional projectors $P_l, Q_m \in L(X)$; $l, m \in \mathbf{N}$. Denote

$$\mathcal{L}_l = R(P_l), \quad \mathcal{M}_m = R(Q_m)$$

and suppose the projectors P_l, Q_m satisfy the following conditions.

Condition 2.9. *For all $x \in X$,*

$$\lim_{l \to \infty} \|x - P_l x\|_X = 0, \quad \lim_{m \to \infty} \|x - Q_m x\|_X = 0. \tag{31}$$

Condition 2.10.

$$\|A(P_l - E)\|_{L(X)} \le \zeta_l, \quad \|(Q_m - E)A\|_{L(X)} \le \eta_m, \tag{32}$$

$$\lim_{l \to \infty} \zeta_l = \lim_{m \to \infty} \eta_m = 0. \tag{33}$$

Conditions (31) mean that the families of subspaces $\{\mathcal{L}_l\}_{l \in \mathbf{N}}$ and $\{\mathcal{M}_m\}_{m \in \mathbf{N}}$ are dense in X. Assumptions of this kind are typical for the theory of projecting methods. Using Banach and Steinhaus' theorem, from (31) we conclude that the norms of operators P_l and Q_m are uniformly bounded:

$$\sup_{l \in \mathbf{N}} \|P_l\|_{L(X)} < \infty, \quad \sup_{m \in \mathbf{N}} \|Q_m\|_{L(X)} < \infty. \tag{34}$$

Conditions (32) and (33) immediately imply that A is a compact operator from X into X. At the same time, compactness of approximate operators A_h is not assumed.

According to the abstract scheme of constructing projection methods for equations (1) with continuously invertible operators, we associate with (1) the family of finite–dimensional equations

$$Q_m A_h P_l x = Q_m f_\delta, \quad x \in X. \tag{35}$$

The equality of dimensions $\dim \mathcal{L}_l = \dim \mathcal{M}_m$ is not assumed here. Let us remark that in our context this equality can't ensure a unique solvability of (35), on the contrary with the classical (regular) case where $A^{-1} \in L(X)$. In other words, in our situation the approximating equation (35) remains ill–posed for all sufficiently large numbers l, m. Applying to equation (35) the regularization procedure (14), we obtain the following finite–dimensional scheme for solving (1):

$$x_{\alpha(h,\delta)}^{(h,\delta)} = \Re_{\alpha(h,\delta)}(A_h, f_\delta) =$$

$$= [E - \Theta(Q_{m(\alpha(h,\delta))} A_h P_{l(\alpha(h,\delta))}, \alpha(h,\delta)) Q_{m(\alpha(h,\delta))} A_h P_{l(\alpha(h,\delta))}]\xi +$$

$$+ \Theta(Q_{m(\alpha(h,\delta))} A_h P_{l(\alpha(h,\delta))}, \alpha(h,\delta)) Q_{m(\alpha(h,\delta))} f_\delta. \tag{36}$$

By (13), (32), and (34), it follows that

$$\|Q_m A_h P_l - A\|_{L(X)} \le c_{21}(h + \zeta_l + \eta_m),$$

$$\|Q_m f_\delta - f\|_X \le \|Q_m f_\delta - Q_m f\|_X + \|Q_m A x^* - A x^*\|_X \le$$
$$\le c_{21}(\delta + \eta_m).$$

The next result is an immediate consequence of Theorems 2.11 and 2.13.

Theorem 2.14. *1) Let A be a compact operator. Suppose conditions 1.1 and 2.6–2.10 are satisfied, the initial discrepancy $x^* - \xi$ has the sourcewise representation (5), and the regularization parameter $\alpha = \alpha(h,\delta)$ and the numbers $l = l(\alpha(h,\delta)), m = m(\alpha(h,\delta))$ in (36) are connected with the error levels h, δ such that (16) is valid and*

$$\lim_{h,\delta \to 0} l(\alpha(h,\delta)) = \lim_{h,\delta \to 0} m(\alpha(h,\delta)) = \infty; \quad \alpha(h,\delta) \in (0, \alpha_0],$$

$$\lim_{h,\delta \to 0} \alpha(h,\delta) = \lim_{h,\delta \to 0} \frac{h + \delta + \zeta_{l(\alpha(h,\delta))} + \eta_{m(\alpha(h,\delta))}}{\alpha(h,\delta)} = 0.$$

Then we have

$$\|x_{\alpha(h,\delta)}^{(h,\delta)} - x^*\|_X \le$$

$$\le c_{22}\left(\frac{h + \delta + \zeta_{l(\alpha(h,\delta))} + \eta_{m(\alpha(h,\delta))}}{\alpha(h,\delta)} + \left(\nu_p(\alpha(h,\delta))h + \alpha^p(h,\delta)\right)\|v\|_X\right).$$

In particular,

$$\lim_{h,\delta\to 0}\|x_{\alpha(h,\delta)}^{(h,\delta)} - x^*\|_X = 0. \tag{37}$$

2) Let A be a compact operator, X a reflexive Banach space, and $x^* \in X_*$ satisfy $x^* - \xi \in \overline{R(A)}$. Suppose Conditions 1.1, 2.6, 2.8–2.10 and inequality (9) with $p = 1$ are satisfied. Then (37) is valid.

Remark 2.7. The assertion of Theorem 2.14 remains valid in the case where $P_l = E$. In this event, the first conditions from (31) and (32) are trivially fulfilled with $\zeta_l = 0$.

Equality (37) implies that the operator $\mathfrak{R}_{\alpha(h,\delta)} : \mathfrak{F} \to X$ specified by (36) defines a finite–dimensional regularization algorithm for the equation (1).

We shall postpone verification of Conditions 2.6–2.8 for concrete generating functions $\Theta(\lambda, \alpha)$ until Section 2.3. Nevertheless, in the succeeding examples the scheme (36) will be specified for some of these functions.

Example 2.10. In the case of generating function (1.15), $x_\alpha^{(h,\delta)}$ with $\alpha = \alpha(h,\delta), l = l(\alpha(h,\delta)), m = m(\alpha(h,\delta))$ can be obtained from the equation

$$(Q_m A_h P_l + \alpha E)y_\alpha^{(h,\delta)} = Q_m(f_\delta - A_h P_l\xi), \quad y_\alpha^{(h,\delta)} \in \mathcal{M}_m, \tag{38}$$

where $y_\alpha^{(h,\delta)} = x_\alpha^{(h,\delta)} - \xi$. Let, for simplicity, $\dim\mathcal{M}_m = m$. By $\{e_i\}_{i=1}^m$ we denote a basis for the subspace $\mathcal{M}_m$. Suppose

$$y_\alpha^{(h,\delta)} = \sum_{i=1}^m c_i e_i;$$

then from (38) and the expansions

$$Q_m(f_\delta - A_h P_l\xi) = \sum_{i=1}^m \varphi_i e_i; \quad Q_m A_h P_l e_i = \sum_{j=1}^m a_{ij}e_j,$$

$$i = 1, 2, \ldots, m$$

we obtain the following linear system for unknown coefficients $c_i, i = 1, \ldots, m$:

$$\sum_{i=1}^m a_{ij}c_i + \alpha c_j = \varphi_j, \quad j = 1, 2, \ldots, m. \tag{39}$$

If conditions of Theorem 2.14 are satisfied, then by Lemma 2.3, $-\alpha \in \rho(Q_m A_h P_l)$. Therefore the system (39) has a unique solution.

Example 2.11. The scheme (36), (1.23) can be reduced to solving the Cauchy problem

$$\frac{du}{dt} + Q_m A_h P_l u = Q_m f_\delta, \quad u(0) = \xi,$$

so that

$$x_\alpha^{(h,\delta)} = u\left(\frac{1}{\alpha}\right).$$

Setting $u(t) = v(t) + \xi$, we get

$$\frac{dv}{dt} + Q_m A_h P_l v = Q_m(f_\delta - A_h P_l \xi), \quad v(0) = 0.$$

Each solution of this problem can be written as

$$v(t) = \sum_{i=1}^{m} c_i(t) e_i.$$

For the unknown functions $c_i(t), i = 1, \ldots, m$ we have the system of linear ordinary differential equations with constant coefficients

$$\frac{dc_j}{dt} + \sum_{i=1}^{m} a_{ij} c_i = \varphi_j, \quad c_j(0) = 0, \quad j = 1, \ldots, m.$$

Finally, we obtain

$$x_\alpha^{(h,\delta)} = \sum_{i=1}^{m} c_i\left(\frac{1}{\alpha}\right) e_i + \xi.$$

Example 2.12. The explicit iterative scheme (1.31) is equivalent to the iteration

$$x^{(k+1)} = x^{(k)} - \mu_0(Q_m A_h P_l x^{(k)} - Q_m f_\delta), \quad k = 0, \ldots, n-1$$

with $x^{(0)} = \xi$ and sufficiently small $\mu_0 > 0$. Here $\alpha = 1/n$ and

$$x_\alpha^{(h,\delta)} = x^{(n)}.$$

Using the representation

$$x^{(k)} = y^{(k)} + \xi, \quad y^{(k)} = \sum_{i=1}^{m} c_i^{(k)} e_i,$$

we can rewrite this iterative process as

$$c_j^{(k+1)} = c_j^{(k)} - \mu_0 \left(\sum_{i=1}^m a_{ij} c_i^{(k)} - \varphi_j \right), \quad c_j^{(0)} = 0, \quad j = 1, \ldots, m.$$

Therefore $x_\alpha^{(h,\delta)}$ takes the form

$$x_\alpha^{(h,\delta)} = \sum_{i=1}^m c_i^{(n)} e_i + \xi.$$

In case of $P_l = E$ the computational schemes of Examples 2.10–2.12 are simplified. Let us make more precise the coordination conditions obtained in Theorem 2.14.

Example 2.13. Consider the integral equation

$$(Ax)(t) \equiv \int_0^{2\pi} K(t,s)x(s)ds = f(t), \quad t \in (0, 2\pi)$$

and put $X = L_p(0, 2\pi), p \in (1, \infty)$. Suppose

$$K, \quad \frac{\partial K}{\partial s}, \quad \frac{\partial K}{\partial t} \in C([0, 2\pi] \times [0, 2\pi]);$$

then A is a compact operator in $L_p(0, 2\pi)$. As P_l and Q_m we take the Fourier projectors onto the subspaces of trigonometric polynomials:

$$(P_m x)(t) = (Q_m x)(t) = \frac{1}{\sqrt{2\pi}} \sum_{|k| \le m} c_k(x) \exp(ikt), \quad t \in (0, 2\pi), \quad m \ge 1,$$

where

$$c_k(x) = \frac{1}{\sqrt{2\pi}} \int_0^{2\pi} \exp(-ikt)x(t)dt, \quad k \in \mathbf{Z}.$$

Then (31) is true; inequalities (34) follow directly by Riesz's theorem ([41]). Let us derive explicit expressions for ζ_l and η_m in (32) and (33). We have

$$\|A(P_l - E)\|_{L(L_p(0,2\pi))} =$$

$$= \sup_{\|x\|_{L_p(0,2\pi)} \le 1} \left(\int_0^{2\pi} \left| \frac{1}{\sqrt{2\pi}} \sum_{|k|>l} c_k(x) \int_0^{2\pi} K(t,s) \exp(iks)ds \right|^p dt \right)^{1/p} =$$

$$= \sup_{\|x\|_{L_p(0,2\pi)}\leq 1} \left(\int_0^{2\pi} \left| \sum_{|k|>l} c_k(x)c_{-k}(K(t,\cdot)) \right|^p dt \right)^{1/p}, \quad \frac{1}{p} + \frac{1}{q} = 1. \quad (40)$$

Recall the estimate ([149])

$$\left| \sum_{|k|>l} c_k(x)c_{-k}(K(t,\cdot)) \right| \leq c_{23}E_l(x)_p E_l(K(t,\cdot))_q, \quad t \in (0,2\pi), \quad (41)$$

where

$$E_l(x)_p = \inf_{y\in L_l} \|x - y\|_{L_p(0,2\pi)}.$$

Obviously,

$$E_l(x)_p \leq \|x\|_{L_p(0,2\pi)} \quad \forall x \in L_p(0,2\pi). \quad (42)$$

The following estimate is well known:

$$E_l(K(t,\cdot))_q \leq \frac{c_{24}}{l} \left\| \frac{\partial K}{\partial s}(t,\cdot) \right\|_{L_q(0,2\pi)}, \quad t \in (0,2\pi). \quad (43)$$

Combining (40)–(43), we get

$$\|A(P_l - E)\|_{L(L_p(0,2\pi))} \leq \frac{c_{25}}{l} \left(\int_0^{2\pi} \left\| \frac{\partial K}{\partial s}(t,\cdot) \right\|_{L_q(0,2\pi)}^p dt \right)^{1/p}.$$

Thus in (32) we can set

$$\zeta_l = \frac{c_{26}}{l}. \quad (44)$$

Analogously to (43),

$$\|(Q_m - E)A\|_{L(L_p(0,2\pi))} =$$

$$= \sup_{\|x\|_{L_p(0,2\pi)}\leq 1} \|(Q_m - E)Ax\|_{L_p(0,2\pi)} \leq$$

$$\leq \frac{c_{26}}{m} \sup_{\|x\|_{L_p(0,2\pi)}\leq 1} \left\| \int_0^{2\pi} \frac{\partial K}{\partial t}(\cdot, s)x(s)ds \right\|_{L_p(0,2\pi)} \leq$$

$$\leq \frac{c_{26}}{m} \left(\int_0^{2\pi} \left\| \frac{\partial K}{\partial t}(t,\cdot) \right\|_{L_q(0,2\pi)}^p dt \right)^{1/p}.$$

Consequently we can take

$$\eta_m = \frac{c_{27}}{m}. \quad (45)$$

From (44) and (45) it follows that the coordination conditions on $l(h, \delta), m(h, \delta)$, and $\alpha(h, \delta)$ given in Theorem 2.14 are satisfied if

$$\lim_{h,\delta \to 0} l(\alpha(h, \delta))\alpha(h, \delta) = \lim_{h,\delta \to 0} m(\alpha(h, \delta))\alpha(h, \delta) = \infty.$$

Example 2.14. Consider the equation

$$(Ax)(t) \equiv \int_0^1 K(t, s)x(s)ds = f(t), \quad t \in [0, 1]$$

in the space $X = C[0, 1]$. Suppose

$$K, \quad \frac{\partial K}{\partial t} \in C([0, 1] \times [0, 1]);$$

then A is a compact operator in $C[0, 1]$. We set $P_l = E$, $\zeta_l = 0$ and define Q_m as an operator of linear spline interpolation on the segment $[0, 1]$:

$$(Q_m x)(t) = \sum_{i=0}^m x(t_i)e_i(t), \quad t \in [0, 1]; \quad m \geq 1.$$

Here $t_i = i/m, i = 0, 1, \ldots, m$. The basis functions $e_i(t), i = 1, \ldots, m$ are linear on the segments $[t_{i-1}, t_i], [t_i, t_{i+1}]$ and $e_i(t) = 0, t \in [0, 1] \backslash [t_{i-1}, t_{i+1}]$; $x(t_{i-1}) = x(t_{i+1}) = 0, x(t_i) = 1$; the functions $e_0(t)$ and $e_m(t)$ are linear on $[t_0, t_1]$ and $[t_{m-1}, t_m]$ respectively, and $e_0(t), e_m(t)$ are equal to zero outside these segments; $e_0(t_0) = e_m(t_m) = 1, e_0(t_1) = e_m(t_{m-1}) = 0$. By

$$\omega(x, \tau) = \max\{|x(t') - x(t'')| : t', t'' \in [0, 1], |t' - t''| \leq \tau\}$$

we denote the modulus of continuity of a function $x \in C[0, 1]$. It is readily seen that

$$\|x - Q_m x\|_{C[0,1]} \leq \omega\left(x, \frac{1}{m}\right), \quad x \in C[0, 1]$$

and $\|Q_m\|_{L(C[0,1])} = 1$; therefore (31) is true. Since

$$\|(Q_m - E)A\|_{L(C[0,1])} \leq$$

$$\leq \sup_{\|x\|_{C[0,1]} \leq 1} \omega\left(\int_0^1 K(\cdot, s)x(s)ds, \frac{1}{m}\right) \leq$$

$$\leq \frac{1}{m} \max_{t,s \in [0,1]} \left|\frac{\partial K}{\partial t}(t, s)\right|,$$

we can take

$$\eta_m = \frac{c_{28}}{m}.$$

2.3 Necessity of Sourcewise Representation for Rate of Convergence Estimates in Banach Space

The aim of this section is to analyze the necessity of representation (2.5) for the power rate of convergence estimate (2.12). For simplicity, we consider only the case where $q = 0$. Below, it will be shown that (2.5) approaches a necessary condition for (2.12) in the sense that (2.12) with $q = 0$ yields

$$x^* - \xi \in R(A^{p-\varepsilon}) \quad \forall \varepsilon \in (0, p), \tag{1}$$

provided $\Theta(\lambda, \alpha)$ satisfies several additional conditions (compare with Theorems 2.4 and 2.5).

It is more convenient for our purposes to deal with functions $\Theta(\lambda, \alpha)$ defined for all positive values of the regularization parameter α, rather than for $\alpha \in (0, \alpha_0]$. Let Condition 1.1 and Condition 2.6 with $\alpha \in (0, \alpha_0]$ replaced by $\alpha \in (0, \infty)$ be satisfied. Consequently the operator

$$\Theta(A, \alpha) = \frac{1}{2\pi i} \int_{\gamma_\alpha} \Theta(\lambda, \alpha) R(\lambda, A) d\lambda \tag{2}$$

is well defined for each $\alpha \in (0, \infty)$. Recall that γ_α is the boundary of the set $K_\alpha(R_0, d_0, \varphi_0)$ defined by (2.3). Here $R_0 > \|A\|_{L(X)}$; d_0 and φ_0 are specified in Condition 2.6 and Condition 1.1 respectively.

Since $Ax^* = f$, (2.2) and (2) yield

$$x_\alpha - x^* = (E - \Theta(A, \alpha)A)(\xi - x^*) =$$
$$= \frac{1}{2\pi i} \int_{\gamma_\alpha} (1 - \Theta(\lambda, \alpha)\lambda) R(\lambda, A)(\xi - x^*) d\lambda, \quad \alpha \in (0, \infty). \tag{3}$$

We denote

$$D(R_0, d_0, \varphi_0) = \{(\lambda, \alpha) \colon \lambda \in K_\alpha(R_0, d_0, \varphi_0), \alpha \in (0, \infty)\}.$$

Suppose $\Theta(\lambda, \alpha)$ satisfies the following technical conditions.

Condition 2.11. *There exists a constant* $\tau_0 = \tau_0(\alpha_0) > 0$ *such that for all* $\tau \in (0, \tau_0]$,

$$\sup_{\alpha \in [\alpha_0, \infty)} \left(\alpha^{-\tau} \int_{\gamma_\alpha} \frac{|1 - \Theta(\lambda, \alpha)\lambda|}{|\lambda|} |d\lambda| \right) < \infty. \tag{4}$$

Condition 2.12. *There exists a constant $\varepsilon^{(0)} > 0$ such that the function $\Theta(\lambda + \varepsilon, \alpha)$ is continuous in (λ, α) on $D(R_0, d_0, \varphi_0)$ whatever $\varepsilon \in (0, \varepsilon^{(0)}]$.*

Condition 2.11 can be considered as a complementary one with respect to Condition 2.7.

In the sequel we shall impose on $\Theta(\lambda, \alpha)$ several additional restrictions. Suppose (2.12) is valid with $q = 0$ and $p \in (0, p_0]$, i.e.,

$$\|x_\alpha - x^*\|_X \le l\alpha^p \quad \forall \alpha \in (0, \alpha_0] \quad (l > 0). \tag{5}$$

Recall that the value p_0 is defined by Condition 2.7. Using Condition 2.12 and formula (2), we conclude that the mapping $\alpha \to x_\alpha, \alpha \in (0, \infty)$ is continuous from $\mathbf{R}$ into X. Besides, in view of (1.1.8) and (3)–(5), for all

$$0 < \kappa < \min\left\{\frac{2}{3}p, 2\tau_0\right\}$$

we have

$$\int_0^\infty \alpha^{-p-1+\kappa}\|x_\alpha - x^*\|_X d\alpha = \int_0^{\alpha_0} \alpha^{-p-1+\kappa}\|x_\alpha - x^*\|_X d\alpha +$$

$$+ \int_{\alpha_0}^\infty \alpha^{-p-1+\kappa}\|x_\alpha - x^*\|_X d\alpha \le l \int_0^{\alpha_0} \alpha^{-1+\kappa} d\alpha +$$

$$+ \int_{\alpha_0}^\infty \alpha^{-p-1+\kappa}\|E - \Theta(A, \alpha)A\|_{L(X)}\|x^* - \xi\|_X d\alpha \le l\kappa^{-1}\alpha_0^\kappa + \tag{6}$$

$$+ c_1 \int_{\alpha_0}^\infty \alpha^{-p-1+\frac{3}{2}\kappa}\left(\alpha^{-\frac{\kappa}{2}} \int_{\gamma_\alpha} \frac{|1 - \Theta(\lambda, \alpha)\lambda|}{|\lambda|}|d\lambda|\right) d\alpha < \infty.$$

It now follows that the element

$$w_\kappa = \int_0^\infty \alpha^{-p-1+\kappa}(x^* - x_\alpha)d\alpha =$$

$$\tag{7}$$

$$= \int_0^\infty \alpha^{-p-1+\kappa}(E - \Theta(A, \alpha)A)(x^* - \xi)d\alpha \in X$$

is well defined. The integrals in (7) are understood in the sense of Bochner.

Let us outline in short the plan of subsequent considerations. Along with the original operator A we define its regularization $A_\varepsilon = A + \varepsilon E, \varepsilon > 0$ and introduce the element

$$u_\kappa^{(\varepsilon)} = \int_0^\infty \alpha^{-p-1+\kappa}(\Theta(A,\alpha)A - \Theta(A_\varepsilon,\alpha)A_\varepsilon)(x^* - \xi)d\alpha. \qquad (8)$$

First, we intend to prove that integral (8) exists and to establish an upper estimate for $\|u_\kappa^{(\varepsilon)}\|_X$. We shall prove simultaneously that the integral

$$\int_0^\infty \alpha^{-p-1+\kappa}(E - \Theta(A_\varepsilon,\alpha)A_\varepsilon)(x^* - \xi)d\alpha = u_\kappa^{(\varepsilon)} + w_\kappa = w_\kappa^{(\varepsilon)} \qquad (9)$$

exists and derive an estimate for the norm $\|w_\kappa^{(\varepsilon)} - w_\kappa\|_X = \|u_\kappa^{(\varepsilon)}\|_X$. Next, by direct calculations we shall obtain the equality

$$A_\varepsilon^{p-\kappa} w_\kappa^{(\varepsilon)} = C(p,\kappa)(x^* - \xi) \qquad (10)$$

with an appropriate constant $C(p,\kappa)$. At last, using the mentioned estimate, we shall prove that $A_\varepsilon^{p-\kappa} w_\kappa^{(\varepsilon)} = A^{p-\kappa} w_\kappa$ for sufficiently small $\varepsilon = \varepsilon_n \to 0$. This immediately yields the required inclusion (1).

To implement the presented scheme, we primarily suppose the function $\Theta(\lambda,\alpha)$ satisfies the following condition.

Condition 2.13. *For all $\alpha \in (0,\infty)$,*

$$1 - \Theta(\lambda,\alpha)\lambda \neq 0 \quad \forall \lambda \in K_\alpha(R_0,d_0,\varphi_0).$$

By Condition 2.13 and the spectral mapping theorem, the operator $E - \Theta(A,\alpha)A$ is continuously invertible for all $\alpha \in (0,\infty)$. Let us denote

$$\psi(\lambda,\alpha,\varepsilon) = \frac{\Theta(\lambda,\alpha)\lambda - \Theta(\lambda+\varepsilon,\alpha)(\lambda+\varepsilon)}{1 - \Theta(\lambda,\alpha)\lambda}. \qquad (11)$$

Using Condition 2.13, we get

$$u_\kappa^{(\varepsilon)} = \int_0^\infty \alpha^{-p-1+\kappa}\psi(A,\alpha,\varepsilon)(E - \Theta(A,\alpha)A)(x^* - \xi)d\alpha. \qquad (12)$$

From Conditions 2.12 and 2.13 it follows that for each $\varepsilon \in (0,\varepsilon^{(0)}]$, the function $\psi(\lambda,\alpha,\varepsilon)$ is continuous in (λ,α) on $D(R_0,d_0,\varphi_0)$. Therefore the mapping

$\alpha \to \psi(A, \alpha, \varepsilon), \alpha \in (0, \infty)$ is continuous from $\mathbf{R}$ into $L(X)$. Since

$$\int\limits_0^\infty \alpha^{-p-1+\kappa} \|\psi(A,\alpha,\varepsilon)(E - \Theta(A,\alpha)A)(x^* - \xi)\|_X d\alpha \le$$

$$\le \int\limits_0^\infty \alpha^{-p-1+\kappa} \|\psi(A,\alpha,\varepsilon)\|_{L(X)} \|x_\alpha - x^*\|_X d\alpha,$$

to establish existence of integral (12), it suffices to prove that

$$\int\limits_0^\infty \alpha^{-p-1+\kappa} \|\psi(A,\alpha,\varepsilon)\|_{L(X)} \|x_\alpha - x^*\|_X d\alpha < \infty.$$

From (1.1.8) we get

$$\|\psi(A,\alpha,\varepsilon)\|_{L(X)} \le \frac{1}{2\pi} \int\limits_{\gamma_\alpha} |\psi(\lambda,\alpha,\varepsilon)| \|R(\lambda,A)\|_{L(X)} |d\lambda| \le$$

$$\le \frac{r_0}{2\pi} \int\limits_{\gamma_\alpha} \frac{|\psi(\lambda,\alpha,\varepsilon)|}{|\lambda|} |d\lambda|. \tag{13}$$

In connection with (13), let us introduce one more condition on $\Theta(\lambda, \alpha)$.

Condition 2.14. *There exist constants $\varepsilon^{(1)} \in (0, \varepsilon^{(0)}]$ and $s_0 > 0$ such that*
1) for all $\varepsilon \in (0, \varepsilon^{(1)}]$ and $\alpha \in (0, \infty)$,

$$\int\limits_{\gamma_\alpha} \frac{|\psi(\lambda,\alpha,\varepsilon)|}{|\lambda|} |d\lambda| = M(\alpha, \varepsilon) < \infty; \tag{14}$$

2) for all $\varepsilon \in (0, \varepsilon^{(1)}]$ and $s \in (0, s_0]$,

$$\sup_{\alpha \in (0,\alpha_0]} (\alpha^s M(\alpha, \varepsilon)) + \sup_{\alpha \in [\alpha_0, \infty)} (\alpha^{-s} M(\alpha, \varepsilon)) < \infty, \tag{15}$$

$$\sup_{\alpha \in (0,\alpha_0]} \left(\alpha^s \int\limits_0^{\varepsilon^{(1)}} \frac{M(\alpha, \varepsilon)}{\varepsilon} d\varepsilon \right) + \sup_{\alpha \in [\alpha_0, \infty)} \left(\alpha^{-s} \int\limits_0^{\varepsilon^{(1)}} \frac{M(\alpha, \varepsilon)}{\varepsilon} d\varepsilon \right) < \infty. \tag{16}$$

Since

$$\int\limits_0^\infty \alpha^{-p-1+\kappa} \|\psi(A,\alpha,\varepsilon)\|_{L(X)} \|x_\alpha - x^*\|_X d\alpha \le$$

$$\le \frac{r_0}{2\pi} \left(\int\limits_0^{\alpha_0} \alpha^{-p-1+\kappa} M(\alpha,\varepsilon) \|x_\alpha - x^*\|_X d\alpha + \right.$$

$$\left. + \int\limits_{\alpha_0}^\infty \alpha^{-p-1+\kappa} M(\alpha,\varepsilon) \|x_\alpha - x^*\|_X d\alpha \right) \le$$

$$\le \frac{r_0}{2\pi} \left(l \int\limits_0^{\alpha_0} \alpha^{-1+\frac{\kappa}{2}} (\alpha^{\frac{\kappa}{2}} M(\alpha,\varepsilon)) d\alpha + \right.$$

$$\left. + \int\limits_{\alpha_0}^\infty \alpha^{-p-1+3\kappa/2} (\alpha^{-\frac{\kappa}{2}} M(\alpha,\varepsilon)) \|x_\alpha - x^*\|_X d\alpha \right) < \infty,$$

estimates (4)–(6) and (13)–(15) imply that the integral (12) exists with an arbitrary $\varepsilon \in (0, \varepsilon^{(1)}]$ and $0 < \kappa < \min\{\frac{2}{3}p, 2\tau_0, 2s_0\}$. Therefore for each $\varepsilon \in (0, \varepsilon^{(1)}]$, the element $w_\kappa^{(\varepsilon)} \in X$ is well defined.

Using (8), (9), and (12)–(14), we obtain

$$\int\limits_0^{\varepsilon^{(1)}} \frac{\|w_\kappa^{(\varepsilon)} - w_\kappa\|_X}{\varepsilon} d\varepsilon \le$$

$$\le \int\limits_0^{\varepsilon^{(1)}} \int\limits_0^\infty \alpha^{-p-1+\kappa} \frac{\|\psi(A,\alpha,\varepsilon)\|_{L(X)}}{\varepsilon} \|x_\alpha - x^*\|_X d\alpha d\varepsilon \le$$

$$\le \frac{r_0}{2\pi} \left(\int\limits_0^{\alpha_0} \alpha^{-p-1+\frac{\kappa}{2}} \|x_\alpha - x^*\|_X \left(\alpha^{\frac{\kappa}{2}} \int\limits_0^{\varepsilon^{(1)}} \frac{M(\alpha,\varepsilon)}{\varepsilon} d\varepsilon \right) d\alpha + \right.$$

$$\left. + \int\limits_{\alpha_0}^\infty \alpha^{-p-1+\frac{3}{2}\kappa} \|x_\alpha - x^*\|_X \left(\alpha^{-\frac{\kappa}{2}} \int\limits_0^{\varepsilon^{(1)}} \frac{M(\alpha,\varepsilon)}{\varepsilon} d\varepsilon \right) d\alpha \right),$$

$$0 < \kappa < \min\left\{\frac{2}{3}p, 2\tau_0, 2s_0\right\}.$$

Now, by (5), (6), and (16),

$$\int_0^{\varepsilon^{(1)}} \frac{\|w_\kappa^{(\varepsilon)} - w_\kappa\|_X}{\varepsilon} d\varepsilon < \infty. \tag{17}$$

From (17) it follows that there exists a sequence $\{\varepsilon_n\}$ such that $\varepsilon_n > 0$, $\varepsilon_n \to 0$, and $\lim_{n\to\infty} \|w_\kappa^{(\varepsilon_n)} - w_\kappa\|_X = 0$. In fact, assume the converse. Then for some constants $c_2 > 0$ and $0 < \varepsilon^{(2)} \leq \varepsilon^{(1)}$ we have

$$\|w_\kappa^{(\varepsilon)} - w_\kappa\|_X \geq c_2 \quad \forall \varepsilon \in (0, \varepsilon^{(2)}].$$

Hence, contrary to (17),

$$\int_0^{\varepsilon^{(1)}} \frac{\|w_\kappa^{(\varepsilon)} - w_\kappa\|_X}{\varepsilon} d\varepsilon \geq \int_0^{\varepsilon^{(2)}} \frac{c_2}{\varepsilon} d\varepsilon = \infty.$$

This contradiction proves our assertion.

Let us now address to the proof of (10).

Suppose $\kappa > 0$ is sufficiently small. Pick $m \in \mathbf{N}$ such that $p - \kappa \in (m, m+1)$. According to (1.1.9),

$$A_\varepsilon^{p-\kappa} = \frac{(-1)^m \sin \pi(p-\kappa)}{\pi} \int_0^\infty t^{p-\kappa-m-1}(tE + A_\varepsilon)^{-1} A_\varepsilon^{m+1} dt.$$

From (9) it follows that

$$A_\varepsilon^{p-\kappa} w_\kappa^{(\varepsilon)} = \frac{(-1)^m \sin \pi(p-\kappa)}{\pi} \int_0^\infty \int_0^\infty t^{p-\kappa-m-1} \alpha^{-p-1+\kappa} \cdot$$
$$\cdot (tE + A_\varepsilon)^{-1} A_\varepsilon^{m+1}(E - \Theta(A_\varepsilon, \alpha) A_\varepsilon)(x^* - \xi) d\alpha dt. \tag{18}$$

Given $r, r_1, r_2 > 0$ and $z, z_1, z_2 \in \mathbf{C}\backslash\{0\}$ such that

$$r_1 \leq r_2, \quad \arg z_1 \leq \arg z_2,$$

we denote

$$\Gamma_r(z_1, z_2) = \{\zeta \in \mathbf{C} : |\zeta| = r, \arg z_1 \leq \arg \zeta \leq \arg z_2\},$$
$$\Gamma_{(r_1, r_2)}(z) = \{\zeta \in \mathbf{C} : r_1 \leq |\zeta| \leq r_2, \arg \zeta = \arg z\},$$
$$\Gamma^{(\varepsilon)}(\varphi) = \Gamma_{\frac{\varepsilon}{2}}(\exp(-i\varphi), \exp(i\varphi)) \cup \Gamma_{R_0}(\exp(-i\varphi), \exp(i\varphi)) \cup$$
$$\cup \Gamma_{(\frac{\varepsilon}{2}, R_0)}(\exp(i\varphi)) \cup \Gamma_{(\frac{\varepsilon}{2}, R_0)}(\exp(-i\varphi)). \tag{19}$$

Consider the function

$$\Upsilon(\lambda, t, \alpha) = (t + \lambda)^{-1}\lambda^{m+1}(1 - \Theta(\lambda, \alpha)\lambda)$$

and the operator

$$\Upsilon(A_\varepsilon, t, \alpha) = (tE + A_\varepsilon)^{-1}A_\varepsilon^{m+1}(E - \Theta(A_\varepsilon, \alpha)A_\varepsilon)$$

involved in (18).

Suppose $\varepsilon^{(3)}$ satisfies

$$0 < \varepsilon^{(3)} < \min\{\varepsilon^{(1)}, R_0 - \|A\|_{L(X)}\};$$

then the contour $\Gamma^{(\varepsilon)}(\varphi_0), \varepsilon \in (0, \varepsilon^{(3)}]$, where φ_0 is specified by Condition 1.1, surrounds the spectrum $\sigma(A_\varepsilon) = \{\lambda + \varepsilon\colon \lambda \in \sigma(A)\}$. Besides, $\sigma(A_\varepsilon)$ lies in a region of analyticity of the function $\Upsilon(\cdot, t, \alpha)$ whatever $t, \alpha \in (0, \infty)$. Hence for $\Upsilon(A_\varepsilon, t, \alpha)$ we can use representation (1.1.6) with $\Gamma = \Gamma^{(\varepsilon)}(\varphi_0)$. Combining (1.1.6) and (18), we conclude that

$$A_\varepsilon^{p-\kappa}w_\kappa^{(\varepsilon)} =$$

$$= D(p, \kappa) \int_0^\infty \int_0^\infty \int_{\Gamma^{(\varepsilon)}(\varphi_0)} \alpha^{-p-1+\kappa}t^{p-\kappa-m-1}\Upsilon(\lambda, t, \alpha)\cdot \qquad (20)$$

$$\cdot R(\lambda, A_\varepsilon)(x^* - \xi)d\lambda d\alpha dt,$$

where

$$D(p, \kappa) = \frac{(-1)^m \sin \pi(p - \kappa)}{2\pi^2 i}.$$

Now let us show that the order of integrals in (20) can be changed. By Fubini's theorem [58], it suffices to prove that

$$J(\varepsilon) \equiv \int_{\Gamma^{(\varepsilon)}(\varphi_0)} \left(\int_0^\infty \int_0^\infty \alpha^{-p-1+\kappa}t^{p-\kappa-m-1}|\Upsilon(\lambda, t, \alpha)|\cdot \right.$$

$$\left. \cdot \|R(\lambda, A_\varepsilon)(x^* - \xi)\|_X d\alpha dt \right) |d\lambda| < \infty.$$

Since for each $\varepsilon \in (0, \varepsilon^{(3)}]$,

$$E(\varepsilon) \equiv \sup_{\lambda \in \Gamma^{(\varepsilon)}(\varphi_0)} \|R(\lambda, A_\varepsilon)(x^* - \xi)\|_X < \infty,$$

we have

$$J(\varepsilon) \le E(\varepsilon) \int\limits_{\Gamma^{(\varepsilon)}(\varphi_0)} |\lambda|^{m+1} \left(\int\limits_0^\infty \alpha^{-p-1+\kappa} |1 - \Theta(\lambda, \alpha)\lambda| d\alpha \right) \cdot$$

$$\cdot \left(\int\limits_0^\infty t^{p-\kappa-m-1} |t + \lambda|^{-1} dt \right) |d\lambda|. \tag{21}$$

Suppose $\Theta(\lambda, \alpha)$ satisfies the following additional condition.

Condition 2.15. *The function*

$$g(\zeta) = 1 - \Theta(\lambda, \lambda\zeta)\lambda$$

is analytic on an open set $D_0 \supset K(\varphi_0)\backslash\{0\}$ and $g(\zeta)$ is independent of λ when $\lambda \in K(\varphi_0)\backslash\{0\}$. Moreover, for all $t \in (0, p)$ and $p \in (0, p_0]$,

$$\sup_{|\varphi|\le\varphi_0} \int\limits_0^\infty \tau^{-t-1} |g(\tau \exp(i\varphi))| d\tau \equiv Z(t) < \infty, \tag{22}$$

$$\lim_{r\to 0+} r^{-t-1} \int\limits_{\Gamma_r(\exp(-i\varphi_0),\exp(i\varphi_0))} |g(\zeta)||d\zeta| = 0, \tag{23}$$

$$\lim_{R\to\infty} R^{-t-1} \int\limits_{\Gamma_R(\exp(-i\varphi_0),\exp(i\varphi_0))} |g(\zeta)||d\zeta| = 0 \tag{24}$$

with p_0 specified in Condition 2.7.

Let us consider in detail the internal integrals in (21). Using the change of variables $\alpha = |\lambda|\tau, t = |\lambda|\tau$ and (22) with $t = p - \kappa$, we obtain

$$\int\limits_0^\infty \alpha^{-p-1+\kappa} |1 - \Theta(\lambda, \alpha)\lambda| d\alpha =$$

$$= |\lambda|^{-p+\kappa} \int\limits_0^\infty \tau^{-p-1+\kappa} |g(\tau \exp(-i\arg\lambda))| d\tau \le \tag{25}$$

$$\le Z(p - \kappa)|\lambda|^{-p+\kappa},$$

$$\int_0^\infty t^{p-\kappa-m-1}|t+\lambda|^{-1}dt =$$

$$= |\lambda|^{p-\kappa-m-1}\int_0^\infty \tau^{p-\kappa-m-1}\left|\tau+|\lambda|^{-1}\lambda\right|^{-1}d\tau \le \tag{26}$$

$$\le P(p,\kappa)|\lambda|^{p-\kappa-m-1} \quad \forall \lambda \in \Gamma^{(\varepsilon)}(\varphi_0),$$

where $P(p,\kappa)$ is a positive constant independent of λ;

$$0 < \kappa < \min\left\{\frac{2}{3}p, 2\tau_0, 2s_0\right\}.$$

The required relation $J(\varepsilon) < \infty$ follows immediately by (21), (25), and (26).

Changing the order of integrals in (20), we get

$$A_\varepsilon^{p-\kappa}w_\kappa^{(\varepsilon)} = D(p,\kappa)\int_{\Gamma^{(\varepsilon)}(\varphi_0)} \lambda^{m+1}\left(\int_0^\infty \alpha^{-p-1+\kappa}(1-\Theta(\lambda,\alpha)\lambda)d\alpha\right)\cdot$$

$$\cdot\left(\int_0^\infty t^{p-\kappa-m-1}(t+\lambda)^{-1}dt\right)R(\lambda,A_\varepsilon)(x^*-\xi)d\lambda, \quad \varepsilon \in (0,\varepsilon^{(3)}].$$

$$\tag{27}$$

By definition, put

$$\Lambda(z) = \{\zeta \in \mathbf{C}: \zeta = tz, t \ge 0\}, \quad z \in \mathbf{C}.$$

With the substitution $\alpha = \lambda\zeta$, the first internal integral in (27) is reduced to the form

$$\int_0^\infty \alpha^{-p-1+\kappa}(1-\Theta(\lambda,\alpha)\lambda)d\alpha = \lambda^{-p+\kappa}\int_{\Lambda(\bar\lambda)} \zeta^{-p-1+\kappa}g(\zeta)d\zeta. \tag{28}$$

Here $\bar\lambda$ is the complex conjugate of $\lambda \in \mathbf{C}$, and the integration over the ray $\Lambda(z)$ is performed from $\zeta = 0$ to $\zeta = \infty$.

We claim that the value $G(p,\kappa)$ of the integral in (28) doesn't depend on $\lambda \in \Gamma^{(\varepsilon)}(\varphi_0)$. Indeed, suppose $\lambda_1, \lambda_2 \in \Gamma^{(\varepsilon)}(\varphi_0)$ satisfy $\arg\bar\lambda_1 < \arg\bar\lambda_2$; then $\Lambda(\bar\lambda_1) \ne \Lambda(\bar\lambda_2)$. Define the contour

$$\Gamma_{(r,R)}(\bar\lambda_1, \bar\lambda_2) = \Gamma_r(\bar\lambda_1, \bar\lambda_2) \cup \Gamma_R(\bar\lambda_1, \bar\lambda_2) \cup \Gamma_{(r,R)}(\bar\lambda_1) \cup \Gamma_{(r,R)}(\bar\lambda_2),$$

where $0 < r < R$. Since the function $g(\zeta)$ is analytic on

$$K(\varphi_0)\backslash\{0\} \supset \Gamma_{(r,R)}(\bar\lambda_1, \bar\lambda_2),$$

by Cauchy's theorem we have

$$\int\limits_{\Gamma_{(r,R)}(\bar\lambda_1,\bar\lambda_2)} \zeta^{-p-1+\kappa}g(\zeta)d\zeta = \int\limits_{\Gamma_r(\bar\lambda_1,\bar\lambda_2)} \zeta^{-p-1+\kappa}g(\zeta)d\zeta +$$

$$+ \int\limits_{\Gamma_R(\bar\lambda_1,\bar\lambda_2)} + \int\limits_{\Gamma_{(r,R)}(\bar\lambda_1)} + \int\limits_{\Gamma_{(r,R)}(\bar\lambda_2)} = 0.$$

Passing to the limits as $r \to 0$ and $R \to \infty$, with the use of (23) and (24) we get

$$\int\limits_{\Lambda(\bar\lambda_1)} \zeta^{-p-1+\kappa}g(\zeta)d\zeta = \int\limits_{\Lambda(\bar\lambda_2)} \zeta^{-p-1+\kappa}g(\zeta)d\zeta.$$

Thus $G(p,\kappa)$ is independent of $\lambda \in \Gamma^{(\varepsilon)}(\varphi_0)$.

In the similar way, for the second internal integral in (27) we obtain

$$\int\limits_0^\infty t^{p-\kappa-m-1}(t+\lambda)^{-1}dt = \lambda^{p-\kappa-m-1} \int\limits_{\Lambda(\bar\lambda)} \zeta^{p-\kappa-m-1}(1+\zeta)^{-1}d\zeta. \quad (29)$$

As above, it can easily be checked that the value

$$H(p,\kappa) = \int\limits_{\Lambda(\bar\lambda)} \zeta^{p-\kappa-m-1}(1+\zeta)^{-1}d\zeta$$

is independent of $\lambda \in \Gamma^{(\varepsilon)}(\varphi_0)$. Finally, combining (27)–(29), we conclude that for each $\varepsilon \in (0, \varepsilon^{(3)}]$,

$$A_\varepsilon^{p-\kappa}w_\kappa^{(\varepsilon)} = \frac{C(p,\kappa)}{2\pi i} \int\limits_{\Gamma^{(\varepsilon)}(\varphi_0)} R(\lambda, A_\varepsilon)(x^* - \xi)d\lambda = C(p,\kappa)(x^* - \xi),$$

where $C(p,\kappa) = 2\pi i D(p,\kappa)G(p,\kappa)H(p,\kappa) \neq 0$. Consequently the equality (10) is valid.

The first main result of this section is as follows.

Theorem 2.15. *Let Conditions 1.1, 2.6, and 2.11–2.15 be satisfied. Suppose estimate (5) is fulfilled. Then inclusion (1) is valid.*

Proof. Let $\{\varepsilon_n\}$ be the sequence specified above, namely,

$$\varepsilon_n > 0, \quad \lim_{n\to\infty} \varepsilon_n = 0, \quad \lim_{n\to\infty} \|w_\kappa^{(\varepsilon_n)} - w_\kappa\|_X = 0.$$

Let us prove that

$$A_{\varepsilon_n}^{p-\kappa} w_\kappa^{(\varepsilon_n)} = A^{p-\kappa} w_\kappa,$$

where the element w_κ is defined by (7). Denote $m = [p - \kappa]$, $\mu = \{p - \kappa\}$, so that $p - \kappa = m + \mu$, $\mu \in (0, 1)$. We have the estimate

$$\|A_{\varepsilon_n}^{p-\kappa} w_\kappa^{(\varepsilon_n)} - A^{p-\kappa} w_\kappa\|_X = \|A_{\varepsilon_n}^{m+\mu} w_\kappa^{(\varepsilon_n)} - A^{m+\mu} w_\kappa\|_X \le$$
$$\le \|A_{\varepsilon_n}^{m+\mu} - A^{m+\mu}\|_{L(X)} \|w_\kappa^{(\varepsilon_n)}\|_X + \|A^{m+\mu}\|_{L(X)} \|w_\kappa^{(\varepsilon_n)} - w_\kappa\|_X. \tag{30}$$

Further,

$$\|A_{\varepsilon_n}^{m+\mu} - A^{m+\mu}\|_{L(X)} \le \|A_{\varepsilon_n}^m - A^m\|_{L(X)} \|A^\mu\|_{L(X)} +$$
$$+ \|A_{\varepsilon_n}\|_{L(X)}^m \|A_{\varepsilon_n}^\mu - A^\mu\|_{L(X)}. \tag{31}$$

It is readily seen that

$$\|A_\varepsilon^m - A^m\|_{L(X)} \le c_3 \varepsilon \quad \forall \varepsilon \in (0, \varepsilon^{(3)}]. \tag{32}$$

The application of Lemma 1.1 yields

$$\|A_\varepsilon^\mu - A^\mu\|_{L(X)} \le c_4 \varepsilon^\mu \quad \forall \varepsilon \in (0, \varepsilon^{(3)}]. \tag{33}$$

Notice that the constants c_3, c_4 in (32) and (33) depend only on p and κ. From (30)–(33) we get

$$\lim_{n \to \infty} \|A_{\varepsilon_n}^{p-\kappa} w_\kappa^{(\varepsilon_n)} - A^{p-\kappa} w_\kappa\|_X = 0. \tag{34}$$

Since by (10), the element $A_\varepsilon^{p-\kappa} w_\kappa^{(\varepsilon)}$ is independent of ε, from (34) it follows that

$$A^{p-\kappa} w_\kappa = C(p, \kappa)(x^* - \xi).$$

Therefore,

$$A^{p-\kappa} v_\kappa = x^* - \xi, \quad v_\kappa = C(p, \kappa)^{-1} w_\kappa.$$

Thus,

$$x^* - \xi \in R(A^{p-\kappa})$$

for all sufficiently small $\kappa > 0$. Since $R(A^{p_1}) \subset R(A^{p_2})$ whatever $0 < p_2 < p_1$, it follows that

$$x^* - \xi \in R(A^{p-\varepsilon})$$

for each $\varepsilon \in (0, p)$.   □

Let us now give some examples of procedures (2.2), to which Theorem 2.15 and results of Section 2.2 are applicable.

We denote

$$l_r(z) = \{\zeta \in \mathbf{C} : |\zeta - z| = r\}.$$

For the function (1.18), it is convenient to check Conditions 2.6–2.8 with the contours

$$\Gamma_\alpha = \Gamma_\alpha^{(1)} \cup \Gamma_\alpha^{(2)}; \quad \Gamma_\alpha^{(1)} = l_{(1-d_0)\alpha}(-\alpha), \quad \Gamma_\alpha^{(2)} = l_{R_0}(0), \quad \alpha \in (0, \alpha_0].$$

Suppose $\alpha_0 > 0$ is sufficiently small, so that

$$\Gamma_\alpha^{(1)} \cap \Gamma_\alpha^{(2)} = \emptyset \quad \forall \alpha \in (0, \alpha_0];$$

then we have $d_0 \alpha_0 < R_0$.

Lemma 2.4. *The function (1.18) satisfies Conditions 2.6–2.8 and 2.11–2.15 for all $\varphi_0 \in (0, \pi)$. Condition 2.7 is fulfilled with $p_0 = N$, Conditions 2.11 and 2.14 are valid for all $\tau_0, s_0 > 0$.*

Proof. It is readily seen that Condition 2.6 is satisfied with $D_\alpha = \mathbf{C} \setminus \{-\alpha\}$. Next, let us consider the integral

$$
\int_{\Gamma_\alpha} |1 - \Theta(\lambda, \alpha)\lambda| |\lambda|^{p-1} |d\lambda| =
$$

$$
= \int_{\Gamma_\alpha^{(1)}} \left| \frac{\alpha}{\lambda + \alpha} \right|^N |\lambda|^{p-1} |d\lambda| + \int_{\Gamma_\alpha^{(2)}} \left| \frac{\alpha}{\lambda + \alpha} \right|^N |\lambda|^{p-1} |d\lambda|.
\tag{35}
$$

Assuming that $p \in [0, N]$, for the summands in (35) we obtain

$$
\int_{\Gamma_\alpha^{(1)}} \left| \frac{\alpha}{\lambda + \alpha} \right|^N |\lambda|^{p-1} |d\lambda| \le \frac{(2 - d_0)^{p-1}}{(1 - d_0)^N} \alpha^{p-1} \int_{\Gamma_\alpha^{(1)}} |d\lambda| =
$$

$$
= \frac{2\pi(2 - d_0)^{p-1}}{(1 - d_0)^{N-1}} \alpha^p,
\tag{36}
$$

$$
\int_{\Gamma_\alpha^{(2)}} \left| \frac{\alpha}{\lambda + \alpha} \right|^N |\lambda|^{p-1} |d\lambda| \le \frac{\alpha^N R_0^{p-1}}{(R_0 - \alpha)^N} \int_{\Gamma_\alpha^{(2)}} |d\lambda| \le
$$

$$
\le \frac{2\pi R_0^p \alpha_0^{N-p}}{(R_0 - \alpha_0)^N} \alpha^p \quad \forall \alpha \in (0, \alpha_0].
\tag{37}
$$

Combining (36) and (37), we conclude that Condition 2.7 is satisfied with $p_0 = N$.

The integral from Condition 2.8 has the form

$$
\int_{\Gamma_\alpha} \frac{|\Theta(\lambda,\alpha)|}{|\lambda|}\,|d\lambda| = \int_{\Gamma_\alpha^{(1)}} |\lambda|^{-2}\left|1 - \left(\frac{\alpha}{\lambda+\alpha}\right)^N\right||d\lambda| +
$$

$$
+ \int_{\Gamma_\alpha^{(2)}} |\lambda|^{-2}\left|1 - \left(\frac{\alpha}{\lambda+\alpha}\right)^N\right||d\lambda| \quad \forall \alpha \in (0,\alpha_0]. \tag{38}
$$

Observe that

$$
\int_{\Gamma_\alpha^{(1)}} |\lambda|^{-2}\left|1 - \left(\frac{\alpha}{\lambda+\alpha}\right)^N\right||d\lambda| \le \frac{1 + (1-d_0)^{-N}}{d_0^2\alpha^2}\int_{\Gamma_\alpha^{(1)}} |d\lambda| =
$$

$$
= \frac{2\pi(1-d_0)(1+(1-d_0)^{-N})}{d_0^2\alpha}, \tag{39}
$$

$$
\int_{\Gamma_\alpha^{(2)}} |\lambda|^{-2}\left|1 - \left(\frac{\alpha}{\lambda+\alpha}\right)^N\right||d\lambda| \le
$$

$$
\le \frac{1}{R_0^2}\left(1 + \left(\frac{\alpha_0}{R_0-\alpha_0}\right)^N\right)\int_{\Gamma_\alpha^{(2)}} |d\lambda| \le \tag{40}
$$

$$
\le \frac{2\pi\alpha_0}{R_0\alpha}\left(1 + \left(\frac{\alpha_0}{R_0-\alpha_0}\right)^N\right) \quad \forall \alpha \in (0,\alpha_0].
$$

From (38)–(40) it follows that Condition 2.8 is satisfied.

Let us analyze Condition 2.11. Notice that for all $\lambda \in \gamma_\alpha$ we have

$$
|\lambda + \alpha| \ge \min\{1 - d_0, \sin\varphi_0\}\alpha.
$$

Consequently for each $\tau > 0$,

$$
\sup_{\alpha \in [\alpha_0,+\infty)}\left(\alpha^{-\tau}\int_{\gamma_\alpha} \frac{|1 - \Theta(\lambda,\alpha)\lambda|}{|\lambda|}|d\lambda|\right) =
$$

$$
= \sup_{\alpha \in [\alpha_0,+\infty)}\left(\alpha^{-\tau}\int_{\gamma_\alpha} \left|\frac{\alpha}{\lambda+\alpha}\right|^N \frac{|d\lambda|}{|\lambda|}\right) \le
$$

$$
\le \sup_{\alpha \in [\alpha_0,+\infty)}\left(\alpha^{-\tau}\frac{1}{d_0\alpha(\min\{1-d_0,\sin\varphi_0\})^N}\int_{\gamma_\alpha} |d\lambda|\right) \le
$$

$$
\le \frac{2(\pi+1)R_0}{d_0\alpha_0^{\tau+1}(\min\{1-d_0,\sin\varphi_0\})^N}.
$$

Thus Condition 2.11 is satisfied for each $\tau_0 > 0$.

The check of Conditions 2.12 and 2.13 is trivial.

The function $\psi(\lambda, \alpha, \varepsilon)$ from Condition 2.14 has the form

$$\psi(\lambda, \alpha, \varepsilon) = \left(1 - \frac{\varepsilon}{\lambda + \alpha + \varepsilon}\right)^N - 1.$$

Observe that for each $\alpha > 0$,

$$\left|\frac{\varepsilon}{\lambda + \alpha + \varepsilon}\right| \le$$

$$\le \frac{\varepsilon}{\min\{(1 - d_0)\alpha + \varepsilon, (\alpha + \varepsilon)\sin\varphi_0\}} < \frac{1}{\sin\varphi_0}$$

$$\forall \varepsilon \in (0, \varepsilon^{(1)}] \quad (\varepsilon^{(1)} > 0).$$

Therefore for all $\varepsilon \in (0, \varepsilon^{(1)}]$ we have

$$\int_{\gamma_\alpha} \frac{|\psi(\lambda, \alpha, \varepsilon)|}{|\lambda|}|d\lambda| = \int_{\gamma_\alpha} \frac{1}{|\lambda|}\left|\left(1 - \frac{\varepsilon}{\lambda + \alpha + \varepsilon}\right)^N - 1\right||d\lambda| =$$

$$= \int_{\gamma_\alpha} \frac{1}{|\lambda|}\left|\frac{\varepsilon}{\lambda + \alpha + \varepsilon}\right|\left|-\binom{N}{1} + \binom{N}{2}\left(\frac{\varepsilon}{\lambda + \alpha + \varepsilon}\right) - \cdots +\right.$$

$$+(-1)^N\left(\frac{\varepsilon}{\lambda + \alpha + \varepsilon}\right)^{N-1}\right||d\lambda| \le$$

$$\le \frac{\varepsilon \sin\varphi_0}{(\alpha + \varepsilon)\min\{1 - d_0, \sin\varphi_0\}}\left[\left(1 + \frac{1}{\sin\varphi_0}\right)^N - 1\right]\int_{\gamma_\alpha} \frac{|d\lambda|}{|\lambda|},$$

where

$$\binom{N}{m} = \frac{N(N - 1)\ldots(N - m + 1)}{1 \cdot 2 \ldots m}.$$

Further,

$$\frac{\varepsilon \sin\varphi_0}{(\alpha + \varepsilon)\min\{1 - d_0, \sin\varphi_0\}}\left[\left(1 + \frac{1}{\sin\varphi_0}\right)^N - 1\right]\int_{\gamma_\alpha} \frac{|d\lambda|}{|\lambda|} =$$

$$= \frac{2\varepsilon \sin\varphi_0}{(\alpha + \varepsilon)\min\{1 - d_0, \sin\varphi_0\}}\left[\left(1 + \frac{1}{\sin\varphi_0}\right)^N - 1\right]\left(\pi + \ln\frac{R_0}{d_0\alpha}\right) \equiv$$

$$\equiv M(\alpha, \varepsilon).$$

It follows that (15) is true for all $s_0 > 0$. Moreover,

$$\int_0^{\varepsilon^{(1)}} \frac{M(\alpha, \varepsilon)}{\varepsilon} d\varepsilon =$$

$$= \frac{2\sin\varphi_0}{\min\{1 - d_0, \sin\varphi_0\}} \left[\left(1 + \frac{1}{\sin\varphi_0}\right)^N - 1\right]\left(\pi + \ln\frac{R_0}{d_0\alpha}\right)\ln\frac{\alpha + \varepsilon^{(1)}}{\alpha}.$$

This yields that (16) is also satisfied for each $s_0 > 0$.

Let us now consider Condition 2.15. It is obvious that the function

$$g(\zeta) = 1 - \Theta(\lambda, \lambda\zeta)\lambda = \left(\frac{\zeta}{1 + \zeta}\right)^N$$

doesn't depend on λ, and $g(\zeta)$ is analytic on

$$\mathbf{C}\backslash\{-1\} \supset K(\varphi_0)\backslash\{0\}.$$

Furthermore,

$$\int_0^\infty \tau^{-t-1}|g(\tau\exp(i\varphi))|d\tau =$$

$$= \int_0^1 \tau^{-t-1}\left|\frac{\tau\exp(i\varphi)}{1 + \tau\exp(i\varphi)}\right|^N d\tau + \int_1^\infty \tau^{-t-1}\left|\frac{\tau\exp(i\varphi)}{1 + \tau\exp(i\varphi)}\right|^N d\tau.$$

$$(41)$$

In the assumption that $t \in (0, p), p \in (0, N]$, we can estimate the summands in (41) as follows:

$$\int_0^1 \tau^{-t-1}\left|\frac{\tau\exp(i\varphi)}{1 + \tau\exp(i\varphi)}\right|^N d\tau \le \frac{1}{(\sin\varphi_0)^N}\int_0^1 \tau^{-1-t+N}d\tau =$$

$$= \frac{1}{(N - t)(\sin\varphi_0)^N};$$

$$(42)$$

$$\int_{1}^{\infty} \tau^{-t-1} \left| \frac{\tau \exp(i\varphi)}{1 + \tau \exp(i\varphi)} \right|^{N} d\tau =$$

$$= \int_{1}^{\infty} \tau^{-t-1} \left| 1 - \frac{1}{1 + \tau \exp(i\varphi)} \right|^{N} d\tau \le \qquad (43)$$

$$\le \left(1 + \frac{1}{\sin \varphi_0} \right)^{N} \int_{1}^{\infty} \tau^{-t-1} d\tau =$$

$$= \frac{1}{t} \left(1 + \frac{1}{\sin \varphi_0} \right)^{N}, \qquad |\varphi| \le \varphi_0.$$

As a function $Z(t)$ we can now take the sum of expressions from the right parts of (42) and (43). Also, observe that for all $p \in (0, N]$ and $t \in (0, p)$,

$$\lim_{r \to 0} r^{-t-1} \int_{\Gamma_r(\exp(-i\varphi_0), \exp(i\varphi_0))} \left| \frac{\varsigma}{1 + \varsigma} \right|^{N} |d\varsigma| \le$$

$$\le \lim_{r \to 0} \frac{r^{-t-1+N}}{(1-r)^{N}} \int_{\Gamma_r(\exp(-i\varphi_0), \exp(i\varphi_0))} |d\varsigma| = \lim_{r \to 0} \frac{2\varphi_0 r^{-t+N}}{(1-r)^{N}} = 0,$$

$$\lim_{R \to \infty} R^{-t-1} \int_{\Gamma_R(\exp(-i\varphi_0), \exp(i\varphi_0))} \left| \frac{\varsigma}{1 + \varsigma} \right|^{N} |d\varsigma| \le$$

$$\le \lim_{R \to \infty} R^{-t-1} \left(\frac{R}{R-1} \right)^{N} \int_{\Gamma_R(\exp(-i\varphi_0), \exp(i\varphi_0))} |d\varsigma| =$$

$$= \lim_{R \to \infty} 2\varphi_0 R^{-t} \left(\frac{R}{R-1} \right)^{N} = 0.$$

Thus Condition 2.15 is satisfied for each $p \in (0, N]$. $\qquad \square$

Lemma 2.5. *Let $\varphi_0 \in (0, \pi/2)$. The function (1.23) satisfies Conditions 2.6–2.8 and 2.11–2.15. Condition 2.7 is valid for each $p_0 > 0$. Conditions 2.11 and 2.14 are satisfied for all $\tau_0, s_0 > 0$.*

Proof. Since the function (1.23) is analytic on the all of **C**, Condition 2.6 is satisfied with $D_\alpha = \mathbf{C}$. We put $\Gamma_\alpha = \gamma_\alpha$, $\alpha \in (0, \alpha_0]$ and denote

$$
\begin{aligned}
\gamma_\alpha^{(1)} &= \Gamma_{d_0\alpha}(\exp(i\varphi_0), \exp(i(2\pi - \varphi_0))), \\
\gamma_\alpha^{(2)} &= \Gamma_{R_0}(\exp(-i\varphi_0), \exp(i\varphi_0)), \\
\gamma_\alpha^{(3)} &= \Gamma_{(d_0\alpha, R_0)}(\exp(i\varphi_0)), \\
\gamma_\alpha^{(4)} &= \Gamma_{(d_0\alpha, R_0)}(\exp(-i\varphi_0))
\end{aligned}
\tag{44}
$$

so that $\Gamma_\alpha = \gamma_\alpha^{(1)} \cup \gamma_\alpha^{(2)} \cup \gamma_\alpha^{(3)} \cup \gamma_\alpha^{(4)}$. For all $p > 0, \alpha > 0$ we have

$$
\int_{\Gamma_\alpha} |1 - \Theta(\lambda, \alpha)\lambda||\lambda|^{p-1}|d\lambda| = \int_{\Gamma_\alpha} |\exp(-\alpha^{-1}\lambda)||\lambda|^{p-1}|d\lambda| \le
$$

$$
\le \int_{\gamma_\alpha^{(1)}} \exp(\alpha^{-1}|\lambda|)|\lambda|^{p-1}|d\lambda| +
$$

$$
+ \int_{\gamma_\alpha^{(2)}} \exp(-\alpha^{-1}|\lambda|\cos\varphi_0)|\lambda|^{p-1}|d\lambda| + \int_{\gamma_\alpha^{(3)}} + \int_{\gamma_\alpha^{(4)}}.
$$

The integrals over $\gamma_\alpha^{(1)}$ and $\gamma_\alpha^{(2)}$ are estimated as follows:

$$
\int_{\gamma_\alpha^{(1)}} \exp(\alpha^{-1}|\lambda|)|\lambda|^{p-1}|d\lambda| = \exp(d_0)d_0^{p-1}\alpha^{p-1} \int_{\gamma_\alpha^{(1)}} |d\lambda| =
$$
$$
= 2(\pi - \varphi_0)\exp(d_0)d_0^p\alpha^p,
\tag{45}
$$

$$
\int_{\gamma_\alpha^{(2)}} \exp(-\alpha^{-1}|\lambda|\cos\varphi_0)|\lambda|^{p-1}|d\lambda| =
$$
$$
= \exp(-\alpha^{-1}R_0\cos\varphi_0)R_0^{p-1} \int_{\gamma_\alpha^{(2)}} |d\lambda| =
\tag{46}
$$
$$
= 2\varphi_0 R_0^p \exp(-\alpha^{-1}R_0\cos\varphi_0) \le c_5\alpha^p;
$$
$$
c_5 = 2\varphi_0 R_0^p \sup_{\alpha>0}\left(\alpha^{-p}\exp(-\alpha^{-1}R_0\cos\varphi_0)\right).
$$

Using the substitution $t = \alpha^{-1}|\lambda|$, we get

$$\int\limits_{\gamma_\alpha^{(3)}} \exp(-\alpha^{-1}|\lambda|\cos\varphi_0)|\lambda|^{p-1}|d\lambda| =$$

$$= \alpha^p \int\limits_{d_0}^{R_0\alpha^{-1}} \exp(-t\cos\varphi_0)t^{p-1}dt \leq \tag{47}$$

$$\leq \frac{\alpha^p}{(\cos\varphi_0)^p} \int\limits_{d_0\cos\varphi_0}^{R_0\alpha^{-1}\cos\varphi_0} \exp(-\tau)\tau^{p-1}d\tau \leq \frac{\Gamma(p)\alpha^p}{(\cos\varphi_0)^p}.$$

Recall that $\Gamma(\cdot)$ denotes the Euler gamma–function. The integral over $\gamma_\alpha^{(4)}$ is estimated analogously to (47). At last, we note that (45)–(47) are satisfied for each $p > 0$. This completes the check of Condition 2.7.

The integral from Condition 2.8 has the form

$$\int\limits_{\Gamma_\alpha} \frac{|\Theta(\lambda,\alpha)|}{|\lambda|}|d\lambda| =$$

$$= \int\limits_{\gamma_\alpha^{(1)}} |\lambda|^{-2}|1 - \exp(-\alpha^{-1}\lambda)||d\lambda| + \int\limits_{\gamma_\alpha^{(2)}} + \int\limits_{\gamma_\alpha^{(3)}} + \int\limits_{\gamma_\alpha^{(4)}}. \tag{48}$$

For all $\alpha \in (0, \alpha_0]$,

$$\int\limits_{\gamma_\alpha^{(1)}} |\lambda|^{-2}|1 - \exp(-\alpha^{-1}\lambda)||d\lambda| \leq$$

$$\leq \frac{1 + \exp(d_0)}{d_0^2\alpha^2} \int\limits_{\gamma_\alpha^{(1)}} |d\lambda| = \frac{2(\pi - \varphi_0)(1 + \exp(d_0))}{d_0\alpha}, \tag{49}$$

$$\int\limits_{\gamma_\alpha^{(2)}} |\lambda|^{-2}|1 - \exp(-\alpha^{-1}\lambda)||d\lambda| \leq \frac{2}{R_0^2} \int\limits_{\gamma_\alpha^{(2)}} |d\lambda| \leq \frac{4\varphi_0\alpha_0}{R_0\alpha}, \tag{50}$$

$$\int\limits_{\gamma_\alpha^{(3)}} |\lambda|^{-2}|1 - \exp(-\alpha^{-1}\lambda)||d\lambda| \leq 2 \int\limits_{d_0\alpha}^{R_0} |\lambda|^{-2}d|\lambda| \leq \frac{2}{d_0\alpha}. \tag{51}$$

An estimate similar to (51) is valid for the integral over $\gamma_\alpha^{(4)}$. From (48)–(51) it follows that Condition 2.8 is satisfied.

Since

$$\sup_{\alpha\in[\alpha_0,+\infty)}\left(\alpha^{-\tau}\int_{\gamma_\alpha}\frac{|1-\Theta(\lambda,\alpha)\lambda|}{|\lambda|}|d\lambda|\right)=$$

$$=\sup_{\alpha\in[\alpha_0,+\infty)}\left(\alpha^{-\tau}\int_{\gamma_\alpha}\frac{|\exp(-\alpha^{-1}\lambda)|}{|\lambda|}|d\lambda|\right)\leq$$

$$\leq\sup_{\alpha\in[\alpha_0,+\infty)}\left(\alpha^{-\tau}\frac{\exp(d_0)}{d_0\alpha}\int_{\gamma_\alpha}|d\lambda|\right)\leq$$

$$\leq\frac{2(\pi+1)R_0\exp(d_0)}{d_0\alpha_0^{\tau+1}}\qquad\forall\tau>0,$$

Condition 2.11 is fulfilled with an arbitrary $\tau_0>0$.

The check of Conditions 2.12 and 2.13 creates no difficulties.

The function $\psi(\lambda,\alpha,\varepsilon)$ from Condition 2.14 has the form

$$\psi(\lambda,\alpha,\varepsilon)=\exp(-\alpha^{-1}\varepsilon)-1.$$

Since $d_0\alpha_0<R_0$, we have

$$\int_{\gamma_\alpha}\frac{|\psi(\lambda,\alpha,\varepsilon)|}{|\lambda|}|d\lambda|=\int_{\gamma_\alpha}\frac{1-\exp(-\alpha^{-1}\varepsilon)}{|\lambda|}|d\lambda|=$$

$$=2\left(\pi+\ln\frac{R_0}{d_0\alpha}\right)(1-\exp(-\alpha^{-1}\varepsilon))\equiv M(\alpha,\varepsilon).$$

Therefore (15) is true for each $s_0>0$. Moreover,

$$\int_0^{\varepsilon^{(1)}}\frac{M(\alpha,\varepsilon)}{\varepsilon}d\varepsilon=2\left(\pi+\ln\frac{R_0}{d_0\alpha}\right)\int_0^{\varepsilon^{(1)}}\frac{1-\exp(-\alpha^{-1}\varepsilon)}{\varepsilon}d\varepsilon\leq$$

$$\leq2\left(\pi+\ln\frac{R_0}{d_0\alpha}\right)\left(\int_0^{\alpha_0^{-1}\varepsilon^{(1)}}\frac{1-\exp(-t)}{t}dt+\int_{\alpha_0^{-1}\varepsilon^{(1)}}^{\alpha^{-1}\varepsilon^{(1)}}\frac{1-\exp(-t)}{t}dt\right)\leq$$

$$\leq2\left(\pi+\ln\frac{R_0}{d_0\alpha}\right)\left(c_6+\ln\frac{\alpha_0}{\alpha}\right),\qquad c_6=\int_0^{\alpha_0^{-1}\varepsilon^{(1)}}\frac{1-\exp(-t)}{t}dt.$$

It follows that (16) is fulfilled for each $s_0>0$. The check of Condition 2.14 is complete.

Let us turn to Condition 2.15. The function

$$g(\zeta) = 1 - \Theta(\lambda, \lambda\zeta)\lambda = \exp(-\zeta^{-1})$$

is analytic on an open set $\operatorname{int} K(\varphi_0') \supset K(\varphi_0)\backslash\{0\}$ for each $\varphi_0' \in (\varphi_0, \pi)$, and $g(\zeta)$ doesn't depend on λ. Let $t \in (0, p)$ with some $p > 0$. We obtain

$$\int_0^\infty \tau^{-t-1}|g(\tau\exp(i\varphi))|d\tau = \int_0^\infty \tau^{-t-1}|\exp(-\tau^{-1}\exp(-i\varphi))|d\tau \leq$$

$$\leq \int_0^\infty \tau^{-t-1}\exp(-\tau^{-1}\cos\varphi_0)d\tau = \int_0^1 \tau^{-t-1}\exp(-\tau^{-1}\cos\varphi_0)d\tau + \quad (52)$$

$$+ \int_1^\infty \tau^{-t-1}\exp(-\tau^{-1}\cos\varphi_0)d\tau.$$

The integrand of the first summand in (52) is bounded and doesn't depend on φ. The second summand is estimated by the integral

$$\int_1^\infty \tau^{-t-1}d\tau < \infty,$$

which is also independent of φ. Hence the inequality (22) is valid. In addition,

$$\lim_{r\to 0} r^{-t-1} \int_{\Gamma_r(\exp(-i\varphi_0),\exp(i\varphi_0))} |g(\zeta)||d\zeta| \leq$$

$$\leq \lim_{r\to 0} r^{-t-1} \int_{\Gamma_r(\exp(-i\varphi_0),\exp(i\varphi_0))} \exp(-|\zeta|^{-1}\cos\varphi_0)|d\zeta| = \quad (53)$$

$$= \lim_{r\to 0} 2\varphi_0 r^{-t}\exp(-r^{-1}\cos\varphi_0) = 0 \quad \forall t \in (0, p).$$

Besides, for each $t \in (0, p)$ we have

$$\lim_{R\to\infty} R^{-t-1} \int_{\Gamma_R(\exp(-i\varphi_0),\exp(i\varphi_0))} |g(\zeta)||d\zeta| \leq$$

$$\leq \lim_{R\to\infty} R^{-t-1} \int_{\Gamma_R(\exp(-i\varphi_0),\exp(i\varphi_0))} \exp(-R^{-1}\cos\varphi_0)|d\zeta| \leq \quad (54)$$

$$\leq \lim_{R\to\infty} 2\varphi_0 R^{-t} = 0.$$

From (53) and (54) it follows that Condition 2.15 is satisfied.   □

The following statements are direct consequences of Theorems 2.10, 2.15 and Lemmas 2.4, 2.5.

Theorem 2.16. *Let (2.5) be satisfied and $\Theta(\lambda, \alpha)$ be the function (1.18) or (1.23). Assume that $p \in (0, N]$ in the case of (1.18) and $p > 0$ in the case of (1.23). Then estimate (5) is valid.*

Theorem 2.17. *Suppose (5) is fulfilled and $\Theta(\lambda, \alpha)$ is the function (1.18) or (1.23). Let $p \in (0, N]$ in the case of (1.18) and $p > 0$ in the case of (1.23). Then (1) is valid.*

Let us proceed to consider examples of generating functions $\Theta(\lambda, \alpha)$. By

$$\Theta_1(\lambda, \alpha) = \frac{1}{\lambda}\left(1 - (1 - \mu_0\lambda)^{1/\alpha}\right)$$

and

$$\Theta_2(\lambda, \alpha) = \frac{1}{\lambda}\left(1 - \left(\frac{\mu_0}{\mu_0 + \lambda}\right)^{1/\alpha}\right)$$

we denote the functions from Example 2.4. At the point $\lambda = 0$ these functions are defined by continuity (see (1.29)). Let us modify Condition 2.6 by setting

$$K_\alpha(R_0, d_0, \varphi_0) = K(R_0, \varphi_0) \cup S(\min\{\overline{\mu}_0, d_0\alpha\}),$$

where $0 < \overline{\mu}_0 < \min\{R_0, \mu_0\}$. It can easily be checked that all the results of Section 2.2 remain valid.

Lemma 2.6. *1) Let $\varphi_0 \in (0, \pi/2)$ and*

$$0 < \mu_0 < \frac{\min\{2\cos\varphi_0, 1\}}{R_0}.$$

Then the function $\Theta_1(\lambda, \alpha)$ satisfies Conditions 2.6–2.8 and 2.11–2.14.
2) Let $\varphi_0 \in (0, \pi/2)$ and $\mu_0 > 0$. Then the function $\Theta_2(\lambda, \alpha)$ satisfies Conditions 2.6–2.8 and 2.11–2.14.
In both cases, Condition 2.7 is valid for each $p_0 > 0$, Conditions 2.11 and 2.14 are satisfied for all $\tau_0, s_0 > 0$.

Proof. The check of Condition 2.6 is obvious.

Let $\Gamma_\alpha = \gamma_\alpha$, $\alpha \in (0, \alpha_0]$, where without loss of generality it can be assumed that $0 < \alpha_0 < d_0^{-1}\overline{\mu}_0$. According to (44),

$$\Gamma_\alpha = \gamma_\alpha^{(1)} \cup \gamma_\alpha^{(2)} \cup \gamma_\alpha^{(3)} \cup \gamma_\alpha^{(4)}.$$

For each $p > 0$ we have

$$\int\limits_{\gamma_\alpha^{(1)}} |1 - \Theta_1(\lambda, \alpha)\lambda||\lambda|^{p-1}|d\lambda| = \int\limits_{\gamma_\alpha^{(1)}} |1 - \mu_0\lambda|^{1/\alpha}|\lambda|^{p-1}|d\lambda| \leq$$

$$\leq \int\limits_{\gamma_\alpha^{(1)}} (1+\mu_0|\lambda|)^{1/\alpha}|\lambda|^{p-1}|d\lambda| = \tag{55}$$

$$= (1 + \mu_0 d_0\alpha)^{1/\alpha}d_0^{p-1}\alpha^{p-1} \int\limits_{\gamma_\alpha^{(1)}} |d\lambda| \leq$$

$$\leq 2(\pi - \varphi_0) \exp(\mu_0 d_0)d_0^p\alpha^p.$$

In conditions of the lemma, for all $\alpha \in (0, \alpha_0]$ there is

$$|1 - \mu_0\lambda| \leq q_1 < 1 \quad \forall \lambda \in \gamma_\alpha^{(2)}. \tag{56}$$

By (56) it follows that

$$\int\limits_{\gamma_\alpha^{(2)}} |1 - \mu_0\lambda|^{1/\alpha}|\lambda|^{p-1}|d\lambda| \leq$$

$$\leq q_1^{1/\alpha} R_0^{p-1} \int\limits_{\gamma_\alpha^{(2)}} |d\lambda| \leq c_7\alpha^p \quad \forall \alpha \in (0, \alpha_0], \tag{57}$$

where

$$c_7 = 2R_0^p\varphi_0 \sup_{\alpha \in (0,\alpha_0]} (q_1^{1/\alpha}\alpha^{-p}).$$

On the other hand,

$$\int\limits_{\gamma_\alpha^{(3)}} |1 - \mu_0\lambda|^{1/\alpha}|\lambda|^{p-1}|d\lambda| =$$

$$= \int\limits_{d_0\alpha}^{R_0} (1 - 2\mu_0|\lambda| \cos \varphi_0 + \mu_0^2|\lambda|^2)^{1/(2\alpha)}|\lambda|^{p-1}d|\lambda| = \tag{58}$$

$$= \alpha^p \int\limits_{d_0}^{R_0\alpha^{-1}} (1 - 2\mu_0\alpha t \cos \varphi_0 + \mu_0^2\alpha^2 t^2)^{1/(2\alpha)}t^{p-1}dt.$$

The second integral in (58) was transformed by the substitution $|\lambda| = t\alpha$. Due to the upper restriction on μ_0, there exists $\mu_1 > 0$ such that for all $t \in [0, R_0\alpha^{-1}]$,

$$1 - 2\mu_0\alpha t \cos\varphi_0 + \mu_0^2\alpha^2 t^2 \leq 1 - \mu_1\alpha t.$$

Therefore the last integral in (58) is estimated by a constant uniformly in $\alpha \in (0, \alpha_0]$:

$$\int\limits_{d_0}^{R_0\alpha^{-1}} (1 - 2\mu_0\alpha t \cos\varphi_0 + \mu_0^2\alpha^2 t^2)^{1/(2\alpha)} t^{p-1} dt \leq$$

$$\leq \int\limits_{d_0}^{R_0\alpha^{-1}} (1 - \mu_1\alpha t)^{1/(2\alpha)} t^{p-1} dt \leq \int\limits_{d_0}^{\infty} \exp(-\mu_1 t/2) t^{p-1} dt < \infty.$$

The integral over $\gamma_\alpha^{(4)}$ has the same estimate. From (55), (57), and (58) it follows that the function $\Theta_1(\lambda, \alpha)$ satisfies Condition 2.7 for each $p_0 > 0$.

Now let us consider the function $\Theta_2(\lambda, \alpha)$. For each $p > 0$ and sufficiently small $\alpha_0 > 0$ we have

$$\int\limits_{\gamma_\alpha^{(1)}} |1 - \Theta_2(\lambda, \alpha)\lambda| |\lambda|^{p-1} |d\lambda| = \int\limits_{\gamma_\alpha^{(1)}} \left|\frac{\mu_0}{\mu_0 + \lambda}\right|^{1/\alpha} |\lambda|^{p-1} |d\lambda| \leq$$

$$\leq \int\limits_{\gamma_\alpha^{(1)}} \left(1 + \frac{d_0\alpha}{\mu_0 - d_0\alpha_0}\right)^{1/\alpha} d_0^{p-1} \alpha^{p-1} |d\lambda| \leq \qquad (59)$$

$$\leq 2(\pi - \varphi_0) \exp\left(\frac{d_0}{\mu_0 - d_0\alpha_0}\right) d_0^p \alpha^p \quad \forall \alpha \in (0, \alpha_0].$$

Besides, there exists a constant q_2 such that

$$\left|\frac{\mu_0}{\mu_0 + \lambda}\right| \leq q_2 < 1 \quad \forall \lambda \in \gamma_\alpha^{(2)} \cup \gamma_\alpha^{(3)} \cup \gamma_\alpha^{(4)} \quad \forall \alpha \in (0, \alpha_0]. \qquad (60)$$

From (60) it follows that the integral of $|1 - \Theta_2(\lambda, \alpha)\lambda| |\lambda|^{p-1}$ over $\gamma_\alpha^{(2)}$ is estimated analogously to (57). For the integrals over $\gamma_\alpha^{(3)}$ and $\gamma_\alpha^{(4)}$ we obtain

$$\int\limits_{\gamma_\alpha^{(k)}} \left|\frac{\mu_0}{\mu_0 + \lambda}\right|^{1/\alpha} |\lambda|^{p-1} |d\lambda| \leq$$

$$\leq \int\limits_{d_0\alpha}^{R_0} \frac{|\lambda|^{p-1}}{(1 + 2\mu_0^{-1}|\lambda| \cos\varphi_0)^{1/(2\alpha)}} d|\lambda|, \quad k = 3, 4.$$

The last integral, as the expression in (58), is of order $O(\alpha^p), \alpha \in (0, \alpha_0]$. Therefore the function $\Theta_2(\lambda, \alpha)$ satisfies Condition 2.7 for each $p_0 > 0$.

Let us remark that in conditions of the lemma,

$$|1 - \mu_0 \lambda| \le 1 \quad \forall \lambda \in \gamma_\alpha^{(2)} \cup \gamma_\alpha^{(3)} \cup \gamma_\alpha^{(4)}.$$

Hence from (60) we get

$$|\Theta_j(\lambda, \alpha)\lambda| < 2 \quad \forall \lambda \in \gamma_\alpha^{(2)} \cup \gamma_\alpha^{(3)} \cup \gamma_\alpha^{(4)} \quad (j = 1, 2). \tag{61}$$

Consequently,

$$\int_{\gamma_\alpha^{(2)} \cup \gamma_\alpha^{(3)} \cup \gamma_\alpha^{(4)}} \frac{|\Theta_j(\lambda, \alpha)|}{|\lambda|} |d\lambda| \le 2 \int_{\gamma_\alpha^{(2)} \cup \gamma_\alpha^{(3)} \cup \gamma_\alpha^{(4)}} \frac{|d\lambda|}{|\lambda|^2}, \quad j = 1, 2.$$

Furthermore,

$$\int_{\gamma_\alpha^{(2)}} \frac{|d\lambda|}{|\lambda|^2} = \frac{2\varphi_0}{R_0} \le \frac{2\varphi_0 \alpha_0}{R_0 \alpha},$$

$$\int_{\gamma_\alpha^{(k)}} \frac{|d\lambda|}{|\lambda|^2} = \int_{d_0\alpha}^{R_0} \frac{d|\lambda|}{|\lambda|^2} \le \frac{1}{d_0 \alpha} \quad \forall \alpha \in (0, \alpha_0] \quad (k = 3, 4).$$

In addition, we obviously have

$$\int_{\gamma_\alpha^{(1)}} \frac{|\Theta_1(\lambda, \alpha)|}{|\lambda|} |d\lambda| \le \int_{\gamma_\alpha^{(1)}} \frac{1 + (1 + \mu_0 |\lambda|)^{1/\alpha}}{|\lambda|^2} |d\lambda| \le$$

$$\le \frac{1 + (1 + \mu_0 d_0 \alpha)^{1/\alpha}}{d_0^2 \alpha^2} \int_{\gamma_\alpha^{(1)}} |d\lambda| \le \frac{2(1 + \exp(\mu_0 d_0))(\pi - \varphi_0)}{d_0 \alpha}$$

and, analogously,

$$\int_{\gamma_\alpha^{(1)}} \frac{|\Theta_2(\lambda, \alpha)|}{|\lambda|} |d\lambda| \le \int_{\gamma_\alpha^{(1)}} \frac{1}{|\lambda|^2} \left(1 + \left| \frac{\mu_0}{\mu_0 + \lambda} \right|^{1/\alpha} \right) |d\lambda| \le$$

$$\le \frac{1}{d_0^2 \alpha^2} \left(1 + \left(1 + \frac{d_0 \alpha}{\mu_0 - d_0 \alpha} \right)^{1/\alpha} \right) \int_{\gamma_\alpha^{(1)}} |d\lambda| \le$$

$$\le \frac{2(\pi - \varphi_0)}{d_0 \alpha} \left(1 + \exp \left(\frac{d_0}{\mu_0 - d_0 \alpha_0} \right) \right).$$

The last two estimates prove that Condition 2.8 is valid for both functions $\Theta_j(\lambda, \alpha), j = 1, 2$.

Using (55)–(61), we get

$$\sup_{\alpha \in [\alpha_0, \infty)} \left(\alpha^{-\tau} \int_{\gamma_\alpha} \frac{|1 - \Theta_j(\lambda, \alpha)\lambda|}{|\lambda|} |d\lambda| \right) < \infty \quad \forall \tau > 0 \quad (j = 1, 2).$$

Hence the functions $\Theta_j(\lambda, \alpha), j = 1, 2$ satisfy Condition 2.11 for all $\tau_0 > 0$.

The check of Conditions 2.12 and 2.13 is trivial.

At last, let us analyze Condition 2.14. According to (11), the functions $\psi(\lambda, \alpha, \varepsilon)$ corresponding to $\Theta(\lambda, \alpha) = \Theta_j(\lambda, \alpha), j = 1, 2$ have the form

$$\psi_1(\lambda, \alpha, \varepsilon) = \left(1 - \frac{\varepsilon}{\mu_0^{-1} - \lambda} \right)^{1/\alpha} - 1, \quad \psi_2(\lambda, \alpha, \varepsilon) = \left(1 - \frac{\varepsilon}{\mu_0 + \lambda + \varepsilon} \right)^{1/\alpha} - 1.$$

We denote

$$z_1(\lambda, \varepsilon) = 1 - \frac{\varepsilon}{\mu_0^{-1} - \lambda}, \quad z_2(\lambda, \varepsilon) = 1 - \frac{\varepsilon}{\mu_0 + \lambda + \varepsilon}$$

and set

$$\rho_j(\lambda, \varepsilon) = |z_j(\lambda, \varepsilon)|, \quad \varphi_j(\lambda, \varepsilon) = \arg z_j(\lambda, \varepsilon), \quad j = 1, 2.$$

Elementary geometric considerations prove that for sufficiently small $\varepsilon^{(1)} > 0$ we have

$$1 - c_8 \varepsilon \le \rho_j(\lambda, \varepsilon) \le 1 - c_9 \varepsilon, \quad |\varphi_j(\lambda, \varepsilon)| \le c_{10} \varepsilon \quad \forall \lambda \in \gamma_\alpha \quad \forall \alpha > 0$$

$$\forall \varepsilon \in (0, \varepsilon^{(1)}] \quad (j = 1, 2)$$

with the constants c_8, c_9, and c_{10} independent of $\lambda, \alpha, \varepsilon, j$. Since $|\sin t| \le |t|$, this yields

$$|\psi_j(\lambda, \alpha, \varepsilon)| = \left| 1 - \left(\rho_j(\lambda, \varepsilon) \exp(i \varphi_j(\lambda, \varepsilon)) \right)^{1/\alpha} \right| =$$

$$= \left(1 - 2\rho_j^{1/\alpha}(\lambda, \varepsilon) \cos \frac{\varphi_j(\lambda, \varepsilon)}{\alpha} + \rho_j^{2/\alpha}(\lambda, \varepsilon) \right)^{1/2} =$$

$$= \left(\left(1 - \rho_j^{1/\alpha}(\lambda, \varepsilon) \right)^2 + 4\rho_j^{1/\alpha}(\lambda, \varepsilon) \sin^2 \frac{\varphi_j(\lambda, \varepsilon)}{2\alpha} \right)^{1/2} \le$$

$$\le 1 - \rho_j^{1/\alpha}(\lambda, \varepsilon) + 2\rho_j^{1/(2\alpha)}(\lambda, \varepsilon) \left| \sin \frac{\varphi_j(\lambda, \varepsilon)}{2\alpha} \right| \le$$

$$\le 1 - (1 - c_8 \varepsilon)^{1/\alpha} + c_{10} \alpha^{-1} \varepsilon (1 - c_9 \varepsilon)^{1/(2\alpha)} \equiv$$

$$\equiv \widetilde{M}(\alpha, \varepsilon), \quad j = 1, 2.$$

Consequently,

$$\int_{\gamma_\alpha} \frac{|\psi_j(\lambda,\alpha,\varepsilon)|}{|\lambda|}|d\lambda| \le \widetilde{M}(\alpha,\varepsilon) \int_{\gamma_\alpha} \frac{|d\lambda|}{|\lambda|} =$$

$$= 2\widetilde{M}(\alpha,\varepsilon)\left(\pi + \ln\frac{R_0}{d_0\alpha}\right) \equiv M(\alpha,\varepsilon); \quad j = 1,2.$$

Therefore condition (15) is satisfied. Furthermore,

$$\int_0^{\varepsilon^{(1)}} \frac{M(\alpha,\varepsilon)}{\varepsilon}d\varepsilon =$$

$$= 2\left(\pi + \ln\frac{R_0}{d_0\alpha}\right)\left(\int_0^{\varepsilon^{(1)}} \frac{1-(1-c_8\varepsilon)^{1/\alpha}}{\varepsilon}d\varepsilon + c_{10}\int_0^{\varepsilon^{(1)}} \frac{(1-c_9\varepsilon)^{1/(2\alpha)}}{\alpha}d\varepsilon\right).$$

Now we need to estimate the integrals in the right part of this expression. Suppose $\varepsilon^{(1)}$ satisfies $1 - c_8\varepsilon^{(1)} > 0$; then setting $t = 1 - c_8\varepsilon$, we get

$$\int_0^{\varepsilon^{(1)}} \frac{1-(1-c_8\varepsilon)^{1/\alpha}}{\varepsilon}d\varepsilon = \int_{1-c_8\varepsilon^{(1)}}^1 \frac{1-t^{1/\alpha}}{1-t}dt \le \int_0^1 \frac{1-t^{1/\alpha}}{1-t}dt \le$$

$$\le c_{11}(1 + |\ln\alpha|),$$

$$\int_0^{\varepsilon^{(1)}} \frac{(1-c_9\varepsilon)^{1/(2\alpha)}}{\alpha}d\varepsilon = \frac{2}{c_9(2\alpha+1)}\left(1 - (1-c_9\varepsilon^{(1)})^{1+1/(2\alpha)}\right) \le$$

$$\le \frac{2}{c_9(2\alpha+1)} \le \frac{2}{c_9} \quad \forall\alpha > 0.$$

Hence Condition 2.14 is satisfied for both functions $\Theta_j(\lambda,\alpha), j = 1,2.$ $\qquad\square$

Since the functions $\Theta_1(\lambda,\alpha)$ and $\Theta_2(\lambda,\alpha)$ don't satisfy Condition 2.15, a proof of possible analogs of Theorem 2.15 for these functions requires additional analysis. We recall that in the case of $\Theta(\lambda,\alpha) = \Theta_j(\lambda,\alpha), j = 1,2$, the approximating elements x_α in (2.2) are evaluated only for $\alpha = \alpha_n = n^{-1}, n \in \mathbf{N}$. According to (1.30)–(1.31),

$$x_\alpha = x^{(n)}, \quad \alpha = \frac{1}{n},$$

where $x^{(0)} = \xi$ and

$$x^{(k+1)} = x^{(k)} - \mu_0(Ax^{(k)} - f) \quad (j = 1), \tag{62}$$

$$(\mu_0 E + A)x^{(k+1)} = \mu_0 x^{(k)} + f \quad (j = 2); \tag{63}$$

$$k = 0, \ldots, n - 1.$$

The following theorem is a direct consequence of Theorem 2.10 and Lemma 2.6.

Theorem 2.18. *Let conditions of Lemma 2.6 be satisfied. Suppose (2.5) is fulfilled and approximations $x^{(n)}$ are defined by (62) or (63). Then there exists a constant $l > 0$ such that*

$$\|x^{(n)} - x^*\|_X \leq \frac{l}{n^p}, \quad n \in \mathbf{N}. \tag{64}$$

We are now in a position to prove a converse result for Theorem 2.18.

Theorem 2.19. *Suppose estimate (64) is fulfilled with $x^{(n)}$ defined by (62) or (63). Assume that Condition 1.1 is satisfied with $\varphi_0 \in (0, \pi/4)$. Let*

$$0 < \mu_0 < \frac{3 - \sqrt{25 - 16\sqrt{2}}}{2R_0}$$

in case of the process (62), and $\mu_0 > 0$ for the process (63). Then (1) is valid.

Proof. First of all, we shall prove that (64) implies inequality (5) for all $\alpha \in (0, \alpha_1](\alpha_1 = 1)$ with p replaced by an arbitrary $\widetilde{p} \in (0, p)$ and with a coefficient $l = \widetilde{l}$ depending on the chosen $\widetilde{p}$. Given $\alpha, \beta > 0$, we denote

$$\chi_j(\lambda, \alpha, \beta) = \frac{1 - \Theta_j(\lambda, \alpha)\lambda}{1 - \Theta_j(\lambda, \beta)\lambda}, \quad \lambda \in \mathbf{C}, \quad j = 1, 2.$$

Pick numbers $\widetilde{p} \in (0, p)$, $n \in \mathbf{N}$, and $\alpha \in (\alpha_{n+1}, \alpha_n]$, where $\alpha_n = n^{-1}$. From (3) we obtain

$$\|x_\alpha - x^*\|_X = \|(E - \Theta_j(A, \alpha)A)(\xi - x^*)\|_X =$$

$$= \|\chi_j(A, \alpha, \alpha_n)(E - \Theta_j(A, \alpha_n)A)(\xi - x^*)\|_X \leq$$

$$\leq \|\chi_j(A, \alpha, \alpha_n)\|_{L(X)}\|x^{(n)} - x^*\|_X \leq$$

$$\leq l2^p \alpha^{\widetilde{p}} \left(\|\chi_j(A, \alpha, \alpha_n)\|_{L(X)}\alpha_n^{p-\widetilde{p}}\right), \quad j = 1, 2.$$

Arguing as in (55)–(60), we conclude that the chain of inequalities

$$\|\chi_j(A, \alpha, \alpha_n)\|_{L(X)} \le \frac{r_0}{2\pi} \int\limits_{\gamma_{\alpha_{n+1}}} \frac{|\chi_j(\lambda, \alpha, \alpha_n)|}{|\lambda|} |d\lambda| \le$$

$$\le c_{12}(1 + |\ln\alpha_n|), \quad j = 1, 2$$

is satisfied uniformly in $\alpha \in (\alpha_{n+1}, \alpha_n], n \in \mathbf{N}$. Thus for all numbers

$$\alpha \in (0, \alpha_1] = \bigcup_{n\in\mathbf{N}} (\alpha_{n+1}, \alpha_n]$$

we have

$$\|x_\alpha - x^*\|_X \le \tilde{l}\alpha^{\tilde{p}}, \quad \tilde{l} = c_{12}l2^{\tilde{p}} \sup_{n\in\mathbf{N}} \left(\alpha_n^{\tilde{p}-\tilde{p}}(1 + |\ln\alpha_n|)\right) < \infty.$$

This proves our assertion.

Lemma 2.6 now implies that in conditions of the present theorem, all the results obtained in the proof of Theorem 2.15 until equality (20) remain valid, and (20) is true with $p = \tilde{p}$. Consider the first internal integral in (27). Using the change of variables $\alpha = \tau^{-1}\lambda$, we get

$$\int\limits_0^\infty \alpha^{-\tilde{p}-1+\kappa}(1-\Theta_j(\lambda, \alpha)\lambda)d\alpha =$$

$$= \lambda^{-\tilde{p}+\kappa} \int\limits_{\Lambda(\lambda)} \tau^{\tilde{p}-1-\kappa} u_j(\lambda)^{\tau/\lambda} d\tau, \quad j = 1, 2, \tag{65}$$

where, by definition,

$$u_1(\lambda) = 1 - \mu_0\lambda, \quad u_2(\lambda) = \frac{\mu_0}{\mu_0 + \lambda};$$

$$u(\lambda)^{\tau/\lambda} = \exp\left(\frac{\tau \ln u(\lambda)}{\lambda}\right),$$

and ln is the branch of the logarithmic function characterized by $\ln 1 = 0$.

Fix a number $\kappa \in (0, \tilde{p})$. We shall prove below that

$$\int\limits_{\Lambda(\lambda)} \tau^{\tilde{p}-1-\kappa} u_j(\lambda)^{\tau/\lambda} d\tau = \left(-\frac{\lambda}{\ln u_j(\lambda)}\right)^{\tilde{p}-\kappa} \Gamma(\tilde{p} - \kappa), \quad j = 1, 2 \tag{66}$$

$$\forall \lambda \in K(R_0, \varphi_0)\backslash\{0\}.$$

First, let us show that for each $\lambda \in K(R_0, \pi/4)\backslash\{0\}$ and $j \in \{1, 2\}$, the integral

$$\int_{\Lambda(\mu)} \tau^{\tilde{p}-1-\kappa} u_j(\lambda)^{\tau/\lambda} d\tau \tag{67}$$

doesn't depend on a choice of $\mu \in K(\pi/4)\backslash\{0\}$. We shall establish simultaneously that the order of integrals in (20) can be changed.

As an auxiliary result, let us prove that

$$\arg\left(-\frac{\ln u_j(\lambda)}{\lambda}\right) \in \left(-\frac{\pi}{4}, \frac{\pi}{4}\right)$$

$$\forall \lambda \in K\left(R_0, \frac{\pi}{4}\right)\backslash\{0\}, \quad j = 1, 2. \tag{68}$$

For arbitrary $\lambda = r \exp(i\varphi) \in K(R_0, \pi/4)\backslash\{0\}$ we have

$$\arg\left(-\frac{\ln u_1(\lambda)}{\lambda}\right) = \arg(-\ln u_1(\lambda)) - \varphi =$$

$$= \arctg\left(-2\left(\ln(1 - 2\mu_0 r \cos\varphi + \mu_0^2 r^2)\right)^{-1} \arctg\frac{\mu_0 r \sin\varphi}{1 - \mu_0 r \cos\varphi}\right) - \varphi.$$

The inclusion (68) will follow by the estimates

$$\operatorname{tg}\left(-\frac{\pi}{4} + \varphi\right) <$$

$$< -2\left(\ln(1 - 2\mu_0 r \cos\varphi + \mu_0^2 r^2)\right)^{-1} \arctg\frac{\mu_0 r \sin\varphi}{1 - \mu_0 r \cos\varphi} < \tag{69}$$

$$< \operatorname{tg}\left(\frac{\pi}{4} + \varphi\right) \quad \forall r \in (0, R_0] \quad \forall \varphi \in \left(-\frac{\pi}{4}, \frac{\pi}{4}\right).$$

Due to the conditions on μ_0,

$$1 - 2\mu_0 r \cos\varphi + \mu_0^2 r^2 < 1 \quad \forall \varphi \in \left(-\frac{\pi}{4}, \frac{\pi}{4}\right).$$

Hence the left inequality in (69) is obvious for $\varphi \in [0, \pi/4)$, and the right one is valid for $\varphi \in (-\pi/4, 0]$. If $\varphi \in (-\pi/4, 0]$, then the left inequality in (69) is equivalent to

$$-2\left(\ln(1 - 2\mu_0 r \cos|\varphi| + \mu_0^2 r^2)\right)^{-1} \arctg\frac{\mu_0 r \sin|\varphi|}{1 - \mu_0 r \cos|\varphi|} < \operatorname{tg}\left(\frac{\pi}{4} + |\varphi|\right).$$

Thus it remains to justify the right inequality in (69) for $\varphi \in (0, \pi/4)$. Taking into account that $\operatorname{arctg} t < t \ \forall t > 0$ and $\ln(1+t) \le t \ \forall t > -1$, we obtain

$$
\begin{aligned}
-2\Big(\ln(1 - 2\mu_0 r \cos\varphi + \mu_0^2 r^2)\Big)^{-1} &\operatorname{arctg} \frac{\mu_0 r \sin\varphi}{1 - \mu_0 r \cos\varphi} < \\
< -2\frac{\mu_0 r \sin\varphi}{1 - \mu_0 r \cos\varphi}&(-2\mu_0 r \cos\varphi + \mu_0^2 r^2)^{-1} = \\
= \Big(1 + \frac{\mu_0^2 r^2}{2} &- \mu_0 r\Big(\cos\varphi + \frac{1}{2\cos\varphi}\Big)\Big)^{-1} \operatorname{tg}\varphi \le \\
\le \Big(1 + \frac{\mu_0^2 R_0^2}{2} &- \mu_0 R_0\Big(\cos\varphi + \frac{1}{2\cos\varphi}\Big)\Big)^{-1} \operatorname{tg}\varphi.
\end{aligned}
\tag{70}
$$

The last inequality in (70) is true, since $\mu_0 r \le \mu_0 R_0 < 1$ and for each $\varphi \in (0, \pi/4)$, the function

$$
\mathfrak{g}(t) = 1 + \frac{t^2}{2} - \Big(\cos\varphi + \frac{1}{2\cos\varphi}\Big)t
$$

is decreasing in $t \in (0, 1)$. Using the upper restriction on μ_0 we deduce

$$
\Big(1 + \frac{\mu_0^2 R_0^2}{2} - \mu_0 R_0\Big(\cos\varphi + \frac{1}{2\cos\varphi}\Big)\Big)^{-1} \operatorname{tg}\varphi \le
$$

$$
\le \Big(1 + \frac{\mu_0^2 R_0^2}{2} - \frac{3}{2}\mu_0 R_0\Big)^{-1} \operatorname{tg}\varphi \le (3 + 2\sqrt{2})\operatorname{tg}\varphi \quad \forall \varphi \in \Big(0, \frac{\pi}{4}\Big).
$$

Direct calculations prove that the last inequality in this chain turns to the equality when $\mu_0 R_0 = (3 - \sqrt{25 - 16\sqrt{2}})/2$. It is readily seen that

$$
(3 + 2\sqrt{2})\operatorname{tg}\varphi \le \frac{1 + \operatorname{tg}\varphi}{1 - \operatorname{tg}\varphi} = \operatorname{tg}\Big(\frac{\pi}{4} + \varphi\Big) \quad \forall \varphi \in \Big(0, \frac{\pi}{4}\Big),
$$

where the coefficient $3 + 2\sqrt{2}$ can't be increased. With (70) this implies (69) and (68) for $j = 1$.

For the function $u_2(\lambda)$ we have

$$
\arg\Big(-\frac{\ln u_2(\lambda)}{\lambda}\Big) = \arg(-\ln u_2(\lambda)) - \varphi =
$$

$$
= \operatorname{arctg}\Big(2\Big(\ln\Big(1 + 2\frac{r}{\mu_0}\cos\varphi + \frac{r^2}{\mu_0^2}\Big)\Big)^{-1} \operatorname{arctg}\frac{r\sin\varphi}{\mu_0 + r\cos\varphi}\Big) - \varphi.
$$

As above, it is sufficient to prove that for all $\varphi \in (0, \pi/4)$,

$$2\left(\ln\left(1 + 2\frac{r}{\mu_0}\cos\varphi + \frac{r^2}{\mu_0^2}\right)\right)^{-1} \operatorname{arctg}\frac{r\sin\varphi}{\mu_0 + r\cos\varphi} < \operatorname{tg}\left(\frac{\pi}{4} + \varphi\right).$$

From the inequality

$$\frac{t}{1+t} \le \ln(1+t) \quad \forall t \ge 0$$

it follows that for each $\mu_0 > 0$,

$$2\left(\ln\left(1 + 2\frac{r}{\mu_0}\cos\varphi + \frac{r^2}{\mu_0^2}\right)\right)^{-1} \operatorname{arctg}\frac{r\sin\varphi}{\mu_0 + r\cos\varphi} <$$

$$< \frac{2r\sin\varphi}{\mu_0 + r\cos\varphi}\left(1 + 2\frac{r}{\mu_0}\cos\varphi + \frac{r^2}{\mu_0^2}\right)\left(2\frac{r}{\mu_0}\cos\varphi + \frac{r^2}{\mu_0^2}\right)^{-1} =$$

$$= 2\left(1 + 2\frac{r}{\mu_0}\cos\varphi + \frac{r^2}{\mu_0^2}\right)\left(2 + 2\frac{r}{\mu_0}\left(\cos\varphi + \frac{1}{2\cos\varphi}\right) + \frac{r^2}{\mu_0^2}\right)^{-1}\operatorname{tg}\varphi \le$$

$$\le \frac{2(1 + r/\mu_0)^2}{2 + 2\sqrt{2}r/\mu_0 + (r/\mu_0)^2}\operatorname{tg}\varphi \le 2\operatorname{tg}\varphi \le \operatorname{tg}\left(\frac{\pi}{4} + \varphi\right) \quad \forall\varphi \in \left(0, \frac{\pi}{4}\right).$$

This completes the proof of (68) for $j = 2$.

It can easily be checked that

$$|u_j(\lambda)^{\tau/\lambda}| = \left|\exp\left(\frac{\tau\ln u_j(\lambda)}{\lambda}\right)\right| = a_j(\lambda, \tau)^{|\tau|},$$

where

$$a_j(\lambda, \tau) = \exp\left(\operatorname{Re}\frac{\ln u_j(\lambda)\exp(i\arg\tau)}{\lambda}\right), \quad j = 1, 2.$$

Since in (67),

$$|\arg\tau| = |\arg\mu| < \frac{\pi}{4},$$

from (68) we get

$$a_j(\lambda, \tau) \le q_3 = q_3(\mu) < 1, \quad j = 1, 2$$

$$\forall\lambda \in K\left(R_0, \frac{\pi}{4}\right)\setminus\{0\} \quad \forall\tau \in \Lambda(\mu), \tag{71}$$

whatever $\mu \in K(\pi/4)\setminus\{0\}$. Indeed,

$$a_j(\lambda, \tau) = \exp\left(\left|\frac{\ln u_j(\lambda)}{\lambda}\right|\cos\arg\left(\frac{\ln u_j(\lambda)\exp(i\arg\tau)}{\lambda}\right)\right),$$

where

$$\inf_{\lambda \in K(R_0, \pi/4) \setminus \{0\}} \left| \frac{\ln u_j(\lambda)}{\lambda} \right| > 0. \quad j = 1, 2. \tag{72}$$

Besides, (68) implies

$$\arg\left(\frac{\ln u_j(\lambda) \exp(i \arg \tau)}{\lambda} \right) = \arg\left(-\frac{\ln u_j(\lambda)}{\lambda} \right) + \pi + \arg \tau \in$$

$$\in \left(\frac{3\pi}{4} - |\arg \mu|, \frac{5\pi}{4} + |\arg \mu| \right) \quad \forall \lambda \in K\left(R_0, \frac{\pi}{4} \right) \setminus \{0\}. \tag{73}$$

Using (72) and (73), we immediately obtain (71). By (71) it follows that for each ε, the integrals in (65) exist uniformly in $\lambda \in \Gamma^{(\varepsilon)}(\varphi_0)$ (see (19)). With (21) and (26), this justifies a change of the order of integrals in (20).

Again turning to integral (67), suppose $\mu \in K(\pi/4) \setminus \{0\}$; then we get

$$\lim_{r \to 0} \left| \int_{\Gamma_r(1,\mu)} \tau^{\tilde{p}-1-\kappa} u_j(\lambda)^{\tau/\lambda} d\tau \right| \leq \lim_{r \to 0} \int_{\Gamma_r(1,\mu)} |\tau|^{\tilde{p}-1-\kappa} a_j(\lambda, \tau)^{|\tau|} |d\tau| \leq$$

$$\leq \lim_{r \to 0} r^{\tilde{p}-1-\kappa} q_3^r \int_{\Gamma_r(1,\mu)} |d\tau| = \lim_{r \to 0} r^{\tilde{p}-\kappa} q_3^r |\arg \mu| = 0, \quad j = 1, 2. \tag{74}$$

In the same way,

$$\lim_{R \to \infty} \left| \int_{\Gamma_R(1,\mu)} \tau^{\tilde{p}-1-\kappa} u_j(\lambda)^{\tau/\lambda} d\tau \right| \leq$$

$$\leq \lim_{R \to \infty} R^{\tilde{p}-1-\kappa} q_3^R \int_{\Gamma_R(1,\mu)} |d\tau| = \tag{75}$$

$$= \lim_{R \to \infty} R^{\tilde{p}-\kappa} q_3^R |\arg \mu| = 0, \quad j = 1, 2.$$

Since the function $\tau^{\tilde{p}-1-\kappa} u_j(\lambda)^{\tau/\lambda}$ is analytic in τ on $K(R_0, \pi/4) \setminus \{0\}$, we have

$$\int_{\Gamma_{(r,R)}(1)} \tau^{\tilde{p}-1-\kappa} u_j(\lambda)^{\tau/\lambda} d\tau + \int_{\Gamma_R(1,\mu)} + \int_{\Gamma_{(r,R)}(\mu)} + \int_{\Gamma_r(1,\mu)} = 0.$$

Passing to the limits as $r \to 0$ and $R \to \infty$, with the use of (74) and (75) we obtain

$$\int_{\Lambda(\mu)} \tau^{\tilde{p}-1-\kappa} u_j(\lambda)^{\tau/\lambda} d\tau = \int_{\Lambda(1)} \tau^{\tilde{p}-1-\kappa} u_j(\lambda)^{\tau/\lambda} d\tau, \quad j = 1, 2.$$

This equality proves our assertion concerning independence of the integral in (67) of $\mu \in K(\pi/4)\backslash\{0\}$.

We denote

$$\phi_j(\lambda) = \arg\left(-\frac{\ln u_j(\lambda)}{\lambda}\right), \quad \lambda \in K(R_0, \varphi_0)\backslash\{0\}, \quad j = 1, 2.$$

According to (68),

$$\phi_j(\lambda) \in \left(-\frac{\pi}{4}, \frac{\pi}{4}\right)$$

and hence there exists $\mu \in K(\pi/4)\backslash\{0\}$ such that

$$\Lambda\big(\exp(-i\phi_j(\lambda))\big) = \Lambda(\mu).$$

Taking into account that the integral (67) doesn't depend on $\mu \in K(\pi/4)\backslash\{0\}$, with the use of substitution

$$t = -\frac{\tau \ln u_j(\lambda)}{\lambda}$$

we obtain

$$\int_{\Lambda(\lambda)} \tau^{\tilde{p}-1-\kappa} u_j(\lambda)^{\tau/\lambda} d\tau =$$

$$= \int_{\Lambda(\lambda)} \tau^{\tilde{p}-1-\kappa} \exp\left(\frac{\ln u_j(\lambda)}{\lambda}\tau\right) d\tau =$$

$$= \int_{\Lambda(\exp(-i\phi_j(\lambda)))} \tau^{\tilde{p}-1-\kappa} \exp\left(-\left(-\frac{\ln u_j(\lambda)}{\lambda}\right)\tau\right) d\tau =$$

$$= \left(-\frac{\lambda}{\ln u_j(\lambda)}\right)^{\tilde{p}-\kappa} \Gamma(\tilde{p}-\kappa) \quad \forall \lambda \in K(R_0, \varphi_0)\backslash\{0\}, \quad j = 1, 2$$

The equality (66) is proved.

Setting in (27) $p = \widetilde{p}$, $\alpha = \tau^{-1}\lambda$, $t = \lambda\zeta$ and using (65), for all $\kappa \in (0, \widetilde{p})$, $j \in \{1, 2\}$ and sufficiently small $\varepsilon > 0$ we get

$$
A_\varepsilon^{\widetilde{p}-\kappa} w_\kappa^{(\varepsilon)} = D(\widetilde{p}, \kappa) \int_{\Gamma^{(\varepsilon)}(\varphi_0)} \lambda^{m+1} \left(\int_0^\infty \alpha^{-\widetilde{p}-1+\kappa} (1 - \Theta_j(\lambda, \alpha)\lambda) d\alpha \right) \cdot
$$

$$
\cdot \left(\int_0^\infty t^{\widetilde{p}-\kappa-m-1} (t + \lambda)^{-1} dt \right) R(\lambda, A_\varepsilon)(x^* - \xi) d\lambda =
$$

$$
= D(\widetilde{p}, \kappa) \int_{\Gamma^{(\varepsilon)}(\varphi_0)} \lambda^{m+1} \left(\lambda^{-\widetilde{p}+\kappa} \int_{\Lambda(\lambda)} \tau^{\widetilde{p}-1-\kappa} u_j(\lambda)^{\tau/\lambda} d\tau \right) \cdot
$$

$$
\cdot \left(\lambda^{\widetilde{p}-\kappa-m-1} \int_{\Lambda(\overline{\lambda})} \zeta^{\widetilde{p}-\kappa-m-1} (1 + \zeta)^{-1} d\zeta \right) R(\lambda, A_\varepsilon)(x^* - \xi) d\lambda =
$$

$$
= D(\widetilde{p}, \kappa) H(\widetilde{p}, \kappa) \int_{\Gamma^{(\varepsilon)}(\varphi_0)} h_j(\lambda) R(\lambda, A_\varepsilon)(x^* - \xi) d\lambda.
$$

Therefore,

$$
A_\varepsilon^{\widetilde{p}-\kappa} w_\kappa^{(\varepsilon)} = E(\widetilde{p}, \kappa) h_j(A_\varepsilon)(x^* - \xi). \tag{76}
$$

Here, we have introduced the following notation

$$
E(\widetilde{p}, \kappa) = 2\pi i D(\widetilde{p}, \kappa) H(\widetilde{p}, \kappa),
$$

$$
h_j(\lambda) = \int_{\Lambda(\lambda)} \tau^{\widetilde{p}-1-\kappa} u_j(\lambda)^{\tau/\lambda} d\tau, \quad j = 1, 2.
$$

According to (66),

$$
h_j(\lambda) = \left(-\frac{\lambda}{\ln u_j(\lambda)} \right)^{\widetilde{p}-\kappa} \Gamma(\widetilde{p} - \kappa), \quad j = 1, 2; \quad \lambda \neq 0.
$$

Let us extend the functions $h_j(\lambda)$, $j = 1, 2$ to $\lambda = 0$ by continuity putting $h_j(0) = \lim_{\lambda \to 0} h_j(\lambda) \neq 0$. If $\delta_0 > 0$ is sufficiently small, then the extended functions $h_j(\lambda)$, $j = 1, 2$ are analytic and have no zeros in a δ_0–neighborhood

$$
O_{\delta_0} = \{z \in \mathbf{C} : \text{dist}(z, K(R_0, \varphi_0)) < \delta_0\}
$$

of the sector $K(R_0, \varphi_0)$. Let σ_0 be a contour that lies in this neighborhood and surrounds the neighborhood $O_{\delta_0/2}$ of the same sector. Since

$$
h_j(\lambda) \neq 0 \quad \forall \lambda \in O_{\delta_0},
$$

the functions $1/h_j(\lambda)$, $j = 1, 2$ are analytic on O_{δ_0}. From $\sigma(A_\varepsilon) \subset K(R_0, \varphi_0)$ we conclude that the operators

$$h_j(A_\varepsilon)^{-1} = \frac{1}{2\pi i} \int_{\sigma_0} h_j(\lambda)^{-1} R(\lambda, A_\varepsilon) d\lambda \in L(X), \quad j = 1, 2$$

are well defined. By (76), using (1.1.5) and (1.1.9), we get

$$A_\varepsilon^{\tilde{p}-\kappa} h_j(A_\varepsilon)^{-1} w_\kappa^{(\varepsilon)} = h_j(A_\varepsilon)^{-1} A_\varepsilon^{\tilde{p}-\kappa} w_\kappa^{(\varepsilon)} = E(\tilde{p}, \kappa)(x^* - \xi). \tag{77}$$

Following the scheme of the proof of Theorem 2.15, consider a sequence $\{\varepsilon_n\}$ such that

$$\varepsilon_n \leq \frac{\delta_0}{4}, \quad \lim_{n \to \infty} \varepsilon_n = 0, \quad \lim_{n \to \infty} \|w_\kappa^{(\varepsilon_n)} - w_\kappa\|_X = 0.$$

Let us prove that

$$\lim_{n \to \infty} \|A_{\varepsilon_n}^{\tilde{p}-\kappa} h_j(A_{\varepsilon_n})^{-1} w_\kappa^{(\varepsilon_n)} - A^{\tilde{p}-\kappa} h_j(A)^{-1} w_\kappa\|_X = 0, \quad j = 1, 2. \tag{78}$$

We have the chain of inequalities

$$\|A_{\varepsilon_n}^{\tilde{p}-\kappa} h_j(A_{\varepsilon_n})^{-1} w_\kappa^{(\varepsilon_n)} - A^{\tilde{p}-\kappa} h_j(A)^{-1} w_\kappa\|_X \leq$$
$$\leq \|A_{\varepsilon_n}^{\tilde{p}-\kappa} h_j(A_{\varepsilon_n})^{-1}\|_{L(X)} \|w_\kappa^{(\varepsilon_n)} - w_\kappa\|_X +$$
$$+ \|A_{\varepsilon_n}^{\tilde{p}-\kappa} h_j(A_{\varepsilon_n})^{-1} - A^{\tilde{p}-\kappa} h_j(A)^{-1}\|_{L(X)} \|w_\kappa\|_X \leq$$
$$\leq \|A_{\varepsilon_n}^{\tilde{p}-\kappa}\|_{L(X)} \left(\|h_j(A_{\varepsilon_n})^{-1} - h_j(A)^{-1}\|_{L(X)} + \right. \tag{79}$$
$$+ \|h_j(A)^{-1}\|_{L(X)} \right) \|w_\kappa^{(\varepsilon_n)} - w_\kappa\|_X +$$
$$+ \|A_{\varepsilon_n}^{\tilde{p}-\kappa}\|_{L(X)} \|h_j(A_{\varepsilon_n})^{-1} - h_j(A)^{-1}\|_{L(X)} \|w_\kappa\|_X +$$
$$+ \|A_{\varepsilon_n}^{\tilde{p}-\kappa} - A^{\tilde{p}-\kappa}\|_{L(X)} \|h_j(A)^{-1}\|_{L(X)} \|w_\kappa\|_X.$$

In (79), the multiplier $\|A_{\varepsilon_n}^{\tilde{p}-\kappa}\|_{L(X)}$ is bounded as $n \to \infty$. Besides, by Lemma 1.1,

$$\lim_{n \to \infty} \|A_{\varepsilon_n}^{\tilde{p}-\kappa} - A^{\tilde{p}-\kappa}\|_{L(X)} = 0.$$

To conclude the proof of (78) it suffices to establish the equality

$$\lim_{n \to \infty} \|h_j(A_{\varepsilon_n})^{-1} - h_j(A)^{-1}\|_{L(X)} = 0, \quad j = 1, 2. \tag{80}$$

Since $\varepsilon_n \leq \delta_0/4$, the points $\lambda, \lambda - \varepsilon_n$ lie outside $K(R_0, \varphi_0)$ when $\lambda \in \sigma_0$. From Condition 1.1 we then obtain

$$\|R(\lambda, A)\|_{L(X)} \leq \frac{c_{13}}{|\lambda|}, \quad \|R(\lambda - \varepsilon_n, A)\|_{L(X)} \leq \frac{c_{13}}{|\lambda - \varepsilon_n|} \quad \forall \lambda \in \sigma_0.$$

Furthermore,

$$|\lambda| \geq \frac{\delta_0}{2}, \quad |\lambda - \varepsilon_n| > \frac{\delta_0}{4} \quad \forall \lambda \in \sigma_0,$$

and the compactness of σ_0 implies $|h_j(\lambda)| \geq c_{14} > 0 \quad \forall \lambda \in \sigma_0$. Consequently,

$$\|h_j(A_{\varepsilon_n})^{-1} - h_j(A)^{-1}\|_{L(X)} \leq$$

$$\leq \frac{1}{2\pi} \int_{\sigma_0} |h_j(\lambda)|^{-1} \|R(\lambda, A_{\varepsilon_n}) - R(\lambda, A)\|_{L(X)} |d\lambda| =$$

$$= \frac{1}{2\pi} \int_{\sigma_0} |h_j(\lambda)|^{-1} \|R(\lambda - \varepsilon_n, A) - R(\lambda, A)\|_{L(X)} |d\lambda| =$$

$$= \frac{\varepsilon_n}{2\pi} \int_{\sigma_0} |h_j(\lambda)|^{-1} \|R(\lambda - \varepsilon_n, A)R(\lambda, A)\|_{L(X)} |d\lambda| \leq$$

$$\leq c_{15}\varepsilon_n \int_{\sigma_0} \frac{|d\lambda|}{|\lambda||\lambda - \varepsilon_n||h_j(\lambda)|} \leq c_{16}\varepsilon_n \int_{\sigma_0} |d\lambda| = c_{17}\varepsilon_n,$$

$$j = 1, 2.$$

Therefore the equalities (80) and (78) are valid. Setting in (77) $\varepsilon = \varepsilon_n$ and passing to the limit as $\varepsilon_n \to 0$, due to (78) we get

$$A^{\tilde{p}-\kappa} h_j(A)^{-1} w_\kappa = E(\tilde{p}, \kappa)(x^* - \xi).$$

Thus with the notation

$$v_{\kappa,j} = E(\tilde{p}, \kappa)^{-1} h_j(A)^{-1} w_\kappa, \quad j = 1, 2,$$

we obtain

$$A^{\tilde{p}-\kappa} v_{\kappa,j} = x^* - \xi. \tag{81}$$

Since $\kappa \in (0, \tilde{p})$ can be chosen arbitrarily small and $\tilde{p} \in (0, p)$ can be taken arbitrarily close to p, the equality (81) completes the proof. $\square$

Remark 2.8. The coefficient of R_0^{-1} in the formulation of Theorem 2.19 equals approximately $0.729 \ldots$. Therefore the conditions of this converse theorem are slightly stronger than those of the original Theorem 2.18 regarding both the angle φ_0 and the stepsize μ_0.

Chapter 3

PARAMETRIC
APPROXIMATIONS OF SOLUTIONS
TO NONLINEAR OPERATOR EQUATIONS

In the contemporary theory of ill–posed problems, the linearization formalism is an extensively used tool popular both in itself, as an effective instrument of theoretical analysis of mathematical models, and as a basis for various numerical methods. From the formal point of view, linearization procedures can be used to the same extent in regular and irregular cases; the linearization technique only requires that operators of the problems be sufficiently smooth. However in the irregular situation, unlike the regular case, it is in general impossible to justify the closeness of approximations generated by the linearization procedures to a solution of the original nonlinear equation. Besides, a substantial impediment to the use of classical linearization schemes in the irregular case is that solvability of obtained linearized equations can't be guaranteed. In this chapter we present and study several modifications of the traditional linearization formalism and show that the modified schemes are applicable to nonlinear irregular equations. For this purpose we invoke the technique of parametric approximations of solutions to ill–posed linear equations developed in Chapter 2. We emphasize that our approach doesn't require the solvability of linearized equations. Section 3.1 is devoted to a brief description of the classical linearization technique when applied to smooth nonlinear operator equations in Hilbert and Banach spaces. In Section 3.2 we present several modifications of the classical linearization schemes and establish error estimates for proposed parametric approximations of nonlinear irregular equations. In Section 3.3 we address to a characteristic example of a smooth irregular problem, for which the previously developed formalism is well suited. This is a 3D inverse acoustic scattering problem serving as a model for more complicated and realistic inverse problems of this kind.

3.1   Classical Linearization Schemes

Let us consider a nonlinear operator $F : X_1 \to X_2$, where X_1 and X_2 are Hilbert or Banach spaces. We denote

$$\Omega_R(x_0) = \{x \in X_1 : \|x - x_0\|_{X_1} \le R\} \quad (x_0 \in X_1)$$

and suppose $F(x)$ is Fréchet differentiable on $\Omega_R(x_0)$ with the Lipschitz continuous derivative $F'(x)$ so that

$$\|F'(x) - F'(y)\|_{L(X_1, X_2)} \le L\|x - y\|_{X_1} \quad \forall x, y \in \Omega_R(x_0). \tag{1}$$

The object of our study is the problem of finding a solution to an operator equation

$$F(x) = 0, \quad x \in \Omega_R(x_0) \tag{2}$$

in a small neighborhood $\Omega_R(x_0)$ of the point $x_0 \in X_1$. Let $x^* \in \Omega_R(x_0)$ be a solution to (2). According to Taylor's formula,

$$0 = F(x^*) \approx F(x_0) + F'(x_0)(x^* - x_0).$$

Therefore it is natural to seek an approximation u of the unknown additive correction $x^* - x_0$ for x_0 solving the linear equation

$$F(x_0) + F'(x_0)u = 0, \quad u \in X_1. \tag{3}$$

Assume that the operator $F'(x_0) : X_1 \to X_2$ has a continuous inverse $F'(x_0)^{-1}$ defined on the all of X_2. This is just the case where the original equation (1) is regular, and the linearized problem (3) is well–posed. From (3) we then obtain

$$u = -F'(x_0)^{-1}F(x_0)$$

and come to the following scheme of approximating the unknown solution x^* to (2):

$$x^* \approx x_0 - F'(x_0)^{-1}F(x_0). \tag{4}$$

The error of this approximation method $\|(x_0 + u) - x^*\|_{X_1}$ has the order $O(\|x^* - x_0\|_{X_1}^2)$. Indeed, with the use of (1) and (1.1.14) we get

$$\|(x_0 - F'(x_0)^{-1}F(x_0)) - x^*\|_{X_1} =$$
$$= \|x^* - x_0 + F'(x_0)^{-1}(F(x_0) - F(x^*))\|_{X_1} =$$
$$= \|F'(x_0)^{-1}[F(x_0) - F(x^*) - F'(x_0)(x_0 - x^*)]\|_{X_1} \le \tag{5}$$
$$\le \frac{1}{2}L\|F'(x_0)^{-1}\|_{L(X_2, X_1)}\|x_0 - x^*\|_{X_1}^2.$$

The above arguing can be extended to the case where the linearized equation (3) is well–posed in the least squares sense. In fact, suppose X_1, X_2 are Hilbert spaces and the operator $F'^*(x_0)F'(x_0) : X_1 \to X_1$ has a continuous inverse $(F'^*(x_0)F'(x_0))^{-1}$ defined on the all of X_1. Then (1) is a regular equation and an appropriate correction u for x_0 can be found from the well–posed least squares problem related to (3):

$$\min\{\|F(x_0) + F'(x_0)u\|_{X_2}^2 : u \in X_1\}. \tag{6}$$

In this event, instead of (4) we obtain the following approximation scheme:

$$x^* \approx x_0 - (F'^*(x_0)F'(x_0))^{-1}F'^*(x_0)F(x_0). \tag{7}$$

The error estimate now takes a form similar to (5):

$$\|[x_0-(F'^*(x_0)F'(x_0))^{-1}F'^*(x_0)F(x_0)] - x^*\|_{X_1} \le$$
$$\le \frac{1}{2}L\|(F'^*(x_0)F'(x_0))^{-1}F'^*(x_0)\|_{L(X_2,X_1)}\|x_0 - x^*\|_{X_1}^2. \tag{8}$$

In the irregular situation, when the problems (3) and (6) are ill–posed, the approximations (4) and (7) lose their meaning since the ranges $R(F'(x_0))$ and $R(F'^*(x_0)F'(x_0))$ may be (and in practice actually are) proper not necessarily closed subsets of the spaces X_2 and X_1 respectively. As an important example, we point to the case where the derivative $F'(x_0)$ is a compact operator. On the other hand, compactness of a derivative is a rather typical feature for operators that arise in inverse problems of mathematical physics ([17, 30, 42, 67]). Hence to most of these problems traditional approximation schemes (4) and (7) are not applicable.

It is well known that iteration of formulae (4) and (7) provides a foundation for construction of rapidly converging numerical methods for regular equations (2). Namely, denote the right part of (4) or (7) by x_1; repeating the linearization procedure at the point x_1, in the similar way we determine the next approximation x_2, and so on. In this manner we come to the classical Newton–Kantorovich method

$$x_{n+1} = x_n - F'(x_n)^{-1}F(x_n) \tag{9}$$

and to the Gauss–Newton method

$$x_{n+1} = x_n - (F'^*(x_n)F'(x_n))^{-1}F'^*(x_n)F(x_n). \tag{10}$$

The rapid local convergence of iterative processes (9) and (10) is guaranteed by the continuity of operators $F'(x)^{-1}, (F'^*(x)F'(x))^{-1}, x \in \Omega_R(x_0)$ and by the error estimates (5) and (8). In the irregular case, an effective and even formal implementation of iterations (9) and (10) is impossible for the reason that inclusions $F(x_n) \in R(F'(x_n))$ and $F'^*(x_n)F(x_n) \in R(F'^*(x_n)F'(x_n))$ can't be guaranteed. Below we shall see that in general irregular situation, the technique of approximation of inverse operators $F'(x_0)^{-1}$ and $(F'^*(x_0)F'(x_0))^{-1}$

by parametric expressions (2.1.3), (2.1.10), and (2.2.2) may be fruitful. Such an approximation allows

1) to obtain for x^* local parametric approximations having error estimates of order $O(\|x^* - x_0\|_{X_1}^p)$ with $p > 1$; according to (5) and (8), in regular case the error of approximations (4) and (7) has the order $O(\|x^* - x_0\|_{X_1}^2)$;

2) to construct effective iterative processes applicable to irregular nonlinear operator equations with arbitrary degeneration.

3.2   Parametric Approximations of Irregular Equations

Let us turn to equation (1.2). Suppose $F : X_1 \to X_2$ is a Fréchet differentiable operator satisfying (1.1), and X_1, X_2 are Hilbert spaces. From (1.1) it follows that

$$\|F'(x)\|_{L(X_1,X_2)} \leq N_1 \quad \forall x \in \Omega_R(x_0); \quad N_1 = \|F'(x_0)\|_{L(X_1,X_2)} + LR.$$

Let $x^* \in \Omega_R(x_0)$ be a solution to (1.2). Setting in (1.3) $u = z - x_0$, $z \approx x^*$, and applying the parametric approximation scheme (2.1.11) to the linearized equation

$$F(x_0) + F'(x_0)(z - x_0) = 0, \quad z \in X_1,$$

we obtain

$$x^* \approx z_\alpha(\xi) =$$
$$= \xi - \Theta(F'^*(x_0)F'(x_0), \alpha)F'^*(x_0)(F(x_0) - F'(x_0)(x_0 - \xi))). \tag{1}$$

Formula (1) defines the two–parameter approximating family

$$\{z_\alpha(\xi) : \alpha > 0, \xi \in X_1\}.$$

The problem now is to minimize $\|z_\alpha(\xi) - x^*\|_{X_1}$ by variations of the parameters $\alpha > 0$ and $\xi \in X_1$. The following theorem is a starting point for the subsequent analysis.

Theorem 3.1. *Suppose (1.1) is satisfied and there exist a Borel measurable function $\varphi(\lambda), \lambda \geq 0$ and an element $v \in X_1$ such that*

$$x^* - \xi = \varphi(F'^*(x_0)F'(x_0))v. \tag{2}$$

Also, assume that

$$\sup_{\lambda \in [0, N_1^2]} |\Theta(\lambda, \alpha)\sqrt{\lambda}| \leq \frac{c_1}{\sqrt{\alpha}}, \tag{3}$$

$$\sup_{\lambda \in [0, N_1^2]} |1 - \Theta(\lambda, \alpha)\lambda||\varphi(\lambda)| \leq c_2 \alpha^p \quad \forall \alpha > 0, \tag{4}$$

where $p > 0$. Then

$$\|z_\alpha(\xi) - x^*\|_{X_1} \le \frac{c_1 L}{2\sqrt{\alpha}}\|x_0 - x^*\|_{X_1}^2 + c_2\|v\|_{X_1}\alpha^p, \quad \alpha > 0. \quad (5)$$

Proof. Since $F(x^*) = 0$, we have

$$\begin{aligned}
F'^*(x_0)F(x_0) =& F'^*(x_0)F'(x_0)(x_0 - x^*) + \\
& + F'^*(x_0)[F(x_0) - F(x^*) - F'(x_0)(x_0 - x^*)].
\end{aligned} \quad (6)$$

Now, by (1) and (6),

$$\begin{aligned}
z_\alpha(\xi) - x^* =& -[E - \Theta(F'^*(x_0)F'(x_0), \alpha)F'^*(x_0)F'(x_0)](x^* - \xi) - \\
& - \Theta(F'^*(x_0)F'(x_0), \alpha)F'^*(x_0)[F(x_0) - F(x^*) - F'(x_0)(x_0 - x^*)].
\end{aligned} \quad (7)$$

Combining (2) with (7) and using estimates (3), (4), and (1.1.14), we obtain (5). □

From the geometric point of view, representation (2) with $\|v\|_{X_1} \le d$ means that the element ξ is supposed to lie in an ellipsoid centered at x^* and with diameter depending on d. If the operator $\varphi(F'^*(x_0)F'(x_0))$ is not continuously invertible, this ellipsoid has an empty interior. Notice that Theorem 3.1 doesn't require that operators $F'(x_0)$, $F'^*(x_0)F'(x_0)$ be continuously invertible. Hence this statement covers the case of irregular equations (1.2).

Remark 3.1. When the original problem is regular, we can simply set

$$\Theta(\lambda, \alpha) = \frac{1}{\lambda}, \quad \alpha = 1.$$

Then (1) takes the form (1.7), and conditions (2) and (4) are fulfilled with $\varphi(\lambda) = 1$, $c_2 = 0$. Therefore (5) leads to the classical error estimate (1.8).

The sourcewise representations (2) are most commonly considered with the functions

$$\varphi(\lambda) = \lambda^\mu, \quad \mu > 0$$

(see Chapter 2). In that case all generating functions $\Theta(\lambda, \alpha)$ from Examples 2.1–2.4 satisfy (4) with an appropriate exponent p. If, for instance, $\Theta(\lambda, \alpha) = (\lambda + \alpha)^{-1}$, then in the right part of (4) we have $p = \min\{\mu, 1\}$. For the approximating schemes from Examples 2.3 and 2.4, $p = \mu$ whatever $\mu > 0$.

Along with the controlling element $\xi \in X_1$, the scheme (1) involves a free numerical parameter $\alpha > 0$. Minimizing the right part of (5) over $\alpha > 0$, we can determine an error order, potentially accessible by $z_\alpha(\xi)$ in the irregular case. The following assertion gives the optimal value of α and corresponding optimal error order.

Corollary 3.1. *Suppose conditions of Theorem 3.1 are satisfied and*

$$\alpha = \left(\frac{c_1 L}{4pc_2 \|v\|_{X_1}} \right)^{2/(2p+1)} \|x_0 - x^*\|_{X_1}^{4/(2p+1)}.$$

Then we have

$$\|z_\alpha(\xi) - x^*\|_{X_1} = O(\|x_0 - x^*\|_{X_1}^{4p/(2p+1)}). \tag{8}$$

It is readily seen that estimate (8) remains valid for each choice of α such that

$$c_3 \|x_0 - x^*\|_{X_1}^{4/(2p+1)} \le \alpha \le c_4 \|x_0 - x^*\|_{X_1}^{4/(2p+1)}. \tag{9}$$

From (8) it follows that taking $p > 0$ sufficiently large and making a choice of ξ and α subject to (2) and (9), we can approximate x^* by $z_\alpha(\xi)$ with an error of order

$$\|z_\alpha(\xi) - x^*\|_{X_1} = O(\|x_0 - x^*\|_{X_1}^{2-\varepsilon}),$$

where $\varepsilon > 0$ can be chosen arbitrarily small. At the same time, if the exponent p in (4) appears to be less than $1/2$, then Theorem 3.1 doesn't guarantee the error order better than $O(\|x_0 - x^*\|_{X_1})$ whatever α be. Notice that the trivial approximation $x^* \approx x_0$ provides just the same error order. Nevertheless the approximation (1) with an appropriate α can be essentially better than the trivial one, provided the Lipschitz constant L in (1.1) is sufficiently small.

Another way to increase an exponent of the error order is to iterate the scheme (1). As an example, let us discuss the following two–stage approximating process

$$y_\alpha(\xi) = \xi - \Theta(F'^*(x_0)F'(x_0), \alpha)F'^*(x_0)(F(x_0) - F'(x_0)(x_0 - \xi))), \tag{10}$$

$$x^* \approx z_\alpha(\xi) =$$

$$= \xi - \Theta(F'^*(x_0)F'(x_0), \alpha)F'^*(x_0)(F(y_\alpha(\xi)) - F'(x_0)(y_\alpha(\xi) - \xi))). \tag{11}$$

Compare (10)–(11) with the scheme obtained by the formal iteration of (1) where expressions (10) and (11) are evaluated sequentially with x_0 in (11) replaced by $y_\alpha(\xi)$. Simple calculations prove that potentially accessible error estimate then takes the form

$$\|z_\alpha(\xi) - x^*\|_{X_1} = O(\|x_0 - x^*\|_{X_1}^{4-\varepsilon})$$

with arbitrarily small $\varepsilon > 0$. At the same time, the doubling of the exponent in this error estimate, as compared with the original method (1), is connected with the proportional increase of computational complexity. Since in (10) and (11) the same operators $F'(x_0), \Theta(F'^*(x_0)F'(x_0), \alpha)$ are used, the procedure (10)–(11) is less expensive in the computational sense.

Theorem 3.2. *Suppose* $y_\alpha(\xi) \in \Omega_R(x_0)$ *and conditions of Theorem 3.1 are satisfied. Let* $z_\alpha(\xi)$ *be defined by (10)–(11). Then for all* $\alpha > 0$,

$$\|z_\alpha(\xi) - x^*\|_{X_1} \le c_5 \Big(\alpha^{-3/2} \|x_0 - x^*\|_{X_1}^4 + \alpha^{-1} \|x_0 - x^*\|_{X_1}^3 + \\ + \|v\|_{X_1} \alpha^{p-1/2} \|x_0 - x^*\|_{X_1} + \|v\|_{X_1} \alpha^p (1 + \|v\|_{X_1} \alpha^{p-1/2}) \Big). \tag{12}$$

The proof follows by the equality

$$z_\alpha(\xi) - x^* = [E - \Theta(F'^*(x_0)F'(x_0), \alpha)F'^*(x_0)F'(x_0)](\xi - x^*) - \\ - \Theta(F'^*(x_0)F'(x_0), \alpha)F'^*(x_0) \cdot \\ \cdot [F(y_\alpha(\xi)) - F(x^*) - F'(y_\alpha(\xi))(y_\alpha(\xi) - x^*)] - \\ - \Theta(F'^*(x_0)F'(x_0), \alpha)F'^*(x_0)[F'(y_\alpha(\xi)) - F'(x_0)](y_\alpha(\xi) - x^*)$$

and the estimate

$$\|F'(y_\alpha(\xi)) - F'(x_0)\|_{L(X_1, X_2)} \le L\|y_\alpha(\xi) - x_0\|_{X_1} \le \\ \le L(\|y_\alpha(\xi) - x^*\|_{X_1} + \|x_0 - x^*\|_{X_1}).$$

Minimizing the right part of (12) over $\alpha > 0$, we obtain

Corollary 3.2. *Suppose conditions of Theorem 3.2 are satisfied with* $p \ge 1/2$. *Let*

$$c_6 \|x_0 - x^*\|_{X_1}^{3/(p+1)} \le \alpha \le c_7 \|x_0 - x^*\|_{X_1}^{3/(p+1)}.$$

Then

$$\|z_\alpha(\xi) - x^*\|_{X_1} = O(\|x_0 - x^*\|_{X_1}^{3p/(p+1)}). \tag{13}$$

From (13) it follows that the iterated approximating scheme (10)–(11) with an appropriate choice of ξ and α can potentially (as $p \to \infty$) provide the error order

$$\|z_\alpha(\xi) - x^*\|_{X_1} = O(\|x_0 - x^*\|_{X_1}^{3-\varepsilon}), \quad \varepsilon \in (0, 3).$$

Approximating properties of the scheme (1) can also be studied under smoothness conditions on $F(x)$ somewhat different from (1.1).

Theorem 3.3. *Suppose the derivative $F'(x)$, $x \in \Omega_R(x_0)$ is Hölder continuous, i.e., there exist constants $\gamma \in (0, 1]$ and $L > 0$ such that*

$$\|F'(x) - F'(y)\|_{L(X_1, X_2)} \leq L\|x - y\|_{X_1}^\gamma \quad \forall x, y \in \Omega_R(x_0). \tag{14}$$

Let conditions (2)–(4) be satisfied and $z_\alpha(\xi)$ be defined by (1). Then

$$\|z_\alpha(\xi) - x^*\|_{X_1} \leq \frac{c_1 L}{(1 + \gamma)\sqrt{\alpha}}\|x_0 - x^*\|_{X_1}^{1+\gamma} + c_2\|v\|_{X_1}\alpha^p, \quad \alpha > 0. \tag{15}$$

The proof follows by (7) and the estimate

$$\|F(x_0) - F(x^*) - F'(x_0)(x_0 - x^*)\|_{X_2} \leq \frac{L}{1 + \gamma}\|x_0 - x^*\|_{X_1}^{1+\gamma},$$

which can be derived from (14) and (1.1.13).

Now, by (15) with an appropriate $\alpha > 0$ we get

$$\|z_\alpha(\xi) - x^*\|_{X_1} = O(\|x^* - x_0\|_{X_1}^{2(1+\gamma)p/(2p+1)}).$$

Hence the approximation (1) is better than the trivial one if $\gamma p > 1/2$. In the case where $\gamma = 1$, this condition takes the above–mentioned form $p > 1/2$.

An extension of the presented theory to equations (1.2) in a Banach space $X = X_1 = X_2$ is not complicated. In this event, instead of (2.1.4) one should use the parametric approximation (2.2.2). Thus we obtain the following Banach analog of the scheme (1):

$$x^* \approx z_\alpha(\xi) = \xi - \Theta(F'(x_0), \alpha)(F(x_0) - F'(x_0)(x_0 - \xi)), \quad \alpha > 0. \tag{16}$$

The next theorem establishes an error estimate for the approximation (16).

Theorem 3.4. *Suppose (1.1) is satisfied, $\varphi(F'(x_0)) \in L(X)$, and*

$$x^* - \xi = \varphi(F'(x_0))v, \quad v \in X, \tag{17}$$

$$\|[E - \Theta(F'(x_0), \alpha)F'(x_0)]\varphi(F'(x_0))\|_{L(X)} \leq c_8\alpha^p \tag{18}$$
$$\forall \alpha > 0 \quad (p > 0).$$

Moreover, let a function $g(\alpha)$ be given such that

$$\|\Theta(F'(x_0), \alpha)\|_{L(X)} \le g(\alpha) \quad \forall \alpha > 0. \tag{19}$$

Then

$$\|z_\alpha(\xi) - x^*\|_X \le \frac{1}{2} Lg(\alpha)\|x_0 - x^*\|_X^2 + c_8\|v\|_X \alpha^p, \quad \alpha > 0.$$

The proof follows immediately by (17)–(19) and the equality

$$z_\alpha(\xi) - x^* = -\Theta(F'(x_0), \alpha)[F(x_0) - F(x^*) - F'(x_0)(x_0 - x^*)] - \\ -(E - \Theta(F'(x_0), \alpha)F'(x_0))(x^* - \xi).$$

In Section 2.3, for sufficiently wide class of generating functions $\Theta(\lambda, \alpha)$ it was shown that in the case where $\varphi(\lambda) = \lambda^\mu$, $\mu > 0$, and $A = F'(x_0)$ satisfies Condition 1.1, inequalities (18) and (19) are valid with $g(\alpha) = c_9\alpha^{-1}$ and an appropriate $p > 0$. Given a function $g(\alpha)$, it is not difficult to derive an optimal error estimate. If, for instance, $g(\alpha) = c_9\alpha^{-1}$, then with an appropriate choice of α we have

$$\|z_\alpha(\xi) - x^*\|_X = O(\|x^* - x_0\|_X^{2p/(p+1)}). \tag{20}$$

This estimate in nontrivial when $p > 1$. From (20) it follows that potentially accessible error estimate is

$$O(\|x^* - x_0\|_X^{2-\varepsilon}), \quad \varepsilon \in (0, 2).$$

The reader will easily formulate and prove analogs of Theorems 3.2 and 3.3 for the case of a Banach space X.

Let us return to the proof of Theorem 3.1. It is readily seen that all the assumptions on $F(x)$ used in the proof can be replaced by the single inequality (1.1.14):

$$\|F(x_0) - F(x^*) - F'(x_0)(x_0 - x^*)\|_{X_2} \le \frac{1}{2} L\|x_0 - x^*\|_{X_1}^2.$$

This observation allows to extend Theorem 3.1 to operators

$$F : D(F) \subset X_1 \to X_2,$$

not necessarily smooth and everywhere defined.

Theorem 3.5. *Suppose $x_0 \in D(F)$ and an operator $A \in L(X_1, X_2)$ satisfies the conditions*

$$\|A\|_{L(X_1, X_2)} \le L,$$

$$\|F(x_0) - F(x^*) - A(x_0 - x^*)\|_{X_2} \le \frac{1}{2}L\|x_0 - x^*\|_{X_1}^2. \tag{21}$$

Assume that

$$x^* - \xi = \varphi(A^*A)v, \quad v \in X_1.$$

Let

$$z_\alpha(\xi) = \xi - \Theta(A^*A, \alpha)A^*(F(x_0) - A(x_0 - \xi)), \quad \alpha > 0, \tag{22}$$

where the functions $\varphi(\lambda)$ and $\Theta(\lambda, \alpha)$ satisfy conditions of Theorem 3.1. Then estimate (5) is valid.

We stress that the operator $F(x)$ in Theorem 3.5 is not formally constrained by any smoothness conditions.

Concluding this section, we point to the following variant of Theorem 3.1.

Theorem 3.6. *Suppose (1.1) is satisfied and there exist a Borel measurable function $\varphi(\lambda), \lambda \ge 0$ and elements $v, w \in X_1$ such that*

$$x^* - \xi = \varphi(F'^*(x_0)F'(x_0))v + w, \quad \|w\|_{X_1} \le \Delta. \tag{23}$$

Let (3) and (4) be satisfied, $z_\alpha(\xi)$ be defined by (1) and

$$\sup_{\lambda \in [0, N_1^2]} |1 - \Theta(\lambda, \alpha)\lambda| \le c_{10} \quad \forall \alpha > 0.$$

Then

$$\|z_\alpha(\xi) - x^*\|_{X_1} \le \frac{c_1 L}{2\sqrt{\alpha}}\|x_0 - x^*\|_{X_1}^2 + c_2\|v\|_{X_1}\alpha^p + c_{10}\Delta, \quad \alpha > 0.$$

The proof results immediately from (6) and (7). The weakened source-wise representation (23) provides additional possibilities for optimization of the choice of ξ. The optimal error estimate now takes the form

$$O(\|x^* - x_0\|_X^{2-\varepsilon} + \Delta).$$

3.3   Parametric Approximations in Inverse Scattering Problem

In this section we analyze the technique of parametric approximations in application to a 3D inverse scattering problem.

Consider a scalar inhomogeneous medium where sound waves propagate. Assume that a time–harmonic compactly supported radiating source

$$F(r, t) = \text{Re}(f(r)\exp(-i\omega t))$$

induces in the medium a field of acoustic pressure

$$U(r,t) = \text{Re}(u(r)\exp(-i\omega t)), \quad r \in \mathbf{R}^3.$$

Then the complex amplitude $u(r)$ of the field $U(r,t)$ satisfies the equation [17, 30, 120]

$$\Delta u(r) + k^2(r)u(r) = f(r), \quad r \in \mathbf{R}^3; \tag{1}$$

$$\Delta u \equiv \sum_{j=1}^{3} \frac{\partial^2 u}{\partial r_j^2}.$$

Here

$$k(r) = \frac{\omega}{c(r)}; \tag{2}$$

$c(r)$ is a sound speed at the point $r \in \mathbf{R}^3$ and ω a given frequency of the incident source. The functions $c(r)$ and $k(r)$ characterize acoustic properties of the medium. When these functions are constant on the whole of space, that is,

$$c(r) \equiv c_0 > 0, \quad k(r) \equiv k_0 = \frac{\omega}{c_0},$$

we have a homogeneous medium, and (1) becomes the classical Helmholtz equation. The value

$$n(r) = \frac{c_0^2}{c^2(r)} = \frac{k^2(r)}{k_0^2}, \quad r \in \mathbf{R}^3$$

is called the refractive index of the medium. Equation (1) should be accomplished by the Sommerfeld radiation condition

$$\frac{\partial u(r)}{\partial \rho} - ik_0 u(r) = o\left(\frac{1}{\rho}\right), \quad \rho = |r| \to \infty, \tag{3}$$

which guarantees that the wave $U(r,t)$ is outgoing. If amplitudes $u(r)$ are sufficiently small and inhomogeneities vary slowly, then formulae (1)–(3) provide a satisfactory quantitative model of time–harmonic sound waves in inhomogeneous media.

Suppose the sources occupy a bounded domain $X \subset \mathbf{R}^3$ so that

$$f(r) = 0 \quad \forall r \in \mathbf{R}^3 \backslash X,$$

where for simplicity $f \in C(\mathbf{R}^3)$. Also, assume that the inhomogeneity is localized within a domain $R \subset \mathbf{R}^3$ such that $R \cap X = \emptyset$:

$$c(r) = c_0 > 0 \quad \forall r \in \mathbf{R}^3 \backslash R, \quad c(r) \geq c_1 > 0 \quad \forall r \in R.$$

Suppose that $c \in C(\mathbf{R}^3)$. The direct acoustic scattering problem is to find a solution $u(r)$ to (1)–(3) given the sound speeds c_0 and $c(r), r \in R$. In the above conditions, this problem has a unique solution $u = u(r), r \in \mathbf{R}^3$ [30].

Below we shall analyze the following inverse scattering problem related to equation (1): given c_0, the sound speed outside the inhomogeneity R, and the wave field $u(r)$ for r from a region of observation $Y \subset \mathbf{R}^3$ such that $Y \cap R = \emptyset$, determine the sound speed on the inhomogeneity, that is, $c(r)$ for $r \in R$. We remark that the region of sources X and the region of detectors Y may intersect or even coincide. It is known ([43]) that a solution to this inverse problem is not unique, in general. However the above given setting by a minor complication can be transformed into a uniquely solvable problem. According to [43], it is sufficient to consider a family of experiments with different sources

$$f(r) = f(q, r), \quad q = 1, 2, \ldots, \quad r \in X$$

and corresponding observations

$$u(r) = u(q, r), \quad q = 1, 2, \ldots, \quad r \in Y$$

rather than a single source and observation. Another way to obtain a uniquely solvable version of the problem is to make experiments with a single source $f(r)$ but at several different frequencies $\omega = \omega_1, \omega_2, \ldots$. Nevertheless for the brevity sake we restrict ourselves with the simplest setting and assume that the experiment involves a single incident source $f(r)$ radiating at a single frequency ω.

Using the notation

$$\epsilon(r) = \frac{1}{c_0^2} - \frac{1}{c^2(r)} = \frac{1 - n(r)}{c_0^2}, \quad r \in \mathbf{R}^3, \tag{4}$$

we can rewrite (1) as

$$\Delta u(r) + k_0^2 u(r) = f(r) + \omega^2 \epsilon(r) u(r), \quad r \in \mathbf{R}^3. \tag{5}$$

By

$$\Gamma(r, r') = -\frac{\exp\left(ik_0|r - r'|\right)}{4\pi|r - r'|} \tag{6}$$

we denote the Green function for the Helmholtz operator

$$\Delta u(r) + k_0^2 u(r)$$

under condition (3). Taking into account that $\epsilon(r) = 0, r \in \mathbf{R}^3 \backslash R$, from (5) and (6) we get the integral form of the direct problem (1)–(3):

$$u(r) = u_0(r) + \omega^2 \int_R \Gamma(r, r') \epsilon(r') u(r') dr', \quad r \in \mathbf{R}^3, \tag{7}$$

where

$$u_0(r) = \int_X \Gamma(r, r') f(r') dr', \quad r \in \mathbf{R}^3. \tag{8}$$

In order to obtain an integral form of the inverse problem let us write (7) separately for points $r = y \in Y$ and $r \in R$. We denote

$$W(y) = u(y) - u_0(y), \quad y \in Y$$

and from (7)–(8) derive the system of two nonlinear integral equations:

$$\omega^2 \int_R \Gamma(y, r') \epsilon(r') u(r') dr' = W(y), \quad y \in Y; \tag{9}$$

$$u(r) - \omega^2 \int_R \Gamma(r, r') \epsilon(r') u(r') dr' = u_0(r), \quad r \in R. \tag{10}$$

In (9)–(10), the functions $\epsilon(r), r \in R$ and $u(r), r \in R$ are unknown, while $u_0(r), r \in R$ and $W(y), y \in Y$ are given. When $\epsilon(r)$ is determined, by (4) we obtain the unknown sound speed as follows:

$$c(r) = \frac{c_0}{\sqrt{1 - c_0^2 \epsilon(r)}}, \quad r \in R.$$

Now suppose the function $\epsilon(r), r \in R$ is sufficiently small and consider a formal linearization of the system (9)–(10). Equation (10) implies that the field $u(r)$ inside the scatterer R is close to the incident field $u_0(r), r \in R$. Therefore we can put

$$u(r) = u_0(r) + v(r)$$

and drop terms involving the product $\epsilon(r)v(r)$. We obtain the following system of linear integral equations for $\epsilon(r)$ and $v(r)$:

$$\omega^2 \int_R \Gamma(y, r') \epsilon(r') u_0(r') dr' = W(y), \quad y \in Y; \tag{11}$$

$$v(r) - \omega^2 \int_R \Gamma(r, r') \epsilon(r') u_0(r') dr' = 0, \quad r \in R. \tag{12}$$

In physical literature, the system (11)–(12) is known as Born's approximation to the inverse scattering problem. Approximations of this type find wide applications both as a tool of theoretical analysis and as a basis of numerical methods [25, 46, 50, 80, 109, 120, 122, 147]. However in most of the studies where Born's approximations are used for constructing numerical methods, rigorous theoretical justification for replacement of the original system (9)–(10)

or its analogs by corresponding formal linearized problems is not carried out at all, or the authors simply claim that such a replacement is physically realistic.

Let us note that the replacement of (9)–(10) with (11)–(12) is equivalent to the linearization of (9)–(10) at $(\epsilon, u) = (0, u_0)$ by the classical scheme (1.3). It is readily seen that the operator F of system (9)–(10) has a compact derivative F' in usual Lebesgue and Sobolev spaces. Therefore rigorous justification of this approach to solving (9)–(10) is inevitably connected with the difficulties mentioned above in Section 3.1.

Let us now apply to (9)–(10) the parametric approximation technique developed in Section 3.2. To this end, we set

$$X_1 = L_2(R) \times L_2(R), \quad X_2 = L_2(Y) \times L_2(R) \tag{13}$$

and define the operator $F : X_1 \to X_2$ as follows:

$$F(x) = \left(\omega^2 \int_R \Gamma(y, r')\epsilon(r')u(r')dr' - W(y); \right.$$

$$\left. u(r) - \omega^2 \int_R \Gamma(r, r')\epsilon(r')u(r')dr' - u_0(r) \right), \tag{14}$$

$$x = (\epsilon, u) \in X_1.$$

According to the properties of potential type operators [96], operator (14) is well defined on the all of X_1. Theorem 3.5 now gives an alternative way of approximating a solution to the system (9)–(10). Put $x_0 = (0, u_0)$ and $x^* = (\epsilon^*, u^*)$. It can be shown in the usual way that the Fréchet derivative of $F(x)$ at x_0 has the form

$$F'(x_0)h = \left(\omega^2 \int_R \Gamma(y, r')\epsilon(r')u_0(r')dr'; \right.$$

$$\left. v(r) - \omega^2 \int_R \Gamma(r, r')\epsilon(r')u_0(r')dr' \right), \quad h = (\epsilon, v) \in X_1. \tag{15}$$

We see that $F'(x_0)$ coincides with the operator of formally linearized system (11)–(12). Since $u_0 \in L_2(R)$, operator (15) is continuous from X_1 into X_2 [96].

Let us check that inequality (2.21) is satisfied with $A = F'(x_0)$ and an appropriate constant L. We supply X_1 and X_2 with standard norms of Cartesian products. From (1) and (7) it follows that

$$\|F(x_0) - F(x^*) - A(x_0 - x^*)\|_{X_1}^2 =$$

$$= \omega^4 \left\| \int_R \Gamma(\cdot, r')\epsilon^*(r')(u_0(r') - u^*(r'))dr' \right\|_{L_2(Y)}^2 +$$

$$+ \omega^4 \left\| \int_R \Gamma(\cdot, r')\epsilon^*(r')(u_0(r') - u^*(r'))dr' \right\|_{L_2(R)}^2 .$$

It is not hard to prove that the operators

$$(A_1 v)(y) = \int_R \Gamma(y, r')v(r')dr', \quad y \in Y;$$

$$(A_2 v)(r) = \int_R \Gamma(r, r')v(r')dr', \quad r \in R$$

are continuous from $L_1(R)$ into $L_2(Y)$ and from $L_1(R)$ into $L_2(R)$ respectively. Denote by M_1 and M_2 the norms of these operators. From (13) and (16), using the Cauchy and Schwarz's inequality, we obtain

$$\|F(x_0) - F(x^*) - A(x_0 - x^*)\|_{X_2}^2 \leq$$

$$\leq \omega^4(M_1^2 + M_2^2)\|\epsilon^*(u_0 - u^*)\|_{L_1(R)}^2 \leq$$

$$\leq \omega^4(M_1^2 + M_2^2)\|\epsilon^*\|_{L_2(R)}^2 \|u_0 - u^*\|_{L_2(R)}^2 \leq$$

$$\leq \frac{1}{4}\omega^4(M_1^2 + M_2^2)\|x_0 - x^*\|_{X_1}^4 .$$

Consequently inequality (2.21) is valid with $L = \omega^2(M_1^2 + M_2^2)^{1/2}$. Hence to approximate a solution of (9)–(10), we can use the parametric family

$$\{z_\alpha(\xi)\} = \{(\epsilon_\alpha(\xi), u_\alpha(\xi))\}$$

defined by (2.22) with $A = F'(x_0)$. Theorem 3.5 and preceding discussions provide conditions on the controlling parameter $\xi \in L_2(R) \times L_2(R)$ sufficient for (2.5).

In many instances, by Born's approximation to the problem (9)–(10) is meant only the first equation (11) of the system (11)–(12). The reason is that equation (10) allows to express the internal field $u(r), r \in R$ in terms of $\epsilon(r), r \in R$. After substitution of the resulting function $u(r) = u[\epsilon](r)$ into (9) we get a single equation for $\epsilon(r)$. If we assume $\epsilon(r)$ to be small and apply to the obtained equation the formal linearization procedure, then we come to (11) again. Let us now apply to this equation the parametric approximation scheme (2.22). For this purpose, we define the spaces

$$X_1 = L_2(R), \quad X_2 = L_2(Y)$$

and the operator $F : X_1 \to X_2$,

$$F(\epsilon)(y) = \omega^2 \int_R \Gamma(y, r')\epsilon(r')u[\varepsilon](r')dr' - W(y), \quad y \in Y,$$

where the mapping

$$u = u[\epsilon](r), \quad r \in R$$

is specified by equation (10). Recall that $u_0 \in L_2(R)$. Given a function $\epsilon \in L_2(R)$, define the operator

$$B_\epsilon : L_2(R) \to L_2(R);$$

$$(B_\epsilon u)(r) = \int_R \Gamma(r, r')\epsilon(r')u(r')dr', \quad r \in R.$$

It is readily seen that B_ϵ is continuous from $L_2(R)$ into $L_2(R)$, and

$$\|B_\epsilon\|_{L(L_2(R))} \le M_2\|\epsilon\|_{L_2(R)}.$$

Suppose

$$\omega^2 M_2\|\epsilon\|_{L_2(R)} < 1. \tag{17}$$

The application of Lemma 1.2 yields that for each $\epsilon \in L_2(R)$, equation (10) determines a unique $u \in L_2(R)$. Moreover,

$$\|u\|_{L_2(R)} \le \frac{\|u_0\|_{L_2(R)}}{1 - \omega^2 M_2\|\epsilon\|_{L_2(R)}}. \tag{18}$$

Consider the operator $A \in L(X_1, X_2)$:

$$(A\epsilon)(y) = \omega^2 \int_R \Gamma(y, r')\epsilon(r')u_0(r')dr', \quad y \in Y.$$

Let us show that A satisfies condition (2.21) with $\epsilon_0(r) = 0$. According to (9),

$$\|F(0) - F(\epsilon^*) - A(\epsilon_0 - \epsilon^*)\|_{L_2(Y)} \le$$
$$\le \omega^2 M_1\|\epsilon^*(u_0 - u^*)\|_{L_1(R)} \le \tag{19}$$
$$\le \omega^2 M_1\|\epsilon^*\|_{L_2(R)}\|u_0 - u^*\|_{L_2(R)}.$$

By (10) it follows that

$$(u^*(r) - u_0(r)) - \omega^2 \int_R \Gamma(r, r')\epsilon^*(r')(u^*(r') - u_0(r'))dr' =$$

$$= \omega^2 \int_R \Gamma(r, r')\epsilon^*(r')u_0(r')dr'.$$

Suppose the inhomogeneity is so small that

$$\omega^2 M_2 \|\epsilon^*\|_{L_2(R)} < 1. \tag{20}$$

From (17), (18), and (20) we then deduce

$$\|u^* - u_0\|_{L_2(R)} \leq \frac{\omega^2 M_2 \|u_0\|_{L_2(R)} \|\epsilon^*\|_{L_2(R)}}{1 - \omega^2 M_2 \|\epsilon^*\|_{L_2(R)}}. \tag{21}$$

Combining (19) and (21), we obtain the required estimate:

$$\|F(0) - F(\epsilon^*) - A(\epsilon_0 - \epsilon^*)\|_{L_2(Y)} \leq$$

$$\leq \frac{\omega^4 M_1 M_2 \|u_0\|_{L_2(R)} \|\epsilon^*\|^2_{L_2(R)}}{1 - \omega^2 M_2 \|\epsilon^*\|_{L_2(R)}} = O(\|\epsilon^*\|^2_{L_2(R)}).$$

Condition (20) on the measure of inhomogeneity $\|\epsilon^*\|_{L_2(R)}$ is consistent with the fact that parametric, as well as classical, linearization schemes are advantageous to use just in a neighborhood of the solution to be approximated. Theorem 3.5 now establishes sufficient conditions for the approximating family $z_\alpha(\xi) = \epsilon_\alpha(\xi), \xi \in L_2(R)$ to satisfy (2.5).

Chapter 4

ITERATIVE PROCESSES ON THE BASIS OF PARAMETRIC APPROXIMATIONS

This chapter is devoted to applications of the parametric approximations formalism developed in Section 3.2 to constructing iterative processes for non-linear irregular operator equations in Hilbert and Banach spaces. The design of these processes is based on iteration of the schemes (3.2.1) and (3.2.16). In Section 4.1 we propose a unified approach to constructing and studying iterative solution methods for nonlinear equations with smooth operators in a Hilbert space. In the case of exact input data, rate of convergence estimates are established. For equations with approximately given operators we present a family of regularization algorithms characterized by a stopping of the original processes at an appropriate iteration. A sourcewise representation condition of type (3.2.2), with the solution x^* in place of x_0, is of first importance in the analysis of these algorithms. In particular, it is shown that such representation is sufficient for power rate of convergence estimates of the iterative processes. We also establish stability of presented regularization algorithms with respect to errors in the exact sourcewise representation (3.2.2), when a perturbed representation has the form (3.2.23). In Section 4.2 we prove that the power sourcewise representation actually approaches a necessary condition for convergence of the proposed methods at the power rate. Sections 4.3 and 4.4 are devoted to an extension of the previous results to nonlinear irregular equations in a Banach space. A sourcewise representation condition of the form (3.2.17), with x_0 replaced by x^*, plays an important part here. In Section 4.5 we present and study continuous analogs of the iterative methods derived in Sections 4.1 and 4.3.

4.1 Generalized Gauss–Newton Type Methods for Nonlinear Irregular Equations in Hilbert Space

In this section we study a nonlinear equation

$$F(x) = 0, \quad x \in X_1, \tag{1}$$

where the operator $F : X_1 \to X_2$ acts from a Hilbert space X_1 into a Hilbert space X_2. Let x^* be a solution of equation (1). Suppose $F(x)$ is Fréchet differentiable on

$$\Omega_R(x^*) = \{x \in X_1 : \|x - x^*\|_{X_1} \leq R\}$$

and has a Lipschitz continuous derivative, i.e.,

$$\|F'(x) - F'(y)\|_{L(X_1, X_2)} \leq L\|x - y\|_{X_1} \quad \forall x, y \in \Omega_R(x^*). \tag{2}$$

From (2) it follows that there exists a constant N_1 such that

$$\|F'(x)\|_{L(X_1, X_2)} \leq N_1 \quad \forall x \in \Omega_R(x^*).$$

Let us fix a sequence of regularization parameters $\{\alpha_n\}, \alpha_n > 0$ and an initial point $x_0 \in \Omega_R(x^*)$. Using the scheme (3.2.1) with $\alpha = \alpha_0$ and taking the obtained point $z_{\alpha_0}(\xi)$ as a next approximation x_1, we get the following iterative process

$$x_{n+1} = \xi - \Theta(F'^*(x_n)F'(x_n), \alpha_n)F'^*(x_n)(F(x_n) - F'(x_n)(x_n - \xi)). \tag{3}$$

Hereafter, the iterative methods (3) are called the generalized Gauss–Newton methods. The parameter $\xi \in X_1$ provides a tool for control over the process (3). Suppose the complex–valued generating function $\Theta(\lambda, \alpha)$ is analytic in λ on a domain $D_\alpha \subset \mathbf{C}$ such that

$$D_\alpha \supset [0, N_1^2] \quad \forall \alpha \in (0, \alpha_0].$$

This technical assumption doesn't lead to significant lose of generality, as compared with Section 3.2, since generating functions most popular in computational practice arise to be analytic (see Examples 4.1–4.5 below).

Suppose the sequence of regularization parameters $\{\alpha_n\}$ satisfies the following condition.

Condition 4.1.

$$0 < \alpha_{n+1} \leq \alpha_n, \quad n = 0, 1, \ldots; \quad \lim_{n \to \infty} \alpha_n = 0;$$

$$\sup_{n \in \mathbf{N} \cup \{0\}} \frac{\alpha_n}{\alpha_{n+1}} < \infty.$$

For instance, we can set

$$\alpha_n = \frac{\alpha_0}{(n+1)^a}; \quad \alpha_0, a > 0.$$

Let us introduce the notation

$$r = \sup_{n \in \mathbb{N} \cup \{0\}} \frac{\alpha_n}{\alpha_{n+1}}.$$

Subsequent analysis of the process (3) is based on the assumption that $x^* - \xi$ can be presented in the form (3.2.2), where $\varphi(\lambda) = \lambda^p, p \geq 1/2$, and x_0 is replaced by x^*. Thus we shall assume throughout this section that there exists an element $v \in X_1$ for which

$$x^* - \xi = (F'^*(x^*)F'(x^*))^p v, \quad p \geq \frac{1}{2}. \tag{4}$$

We start with the case where the original operator $F(x)$ is accessible without errors. Below, under appropriate additional conditions on $\Theta(\lambda, \alpha)$, we shall prove that the power sourcewise representation (4) implies the power rate of convergence estimate

$$\|x_n - x^*\|_{X_1} \leq l\alpha_n^p, \quad n = 0, 1, \dots \tag{5}$$

with the same exponent p as in (4).

The following conditions in essence agree with (2.1.46), (3.2.3), and (3.2.4).

Condition 4.2. *There exists $c_1 > 0$ such that*

$$\sup_{\lambda \in [0, N_1^2]} |\Theta(\lambda, \alpha)| \leq \frac{c_1}{\alpha} \quad \forall \alpha \in (0, \alpha_0].$$

Condition 4.3. *There exist a constant $p_0 \geq 1/2$ and a nondecreasing function $c_2 = c_2(p), p \geq 0$ such that for each $p \in [0, p_0]$,*

$$\sup_{\lambda \in [0, N_1^2]} |1 - \Theta(\lambda, \alpha)\lambda|\lambda^p \leq c_2 \alpha^p \quad \forall \alpha \in (0, \alpha_0].$$

Remark 4.1. Conditions 4.2 and 4.3 (with $p = 0$) imply

$$\sup_{\lambda \in [0, N_1^2]} |\Theta(\lambda, \alpha)\sqrt{\lambda}| \leq \frac{\tilde{c}_1}{\sqrt{\alpha}} \quad \forall \alpha \in (0, \alpha_0].$$

From here on, we suppose that representation (4) is fulfilled with some exponent $p \in [1/2, p_0]$.

Assuming $x_n \in \Omega_R(x^*)$, let us estimate $\|x_{n+1} - x^*\|_{X_1}$. By (3) and (4), it follows that

$$
\begin{aligned}
x_{n+1} - x^* &= -\Theta(F'^*(x_n)F'(x_n), \alpha_n)F'^*(x_n)\cdot \\
&\cdot [F(x_n) - F(x^*) - F'(x_n)(x_n - x^*)]- \\
&- [E - \Theta(F'^*(x^*)F'(x^*), \alpha_n)F'^*(x^*)F'(x^*)](F'^*(x^*)F'(x^*))^p v- \quad (6) \\
&- [\Theta(F'^*(x^*)F'(x^*), \alpha_n)F'^*(x^*)F'(x^*)- \\
&- \Theta(F'^*(x_n)F'(x_n), \alpha_n)F'^*(x_n)F'(x_n)](F'^*(x^*)F'(x^*))^p v.
\end{aligned}
$$

According to (2) and (1.1.14),

$$
\|F(x_n) - F(x^*) - F'(x_n)(x_n - x^*)\|_{X_2} \leq \frac{1}{2}L\|x_n - x^*\|_{X_1}^2. \quad (7)
$$

Combining (6) with (7) and using Conditions 4.2, 4.3 and Remark 4.1, we obtain

$$
\begin{aligned}
\|x_{n+1} - x^*\|_{X_1} &\leq \frac{\tilde{c}_1 L}{2\sqrt{\alpha_n}}\|x_n - x^*\|_{X_1}^2 + \\
&+ c_2\alpha_n^p\|v\|_{X_1} + \|[\Theta(F'^*(x^*)F'(x^*), \alpha_n)F'^*(x^*)F'(x^*)- \\
&- \Theta(F'^*(x_n)F'(x_n), \alpha_n)F'^*(x_n)F'(x_n)](F'^*(x^*)F'(x^*))^p v\|_{X_1}.
\end{aligned} \quad (8)
$$

To estimate the last norm in (8), we need a family of contours $\{\Gamma_\alpha\}_{\alpha \in (0,\alpha_0]}$ such that $\Gamma_\alpha \subset D_\alpha$, and Γ_α surrounds the segment $[0, N_1^2]$ of the real line. Also, suppose the following condition is satisfied.

Condition 4.4.

$$
\sup_{\alpha \in (0,\alpha_0]} \sup_{\lambda \in \Gamma_\alpha} |\lambda| < \infty, \quad (9)
$$

$$
\sup_{\alpha \in (0,\alpha_0]} \sup_{\substack{\lambda \in \Gamma_\alpha \\ \mu \in [0,N_1^2]}} \frac{|\lambda| + \mu}{|\lambda - \mu|} < \infty. \quad (10)
$$

Denote by M_1 and M_2 the constants from (9) and (10). Using the Riesz–Dunford formula (1.1.6), we get

$$
\begin{aligned}
\|[\Theta(F'^*(x^*)&F'(x^*), \alpha_n)F'^*(x^*)F'(x^*)- \\
&-\Theta(F'^*(x_n)F'(x_n), \alpha_n)F'^*(x_n)F'(x_n)](F'^*(x^*)F'(x^*))^p v\|_{X_1} \leq \\
&\leq \frac{1}{2\pi}\int_{\Gamma_{\alpha n}} |1 - \Theta(\lambda, \alpha_n)\lambda|\|[R(\lambda, F'^*(x^*)F'(x^*))- \quad (11) \\
&- R(\lambda, F'^*(x_n)F'(x_n))](F'^*(x^*)F'(x^*))^p v\|_{X_1}|d\lambda|.
\end{aligned}
$$

Further, we have

$$
\begin{aligned}
&\|[R(\lambda, F'^*(x^*)F'(x^*)) - \\
&- R(\lambda, F'^*(x_n)F'(x_n))](F'^*(x^*)F'(x^*))^p\|_{L(X_1)} \leq \\
&\leq \|R(\lambda, F'^*(x_n)F'(x_n))F'^*(x_n)\|_{L(X_2,X_1)}\|F'(x^*) - F'(x_n)\|_{L(X_1,X_2)} \cdot \\
&\quad \cdot \|R(\lambda, F'^*(x^*)F'(x^*))(F'^*(x^*)F'(x^*))^p\|_{L(X_1)} + \\
&+ \|R(\lambda, F'^*(x_n)F'(x_n))\|_{L(X_1)}\|F'^*(x^*) - F'^*(x_n)\|_{L(X_2,X_1)} \cdot \\
&\quad \cdot \|F'(x^*)R(\lambda, F'^*(x^*)F'(x^*))(F'^*(x^*)F'(x^*))^p\|_{L(X_1,X_2)}.
\end{aligned}
\tag{12}
$$

Since $x^*, x_n \in \Omega_R(x^*)$, from (2) it follows that

$$
\begin{aligned}
\|F'^*(x^*) - F'^*(x_n)\|_{L(X_2,X_1)} &= \\
= \|F'(x^*) - F'(x_n)\|_{L(X_1,X_2)} &\leq L\|x_n - x^*\|_{X_1}.
\end{aligned}
\tag{13}
$$

Using the spectral decomposition technique (see Section 1.1), it is not hard to prove that for each $s \geq 0$,

$$
\|R(\lambda, F'^*(x)F'(x))(F'^*(x)F'(x))^s\|_{L(X_1)} \leq \frac{M_3}{|\lambda|^{1-\min\{s,1\}}},
\tag{14}
$$

$$
\begin{aligned}
&\|F'(x)R(\lambda, F'^*(x)F'(x))(F'^*(x)F'(x))^s\|_{L(X_1,X_2)} \leq \\
&\leq \frac{M_3}{|\lambda|^{1-\min\{s+1/2,1\}}} \quad \forall \lambda \in \Gamma_\alpha \quad \forall \alpha \in (0, \alpha_0] \quad \forall x \in \Omega_R(x^*),
\end{aligned}
\tag{15}
$$

where $M_3 = M_3(s)$. Setting in (14) $s = 1/2, s = p, s = 0$ and in (15) $s = p$, we see that for all $\lambda \in \Gamma_\alpha, \alpha \in (0, \alpha_0]$,

$$
\begin{aligned}
\|R(\lambda, F'^*(x_n)F'(x_n))F'^*(x_n)\|_{L(X_2,X_1)} &= \\
= \|F'(x_n)R(\overline{\lambda}, F'^*(x_n)F'(x_n))\|_{L(X_1,X_2)} &= \\
= \|R(\lambda, F'^*(x_n)F'(x_n))(F'^*(x_n)F'(x_n))^{1/2}\|_{L(X_1)} &\leq \frac{M_3(0)}{\sqrt{|\lambda|}},
\end{aligned}
\tag{16}
$$

$$
\|R(\lambda, F'^*(x^*)F'(x^*))(F'^*(x^*)F'(x^*))^p\|_{L(X_1)} \leq \frac{M_3(p)}{|\lambda|^{1-\min\{p,1\}}},
\tag{17}
$$

$$
\|R(\lambda, F'^*(x_n)F'(x_n))\|_{L(X_1)} \leq \frac{M_3(0)}{|\lambda|},
\tag{18}
$$

$$
\|F'(x^*)R(\lambda, F'^*(x^*)F'(x^*))(F'^*(x^*)F'(x^*))^p\|_{L(X_1,X_2)} \leq M_3(p).
\tag{19}
$$

Let us denote

$$
M_4 = \begin{cases} M_1^{p-1/2} + 1, & p \in [1/2, 1), \\ \sqrt{M_1} + 1, & p \geq 1. \end{cases}
$$

Combining (9), (12), (13), and (16)–(19), we obtain

$$\|[R(\lambda, F'^*(x^*)F'(x^*))-$$
$$-R(\lambda, F'^*(x_n)F'(x_n))](F'^*(x^*)F'(x^*))^p\|_{L(X_1)} \leq$$
$$\leq c_3 M_3 L\|x_n - x^*\|_{X_1}\left(\frac{1}{|\lambda|^{3/2-\min\{p,1\}}} + \frac{1}{|\lambda|}\right) \leq$$
$$\leq \frac{c_3 M_3 M_4 L\|x_n - x^*\|_{X_1}}{|\lambda|}.$$

Inequalities (11) and (12) then yield

$$\|[\Theta(F'^*(x^*)F'(x^*), \alpha_n)F'^*(x^*)F'(x^*)-$$
$$-\Theta(F'^*(x_n)F'(x_n), \alpha_n)F'^*(x_n)F'(x_n)](F'^*(x^*)F'(x^*))^p v\|_{X_1} \leq$$
$$\leq c_4 M_3 M_4 L\|v\|_{X_1}\|x_n - x^*\|_{X_1} \int_{\Gamma_{\alpha_n}} \frac{|1 - \Theta(\lambda, \alpha_n)\lambda|}{|\lambda|}|d\lambda|. \tag{20}$$

In connection with the last estimate, we impose on $\Theta(\lambda, \alpha)$ the following additional condition.

Condition 4.5.

$$\sup_{\alpha\in(0,\alpha_0]} \int_{\Gamma_\alpha} \frac{|1 - \Theta(\lambda, \alpha)\lambda|}{|\lambda|}|d\lambda| < \infty.$$

By (20) and Condition 4.5,

$$\|[\Theta(F'^*(x^*)F'(x^*), \alpha_n)F'^*(x^*)F'(x^*)-$$
$$-\Theta(F'^*(x_n)F'(x_n), \alpha_n)F'^*(x_n)F'(x_n)](F'^*(x^*)F'(x^*))^p v\|_{X_1} \leq \tag{21}$$
$$\leq c_5 M_3 M_4 L\|v\|_{X_1}\|x_n - x^*\|_{X_1}.$$

Combining (8) with (21), we finally obtain

$$\|x_{n+1} - x^*\|_{X_1} \leq \frac{\tilde{c}_1 L}{2\sqrt{\alpha_n}}\|x_n - x^*\|_{X_1}^2 + c_2\alpha_n^p\|v\|_{X_1}+ \tag{22}$$
$$+ c_5 M_3 M_4 L\|v\|_{X_1}\|x_n - x^*\|_{X_1},$$

provided that $x_n \in \Omega_R(x^*)$. The following theorem establishes a rate of convergence estimate for the iteration (3).

Theorem 4.1. *Suppose Conditions 4.1–4.5 are satisfied, and (4) is fulfilled with $p \in [1/2, p_0]$. Assume that*

$$\|x_0 - x^*\|_{X_1} \leq l\alpha_0^p, \tag{23}$$

where

$$0 < l \leq \min\left\{\frac{1}{r^p \widetilde{c}_1 L \alpha_0^{p-1/2}}, \frac{R}{\alpha_0^p}\right\}, \tag{24}$$

$$\|v\|_{X_1} \leq d = \min\left\{\frac{1}{4r^p c_5 M_3 M_4 L}, \frac{l}{4r^p c_2}\right\}. \tag{25}$$

Then estimate (5) is valid.

Proof. The proof in by induction on n. For $n = 0$, there is nothing to prove. Suppose (5) is valid for a number $n \geq 0$, and let $x_n \in \Omega_R(x^*)$. From (23) and (24) it follows that $x_0 \in \Omega_R(x^*)$. By (22), with the use of Condition 4.1 and the induction hypothesis, we get

$$\|x_{n+1} - x^*\|_{X_1} \leq$$

$$\leq \frac{1}{2}\widetilde{c}_1 L l^2 \alpha_n^{2p-1/2} + c_2 \alpha_n^p \|v\|_{X_1} + c_5 M_3 M_4 L l \|v\|_{X_1} \alpha_n^p \leq$$

$$\leq r^p \left(\frac{1}{2}\widetilde{c}_1 L l^2 \alpha_0^{p-1/2} + c_2 \|v\|_{X_1} + c_5 M_3 M_4 L l \|v\|_{X_1}\right) \alpha_{n+1}^p.$$

Further, due to (24) and (25),

$$\frac{1}{2}\widetilde{c}_1 L l^2 \alpha_0^{p-1/2} + c_2 \|v\|_{X_1} \leq \frac{3l}{4r^p}, \quad c_5 M_3 M_4 L l \|v\|_{X_1} \leq \frac{l}{4r^p}.$$

Summing the last two inequalities, we obtain

$$\frac{1}{2}\widetilde{c}_1 L l^2 \alpha_0^{p-1/2} + c_2 \|v\|_{X_1} + c_5 M_3 M_4 L l \|v\|_{X_1} \leq \frac{l}{r^p}$$

and hence,

$$\|x_{n+1} - x^*\|_{X_1} \leq l \alpha_{n+1}^p.$$

Also, note that by (24),

$$\|x_{n+1} - x^*\|_{X_1} \leq l \alpha_{n+1}^p \leq l \alpha_0^p \leq R.$$

Therefore $x_{n+1} \in \Omega_R(x^*)$. Thus estimate (5) is valid for each number $n \in \mathbf{N}$. This completes the proof. $\qquad\square$

Remark 4.2. According to (24) and (25), the constants $l = l(p)$ and $d = d(p)$ can be specified such that

$$\sup_{p \in [a,b]} l(p) < \infty, \quad \inf_{p \in [a,b]} d(p) > 0$$

for each segment $[a, b]$ with $1/2 \leq a < b < \infty$.

From Theorem 4.1 it follows that the least upper bound p^* of possible values p_0 in Condition 4.3 has the meaning of the qualification parameter for (3).

Theorem 4.1 establishes that the sequence $\{x_n\}$ locally converges to x^*, provided the controlling parameter ξ is taken from the ellipsoid

$$\mathbb{M}(x^*) = \{x^* + (F'^*(x^*)F'(x^*))^p v : \|v\|_{X_1} \leq d\} \tag{26}$$

centered at x^*. We stress that in the irregular case, where $0 \in \sigma(F'^*(x^*)F'(x^*))$, the ellipsoid $\mathbb{M}(x^*)$ has an empty interior. Therefore from the standpoint of numerical implementation, the problem of finding an element $\xi \in \mathbb{M}(x^*)$ is rather complicated.

Let us now address to the case where the operator $F(x)$ in equation (1) is available with errors. Assume that instead of $F(x)$ an approximation $\widetilde{F}(x)$ is given, and $\widetilde{F}(x)$ has a Fréchet derivative satisfying (2). Besides, suppose

$$\|\widetilde{F}(x^*)\|_{X_2} \leq \delta; \quad \|\widetilde{F}'(x) - F'(x)\|_{L(X_1,X_2)} \leq \delta \quad \forall x \in \Omega_R(x^*). \tag{27}$$

Without loss of generality we can assume that

$$\|\widetilde{F}'(x)\|_{L(X_1,X_2)} \leq N_1 \quad \forall x \in \Omega_R(x^*).$$

When analyzing processes (3) in the case of approximate data, it is convenient to fix the element x^* and to define the class $\mathfrak{F}$ of all exact and approximate data as follows:

$$\mathfrak{F} = \{F : \|F'(x) - F'(y)\|_{L(X_1,X_2)} \leq L\|x - y\|_{X_1} \; \forall x, y \in \Omega_R(x^*)\}.$$

Let us supply $\mathfrak{F}$ with the metric

$$\rho(F_1, F_2) =$$

$$= \max\left\{\|F_1(x^*) - F_2(x^*)\|_{X_2}, \sup_{x \in \Omega_R(x^*)} \|F_1'(x) - F_2'(x)\|_{L(X_1,X_2)}\right\}$$

and consider the mapping $G : \mathfrak{F} \to X_1$,

$$D(G) = \{F \in \mathfrak{F} : F(x^*) = 0\}, \quad G(F) = x^*.$$

Also, let the mapping $\mathfrak{R}_{1/n}$ take each operator $F \in \mathfrak{F}$ to the element x_n defined by (3). Theorem 4.1 then establishes conditions under which the family $\{\mathfrak{R}_{1/n}\}$ approximates the mapping G in the sense of equality (1.2.6) with $\alpha = 1/n$. If ξ and $\{\alpha_n\}$ are assumed to be fixed then conditions (4), (24), and (25) define a subset of $D(G)$, for which (1.2.6) is true.

Now, given an approximate operator $\widetilde{F}(x)$ and an initial point $x_0 \in \Omega_R(x^*)$, consider the iterative process

$$x_{n+1} = \xi - \Theta(\widetilde{F}'^*(x_n)\widetilde{F}'(x_n), \alpha_n)\widetilde{F}'^*(x_n)(\widetilde{F}(x_n) - \widetilde{F}'(x_n)(x_n - \xi)). \quad (28)$$

Suppose $x^* - \xi$ possesses the approximate sourcewise representation

$$x^* - \xi = (F'^*(x^*)F'(x^*))^p \widetilde{v} + \widetilde{w}, \quad p \geq \frac{1}{2},$$

$$\widetilde{v}, \widetilde{w} \in X_1, \quad \|\widetilde{w}\|_{X_1} \leq \Delta. \quad (29)$$

The value Δ in (29) is an error measure of the exact representation (4). As in the proof of Theorem 4.1, with the use of (27) we obtain

$$\|x_{n+1} - x^*\|_{X_1} \leq \frac{\widetilde{c}_1 L}{2\sqrt{\alpha_n}}\|x_n - x^*\|_{X_1}^2 + c_2\alpha_n^p\|\widetilde{v}\|_{X_1} +$$

$$+ c_6\|\widetilde{v}\|_{X_1}\|x_n - x^*\|_{X_1} + c_7\left(\frac{\delta}{\sqrt{\alpha_n}} + \Delta\right). \quad (30)$$

We observe that the right part of this estimate tends to infinity as the iteration number $n \to \infty$. Hence to obtain an acceptable approximation x_n we should terminate the iteration at an appropriate finite moment $n = N(\delta, \Delta)$ and take $x_{N(\delta,\Delta)}$ as an approximation to x^*.

The following proposition concerning a behavior of the iteration (28) is obtained from (30) by the scheme of the proof of Theorem 4.1.

Theorem 4.2. *Let Conditions 4.1–4.5 be satisfied, and*

$$\|x_0 - x^*\|_{X_1} \leq \widetilde{l}\alpha_0^p,$$

where

$$0 < \widetilde{l} \leq \min\left\{\frac{1}{2r^p\widetilde{c}_1 L\alpha_0^{p-1/2}}, \frac{R}{\alpha_0^p}\right\}.$$

Assume that (29) is fulfilled with

$$p \in \left[\frac{1}{2}, p_0\right], \quad \|v\|_{X_1} \leq \widetilde{d} = \min\left\{\frac{\widetilde{l}}{4r^p c_2}, \frac{1}{4r^p c_6}\right\}. \quad (31)$$

Then for each $0 < \overline{d} < (4r^p c_7)^{-1}\widetilde{l}$ we have

$$\|x_n - x^*\|_{X_1} \leq \widetilde{l}\alpha_n^p, \quad n = 0, 1, \ldots, N(\delta, \Delta), \quad (32)$$

$$N(\delta, \Delta) = \max\left\{n \in \mathbf{N} \cup \{0\} : \frac{\delta}{\alpha_n^{p+1/2}} + \frac{\Delta}{\alpha_n^p} \leq \overline{d}\right\}. \quad (33)$$

It is readily seen that

$$\lim_{\delta,\Delta\to 0} N(\delta,\Delta) = \infty.$$

Given error levels δ, Δ, the number $n = N(\delta,\Delta)$ indicates the stopping point of the process (28) with an approximate operator $\widetilde{F}(x)$.

It is not hard now to establish an estimate for $\|x_{N(\delta,\Delta)} - x^*\|_{X_1}$ in terms of δ, Δ.

Corollary 4.1. *In the conditions of Theorem 4.2, we have*

$$\|x_{N(\delta,\Delta)} - x^*\|_{X_1} \leq r^p \widetilde{l} \left(\frac{\delta + \sqrt{\alpha_{N(\delta,\Delta)}}\Delta}{\overline{d}} \right)^{2p/(2p+1)}. \tag{34}$$

Inequality (34) follows directly by (32) and (33). Indeed,

$$\alpha_{N(\delta,\Delta)} \leq r\alpha_{N(\delta,\Delta)+1} < r\left(\frac{\delta + \sqrt{\alpha_{N(\delta,\Delta)+1}}\Delta}{\overline{d}} \right)^{2/(2p+1)} \leq$$

$$\leq r\left(\frac{\delta + \sqrt{\alpha_{N(\delta,\Delta)}}\Delta}{\overline{d}} \right)^{2/(2p+1)}.$$

According to (34), the equality

$$\lim_{\delta,\Delta\to 0} \|x_{N(\delta,\Delta)} - x^*\|_{X_1} = 0$$

is fulfilled uniformly with respect to a choice of an approximate operator $\widetilde{F}(x)$ subject to (27).

Let us turn again to the special case $\Delta = 0$. Suppose conditions of Theorem 4.2 are satisfied; then from estimate (34) we deduce that the mapping $\mathfrak{R}_{1/N(\delta,0)}$ that takes each operator $\widetilde{F} \in \mathfrak{F}$ to the point $x_{N(\delta,0)}$, defines a regularization algorithm for equation (1). The same result we have in a slightly more general case where $\Delta = \Delta(\delta)$ and $\lim_{\delta\to 0} \Delta(\delta) = 0$. Given ξ and $\{\alpha_n\}$, conditions (29) (with $\widetilde{w} = 0$) and (31) set off a subset of $D(G)$, where the equality (1.2.5) is valid. On the contrary, if (29) and (31) are related by conditions on the parameters ξ and $\{\alpha_n\}$, then (1.2.5) is true on the all of $D(G)$. In general situation where $\Delta \geq 0$, inequality (34) establishes a qualified estimate of the value $\|\mathfrak{R}_{1/N(\delta,\Delta)}(\widetilde{F}) - G(F)\|_{X_1}$ in (1.2.8) on the mentioned subset of $D(G)$. Also, note that when the value $\Delta > 0$ is fixed, the mapping

$$\mathfrak{R}_{1/N(\delta,\Delta)}(\widetilde{F}) = x_{N(\delta,\Delta)}$$

doesn't satisfy (1.2.5) and hence doesn't define a regularization algorithm. At the same time, (34) implies that the operator $\mathfrak{R}_{1/N(\delta,\Delta)}$ is stable with respect to Δ and tends to a regularization mapping $\mathfrak{R}_{1/N(\delta,0)}$ as $\Delta \to 0$. From (29) and (31) it follows that in the case of $\Delta > 0$, an admissible subset of elements ξ is widened, as compared to the case $\Delta = 0$, to a Δ–neighborhood

$$\widetilde{\mathbb{M}}_\Delta(x^*) = \{x^* + (F'^*(x^*)F'(x^*))^p v + w : \|v\|_{X_1} \leq \widetilde{d}, \|w\|_{X_1} \leq \Delta\}$$

of the ellipsoid

$$\widetilde{\mathbb{M}}(x^*) = \{x^* + (F'^*(x^*)F'(x^*))^p v : \|v\|_{X_1} \leq \widetilde{d}\}.$$

The neighborhood $\widetilde{\mathbb{M}}_\Delta(x^*)$, as opposite to $\mathbb{M}(x^*)$ (see (26)) and $\widetilde{\mathbb{M}}(x^*)$, has a nonempty interior. This facilitates a choice of ξ when $\Delta > 0$.

Let us assume that a priori analysis of equation (1) allows to specify a subset $Q \subset X_1$ containing the solution x^*. It is well known that proper accounting of such information offers strong possibilities of refining present methods and constructing new effective computational algorithms for solving (1) [138, 146]. Suppose Q is a closed convex subset of X_1 and consider the iterative process

$$x_{n+1} = P_Q\{\xi - \Theta(\widetilde{F}'^*(x_n)\widetilde{F}'(x_n), \alpha_n)\widetilde{F}'^*(x_n)\cdot$$
$$\cdot (\widetilde{F}(x_n) - \widetilde{F}'(x_n)(x_n - \xi))\}. \tag{35}$$

In the case where $Q = X_1$, (35) takes the form (28). The use of projecting in (35) allows to get iterative points x_n in the subset Q, to which the unknown solution x^* belongs.

Since $x^* \in Q$, from (35) and (1.1.16) it follows that

$$\|x_{n+1} - x^*\|_{X_1} = \|P_Q\{\xi - \Theta(\widetilde{F}'^*(x_n)\widetilde{F}'(x_n), \alpha_n)\widetilde{F}'^*(x_n)\cdot$$
$$\cdot (\widetilde{F}(x_n) - \widetilde{F}'(x_n)(x_n - \xi))\} - P_Q(x^*)\|_{X_1} \leq$$
$$\leq \|\xi - \Theta(\widetilde{F}'^*(x_n)\widetilde{F}'(x_n), \alpha_n)\widetilde{F}'^*(x_n)\widetilde{F}(x_n) - \widetilde{F}'(x_n)(x_n - \xi)) - x^*\|_{X_1}.$$

Now, repeating without any changes the proofs of Theorems 4.1 and 4.2, we obtain analogs of these statements related to the process (35) in cases of exact and approximate data. In particular, the stopping rule (33) provides for the process (35) the same error estimate

$$\|x_{N(\delta,\Delta)} - x^*\|_{X_1} = O((\delta + \Delta)^{2p/(2p+1)}). \tag{36}$$

Practical effectiveness of the scheme (35) evidently depends on the complexity of numerical implementation of the projecting operator P_Q for a chosen subset Q. In [32, 145] one can find several examples of spaces X_1 and subsets

$Q \subset X_1$ such that the projection $P_Q(y)$, $y \in X_1$ can be expressed in an explicit form.

Concluding this section, let us give some examples of generating functions $\Theta(\lambda, \alpha)$ satisfying Conditions 4.2, 4.3, and 4.5. Define the family of contours

$$\Gamma_\alpha = \Gamma_\alpha^{(1)} \cup \Gamma_\alpha^{(2)} \cup \Gamma_\alpha^{(3)} \cup \Gamma_\alpha^{(4)}, \quad \alpha \in (0, \alpha_0], \tag{37}$$

where in the notation of (2.3.19),

$$\Gamma_\alpha^{(1)} = \Gamma_{\alpha/2}(\exp(i\varphi_0), \exp(i(2\pi - \varphi_0))), \quad \varphi_0 \in \left(0, \frac{\pi}{2}\right),$$
$$\Gamma_\alpha^{(2)} = \Gamma_{R_0}(\exp(-i\varphi_0), \exp(i\varphi_0)), \quad R_0 > N_1^2,$$
$$\Gamma_\alpha^{(3)} = \Gamma_{(\alpha/2, R_0)}(\exp(i\varphi_0)),$$
$$\Gamma_\alpha^{(4)} = \Gamma_{(\alpha/2, R_0)}(\exp(-i\varphi_0)).$$

For the family $\{\Gamma_\alpha\}_{\alpha \in (0, \alpha_0]}$ defined by (37), inequality (9) is trivial, and condition (10) follows by the cosine theorem applied to the triangle $O\lambda\mu$. Therefore the contours (37) satisfy Condition 4.4. Simple calculations prove that all generating functions $\Theta(\lambda, \alpha)$ in the following Examples 4.1–4.5 satisfy Condition 4.5 with the contours (37). Conditions 4.2 and 4.3 are also valid.

Example 4.1. In the case of generating function (2.1.15), the process (3) takes the form

$$(F'^*(x_n)F'(x_n) + \alpha_n E)x_{n+1} =$$
$$= \alpha_n \xi + F'^*(x_n)(F'(x_n)x_n - F(x_n)). \tag{38}$$

Condition 4.3 for the iteration (38) is satisfied with $p_0 = 1$.

Example 4.2. An iterative step of the method (3), (2.1.18) can be implemented as the finite iterative process:

$$x_{n+1} = x_{n+1}^{(N)},$$

where

$$x_{n+1}^{(0)} = \xi, \quad (F'^*(x_n)F'(x_n) + \alpha_n E)x_{n+1}^{(k+1)} =$$
$$= \alpha_n x_{n+1}^{(k)} + F'^*(x_n)(F'(x_n)x_n - F(x_n)), \quad k = 0, 1, \ldots, N-1. \tag{39}$$

In this case, Condition 4.3 is fulfilled with $p_0 = N$.

Example 4.3. An iteration of the method (3), (2.1.23) has the following form:

$$x_{n+1} = u\left(\frac{1}{\alpha_n}\right),$$

where $u = u(t)$ is a solution to the Cauchy problem

$$\frac{du(t)}{dt} + F'^*(x_n)F'(x_n)u(t) =$$
$$= F'^*(x_n)(F'(x_n)x_n - F(x_n)), \quad u(0) = \xi. \tag{40}$$

Condition 4.3 is satisfied with an arbitrary $p_0 > 0$.

Example 4.4. Consider the generating function (2.1.29) with

$$g(\lambda) \equiv \mu_0, \quad 0 < \mu_0 < \frac{1}{N_1^2}.$$

The regularization parameter α takes the values $1, 1/2, 1/3, \dots$. An iteration of the method (3), (2.1.29) with $\alpha_n = n^{-1}$ can be implemented as follows:

$$x_{n+1} = x_{n+1}^{(n)},$$

where

$$x_{n+1}^{(0)} = \xi, \quad x_{n+1}^{(k+1)} = (E - \mu_0 F'^*(x_n)F'(x_n))x_{n+1}^{(k)} -$$
$$- F'^*(x_n)(F'(x_n)x_n - F(x_n)), \quad k = 0, 1, \dots, n-1. \tag{41}$$

Example 4.5. For the function (2.1.29) with

$$g(\lambda) = \frac{1}{\lambda + \mu_0}, \quad \mu_0 > 0$$

and $\alpha_n = n^{-1}$, an iterative step of the method (3) takes the form:

$$x_{n+1} = x_{n+1}^{(n)};$$

$$x_{n+1}^{(0)} = \xi, \quad (\mu_0 E + F'^*(x_n)F'(x_n))x_{n+1}^{(k+1)} =$$
$$= \mu_0 x_{n+1}^{(k)} + F'^*(x_n)(F'(x_n)x_n - F(x_n)), \quad k = 0, 1, \dots, n-1. \tag{42}$$

Since the functions $\Theta(\lambda, \alpha)$ from Examples 4.3–4.5 satisfy Condition 4.4 with an arbitrary $p_0 > 0$, the processes (40)–(42) are saturation free.

4.2 Necessity of Sourcewise Representation for Rate of Convergence Estimates of Gauss–Newton Type Methods

In this section, under appropriate conditions on the generating function $\Theta(\lambda, \alpha)$ we prove that the sourcewise representation (1.4) is close to a necessary condition for the power rate of convergence estimate (1.5). Namely, it will be shown that the estimate

$$\|x_n - x^*\|_{X_1} \le l\alpha_n^p, \quad n = 0, 1, \ldots \qquad \left(p > \frac{1}{2}\right) \qquad (1)$$

implies

$$x^* - \xi \in R((F'^*(x^*)F'(x^*))^{p-\varepsilon}) \quad \forall \varepsilon \in (0, p). \qquad (2)$$

This result is similar to assertions of Theorems 2.4 and 2.5.

Suppose (1) is true, and Conditions 4.1, 4.2, 4.3 (with $p = 0$), and 4.4 are satisfied. Without loss of generality we can assume that $l\alpha_0^p \le R$. Therefore

$$x_n \in \Omega_R(x^*), \quad n = 0, 1, \ldots.$$

From (1.6), (1.7), and (1) it follows that

$$\|[E - \Theta(F'^*(x^*)F'(x^*), \alpha_n)F'^*(x^*)F'(x^*)](x^* - \xi)\|_{X_1} \le$$
$$\le l\alpha_{n+1}^p + c_1 l^2 L\alpha_n^{2p-1/2} +$$
$$+ \|[\Theta(F'^*(x^*)F'(x^*), \alpha_n)F'^*(x^*)F'(x^*) -$$
$$- \Theta(F'^*(x_n)F'(x_n), \alpha_n)F'^*(x_n)F'(x_n)](x^* - \xi)\|_{X_1} \le \qquad (3)$$
$$\le c_2 \Big(\alpha_n^p + \|\Theta(F'^*(x^*)F'(x^*), \alpha_n)F'^*(x^*)F'(x^*) -$$
$$- \Theta(F'^*(x_n)F'(x_n), \alpha_n)F'^*(x_n)F'(x_n)\|_{L(X_1)} \Big).$$

In view of (1.11)–(1.13) (with $p = 0$), (1.16), and (1.17), for the last norm in (3) we have the estimate

$$\|\Theta(F'^*(x^*)F'(x^*), \alpha_n)F'^*(x^*)F'(x^*) -$$
$$- \Theta(F'^*(x_n)F'(x_n), \alpha_n)F'^*(x_n)F'(x_n)\|_{L(X_1)} \le$$
$$\le c_3 \alpha_n^p \int_{\Gamma_{\alpha_n}} \frac{|1 - \Theta(\lambda, \alpha_n)\lambda|}{|\lambda|^{3/2}} |d\lambda|, \quad n = 0, 1, \ldots. \qquad (4)$$

The following condition on $\Theta(\lambda, \alpha)$ generalizes Condition 4.5.

Condition 4.6. *For each $s \in [1, 3/2]$,*

$$\sup_{\alpha \in (0, \alpha_0]} \alpha^{s-1} \int_{\Gamma_\alpha} \frac{|1 - \Theta(\lambda, \alpha)\lambda|}{|\lambda|^s} |d\lambda| < \infty.$$

From (4) and Condition 4.6 we get

$$\|\Theta(F'^*(x^*)F'(x^*), \alpha_n)F'^*(x^*)F'(x^*) -$$
$$- \Theta(F'^*(x_n)F'(x_n), \alpha_n)F'^*(x_n)F'(x_n)\|_{L(X_1)} \le c_4 \alpha_n^{p-1/2}. \qquad (5)$$

Given $\alpha \in (0, \alpha_0]$, we denote

$$\Phi(\alpha) = \|[E - \Theta(F'^*(x^*)F'(x^*), \alpha)F'^*(x^*)F'(x^*)](x^* - \xi)\|_{X_1}. \quad (6)$$

According to (3), (5), and (6),

$$\Phi(\alpha_n) \leq c_5 \alpha_n^{p-1/2}. \quad (7)$$

Suppose the following condition is satisfied.

Condition 4.7. *For each $\lambda \in [0, N_1^2]$, the function*

$$\eta(\lambda, \alpha) = |1 - \Theta(\lambda, \alpha)\lambda|$$

is nondecreasing in $\alpha \in (0, \alpha_0]$.

Let $\{E_\lambda\}_{\lambda \in [0, N_1^2]}$ be the spectral family of $F'^*(x^*)F'(x^*)$. Since

$$\Phi(\alpha) = \left(\int\limits_{0-0}^{N_1^2} \eta^2(\lambda, \alpha) d\|E_\lambda(x^* - \xi)\|_{X_1}^2 \right)^{1/2},$$

by Condition 4.7 it follows that the function $\Phi(\alpha)$ is nondecreasing on the segment $(0, \alpha_0]$. According to (7), for all $\alpha \in (\alpha_{n+1}, \alpha_n], n = 0, 1, \ldots$ we have

$$\frac{\Phi(\alpha)}{\alpha^{p-1/2}} \leq \frac{\Phi(\alpha_n)}{\alpha^{p-1/2}} \leq \frac{c_5 \alpha_n^{p-1/2}}{\alpha_{n+1}^{p-1/2}} \leq c_5 r^{p-1/2} < \infty.$$

Therefore,

$$\Phi(\alpha) \leq c_5 r^{p-1/2} \alpha^{p-1/2} \quad \forall \alpha \in (0, \alpha_0]. \quad (8)$$

Since $p > 1/2$, from (8) we conclude that $\lim\limits_{\alpha \to 0} \Phi(\alpha) = 0$. On the other hand, the integral representation for $\Phi(\alpha)$ yields

$$\lim_{\alpha \to 0} \Phi(\alpha) \geq \left(\int\limits_{\{0\}} d\|E_\lambda(x^* - \xi)\|_{X_1}^2 \right)^{1/2} =$$

$$= \|P_{N(F'^*(x^*)F'(x^*))}(x^* - \xi)\|_{X_1}.$$

It now follows that

$$x^* - \xi \perp N(F'^*(x^*)F'(x^*)).$$

Let us apply Theorems 2.4 and 2.5, with (2.1.38) and (2.1.43) replaced by (2.1.46) (see [42]), to the equation

$$F'^*(x^*)F'(x^*)(x - x^*) = 0, \quad x \in X_1,$$

for which $x = x^*$ is a solution nearest to ξ. By (8) we then obtain

$$x^* - \xi \in R((F'^*(x^*)F'(x^*))^{p-1/2-\varepsilon_1}) \quad \forall \varepsilon_1 \in \left(0, p - \frac{1}{2}\right). \qquad (9)$$

We denote $p_1 = p - 1/2 - \varepsilon_1$. From (9) it follows that there exists an element $v^{(1)} \in X_1$ such that

$$x^* - \xi = (F'^*(x^*)F'(x^*))^{p_1} v^{(1)}. \qquad (10)$$

Let us now improve the estimate (7). Substituting (10) into (3), in place of (7) we get

$$\Phi(\alpha_n) =$$
$$= \|[E - \Theta(F'^*(x^*)F'(x^*), \alpha_n)F'^*(x^*)F'(x^*)](x^* - \xi)\|_{X_1} \le$$
$$\le c_6 \left(\alpha_n^p + \int_{\Gamma_{\alpha_n}} |1 - \Theta(\lambda, \alpha_n)\lambda| \|[R(\lambda, F'^*(x^*)F'(x^*)) - \right. \qquad (11)$$
$$\left. - R(\lambda, F'^*(x_n)F'(x_n))](F'^*(x^*)F'(x^*))^{p_1} \|_{L(X_1)} |d\lambda| \right).$$

Analogously to (1.12), for all $\lambda \in \Gamma_{\alpha_n}$ we have

$$\|[R(\lambda, F'^*(x^*)F'(x^*)) -$$
$$- R(\lambda, F'^*(x_n)F'(x_n))](F'^*(x^*)F'(x^*))^{p_1} \|_{L(X_1)} \le$$
$$\le \|R(\lambda, F'^*(x_n)F'(x_n))F'^*(x_n)\|_{L(X_2, X_1)} \|F'(x^*) - F'(x_n)\|_{L(X_1, X_2)} \cdot$$
$$\cdot \|R(\lambda, F'^*(x^*)F'(x^*))(F'^*(x^*)F'(x^*))^{p_1} \|_{L(X_1)} +$$
$$+ \|R(\lambda, F'^*(x_n)F'(x_n))\|_{L(X_1)} \|F'^*(x^*) - F'^*(x_n)\|_{L(X_2, X_1)} \cdot$$
$$\cdot \|F'(x^*)R(\lambda, F'^*(x^*)F'(x^*))(F'^*(x^*)F'(x^*))^{p_1} \|_{L(X_1, X_2)}. \qquad (12)$$

It follows by (1.2) and (1) that

$$\|F'^*(x^*) - F'^*(x_n)\|_{L(X_2, X_1)} =$$
$$= \|F'(x^*) - F'(x_n)\|_{L(X_1, X_2)} \le c_7 \alpha_n^p. \qquad (13)$$

After setting in (1.14) and (1.15) $s = p_1$, we obtain

$$\|R(\lambda, F'^*(x)F'(x))(F'^*(x)F'(x))^{p_1} \|_{L(X_1)} \le \frac{c_8}{|\lambda|^{1-\min\{p_1, 1\}}}, \qquad (14)$$

$$\|F'(x)R(\lambda, F'^*(x)F'(x))(F'^*(x)F'(x))^{p_1} \|_{L(X_1, X_2)} \le$$
$$\le \frac{c_8}{|\lambda|^{1-\min\{p_1+1/2, 1\}}} \quad \forall \lambda \in \Gamma_{\alpha_n}. \qquad (15)$$

According to (12)–(15),

$$\|[R(\lambda, F'^*(x^*)F'(x^*)) - R(\lambda, F'^*(x_n)F'(x_n))](F'^*(x^*)F'(x^*))^{p_1}\|_{L(X_1)} \leq$$

$$\leq \frac{c_9 \alpha_n^p}{|\lambda|^{2-\min\{p_1+1/2,1\}}} \quad \forall \lambda \in \Gamma_{\alpha_n}.$$

Therefore from (11) we get the following refined estimate for $\Phi(\alpha_n)$:

$$\Phi(\alpha_n) \leq c_{10}\alpha_n^p\left(1 + \int_{\Gamma_{\alpha_n}} \frac{|1 - \Theta(\lambda, \alpha_n)\lambda|}{|\lambda|^{2-\min\{p_1+1/2,1\}}}|d\lambda|\right), \quad n = 0, 1, \dots. \quad (16)$$

Let us consider two cases.

1) Suppose $p_1 \geq 1/2$; then Condition 4.6 and (16) yield

$$\Phi(\alpha_n) \leq c_{11}\alpha_n^p.$$

Therefore for each $\varepsilon \in (0, p)$ there exists an element $\tilde{v} = \tilde{v}(\varepsilon) \in X_1$ such that

$$x^* - \xi = (F'^*(x^*)F'(x^*))^{p-\varepsilon}\tilde{v}. \quad (17)$$

Thus (2) is valid.

2) Let $p_1 \in (0, 1/2)$. Then from (4) and (16) we obtain

$$\Phi(\alpha_n) \leq c_{12}\alpha_n^{p-(1/2-p_1)}. \quad (18)$$

Hence for each $\varepsilon_2 \in (0, p - (1/2 - p_1))$ there exists an element $v^{(2)} \in X_1$ such that

$$x^* - \xi = (F'^*(x^*)F'(x^*))^{p-(1/2-p_1)-\varepsilon_2}v^{(2)}. \quad (19)$$

Now we can use (19) to improve estimate (18) by substituting (19) into (3). Continuing the process of iterating estimates for $\Phi(\alpha_n)$, we get sequences $\{p_k\}$, $\{\varepsilon_k\}$, and $\{v^{(k)}\}$ such that

$$p_{k+1} = p - \left(\frac{1}{2} - p_k\right) - \varepsilon_{k+1}, \quad \varepsilon_{k+1} \in \left(0, p - \left(\frac{1}{2} - p_k\right)\right);$$

$$x^* - \xi = (F'^*(x^*)F'(x^*))^{p_{k+1}}v^{(k+1)}, \quad k \in \mathbf{N}. \quad (20)$$

Without loss of generality we can assume that

$$\lim_{k \to \infty} \varepsilon_k = 0, \quad \sup_{k \in \mathbf{N}} \varepsilon_k < p - \frac{1}{2}. \quad (21)$$

If on the k-th iteration we have $p_k \geq 1/2$, then (17) and (2) are valid. We claim that there exists a number $k = k_0$ for which $p_{k_0} \geq 1/2$. Indeed, suppose the contrary, i.e., $p_k < 1/2$ for all numbers $k \in \mathbf{N}$. By (20) and (21) we get

$p_k < p_{k+1} < p$. Since the sequence $\{p_k\}$ is monotone and bounded, there exists a finite limit $\lim\limits_{k\to\infty} p_k = \widetilde{p} \le p$. Passing to limits in both parts of equality (20), with the use of (21) we obtain $p = 1/2$. This contradicts the assumption $p > 1/2$ (see (1)). Thus (2) is valid again. The following statement summarizes the preceding discussion.

Theorem 4.3. *Suppose Conditions 4.1, 4.2, 4.3 (with $p = 0$), 4.4, 4.6, and 4.7 are satisfied and estimate (1) is fulfilled. Then for each $\varepsilon \in (0, p)$,*

$$x^* - \xi \in R((F'^*(x^*)F'(x^*))^{p-\varepsilon}).$$

Note that all generating functions from Examples 4.1–4.5 satisfy Conditions 4.6 and 4.7. As $\{\Gamma_\alpha\}_{\alpha \in (0, \alpha_0]}$, the family of contours (1.37) may be taken.

In a conclusion of this section we shall demonstrate an application of the technique derived above to Tikhonov's method [17, 42, 137]. Suppose the operator $F(x)$ in (1.1) is available without errors. According to the basic scheme of Tikhonov's method, a solution x_α to the regularized problem

$$\min_{x \in X_1} \Psi_\alpha(x); \quad \Psi_\alpha(x) = \frac{1}{2}\|F(x)\|^2_{X_2} + \frac{1}{2}\alpha\|x - \xi\|^2_{X_1}, \quad \alpha > 0 \quad (22)$$

is taken as an approximation of x^*. Here α is a regularization parameter and $\xi \in X_1$ an estimate of x^*.

Suppose (1.2) is satisfied and the operator $F : X_1 \to X_2$ is weakly continuous, i.e., the weak convergence $z_n \to z_0 (n \to \infty)$ in X_1 implies the weak convergence $F(z_n) \to F(z_0)$ in X_2. The last assumption ensures the solvability of (22) for each $\alpha > 0$.

In [111], under the condition (1.4) with $p \in [1/2, 1]$, the following rate of convergence estimate for $\{x_\alpha\}$ was established:

$$\|x_\alpha - x^*\|_{X_1} \le l\alpha^p \quad \forall \alpha \in (0, \alpha_0]. \quad (23)$$

We shall prove below that the sourcewise representation (1.4), sufficient for the estimate (23), is actually close to a necessary condition for (23). Thus let the family $\{x_\alpha\}$ be defined by (22), and (23) be fulfilled with some exponent $p \in (1/2, 1]$. Since x_α is a minimizer of the functional $\Psi_\alpha(x)$, we have

$$\Psi'_\alpha(x_\alpha) = F'^*(x_\alpha)F(x_\alpha) + \alpha(x_\alpha - \xi) = 0.$$

Consequently,

$$F'^*(x_\alpha)[F'(x_\alpha)(x_\alpha - x^*) + G(x_\alpha, x^*)] + \alpha(x_\alpha - \xi) = 0, \quad (24)$$

where
$$G(x_\alpha, x^*) = F(x_\alpha) - F(x^*) - F'(x_\alpha)(x_\alpha - x^*).$$

By analogy with (1.7),

$$\|G(x_\alpha, x^*)\|_{X_2} \leq \frac{1}{2}L\|x_\alpha - x^*\|_{X_1}^2.$$

From (24) it follows that

$$(F'^*(x_\alpha)F'(x_\alpha) + \alpha E)(x_\alpha - x^*) +$$
$$+ F'^*(x_\alpha)G(x_\alpha, x^*) + \alpha(x^* - \xi) = 0.$$

Hence,

$$(F'^*(x_\alpha)F'(x_\alpha) + \alpha E)^{-1}F'^*(x_\alpha)G(x_\alpha, x^*) +$$
$$+ (x_\alpha - x^*) + \alpha[(F'^*(x_\alpha)F'(x_\alpha) + \alpha E)^{-1} -$$
$$- (F'^*(x^*)F'(x^*) + \alpha E)^{-1}](x^* - \xi) +$$
$$+ \alpha(F'^*(x^*)F'(x^*) + \alpha E)^{-1}(x^* - \xi) = 0. \tag{25}$$

The last summand in (25) can be rewritten as

$$\alpha(F'^*(x^*)F'(x^*) + \alpha E)^{-1}(x^* - \xi) =$$
$$= (E - \Theta(F'^*(x^*)F'(x^*), \alpha)F'^*(x^*)F'(x^*))(x^* - \xi), \tag{26}$$

where $\Theta(\lambda, \alpha) = (\lambda + \alpha)^{-1}$. According to (26), by (6), (23), and (25) we get

$$\Phi(\alpha) \leq \|x_\alpha - x^*\|_{X_1} +$$
$$+ \|(F'^*(x_\alpha)F'(x_\alpha) + \alpha E)^{-1}F'^*(x_\alpha)G(x_\alpha, x^*)\|_{X_1} +$$
$$+ \alpha\|[(F'^*(x_\alpha)F'(x_\alpha) + \alpha E)^{-1} -$$
$$- (F'^*(x^*)F'(x^*) + \alpha E)^{-1}](x^* - \xi)\|_{X_1}.$$

Therefore,

$$\Phi(\alpha) \leq \|x_\alpha - x^*\|_{X_1} +$$
$$+ c_{13}\frac{\|x_\alpha - x^*\|_{X_1}^2}{\sqrt{\alpha}} + \alpha\|[(F'^*(x_\alpha)F'(x_\alpha) + \alpha E)^{-1} -$$
$$- (F'^*(x^*)F'(x^*) + \alpha E)^{-1}](x^* - \xi)\|_{X_1} \leq$$
$$\leq c_{14}\alpha^p + \alpha\|[R(-\alpha, F'^*(x_\alpha)F'(x_\alpha)) -$$
$$- R(-\alpha, F'^*(x^*)F'(x^*))](x^* - \xi)\|_{X_1}. \tag{27}$$

Due to (23) and (1.12)–(1.15), for the last norm in (27) we have

$$\|[R(-\alpha, F'^*(x_\alpha)F'(x_\alpha)) -$$
$$- R(-\alpha, F'^*(x^*)F'(x^*))](x^* - \xi)\|_{X_1} \leq c_{15}\alpha^{p-3/2}. \tag{28}$$

Thus (27) and (28) yield

$$\Phi(\alpha) \le c_{16}(\alpha^p + \alpha^{p-1/2}) \le c_{17}\alpha^{p-1/2} \quad \forall \alpha \in (0, \alpha_0]. \qquad (29)$$

Estimate (29) is similar to inequality (7), which was derived for the process (1.3). By the same argument as in the proof of Theorem 4.3, we then obtain representation (10). Next, using (10) for improvement of (28), we get (20) and (21). Continuing this line of reasoning, we come to the following result.

Theorem 4.4. *Suppose (1.2) is satisfied and $F(x)$ is a weakly continuous operator. Let (23) be fulfilled with $p \in (1/2, 1]$. Then for each $\varepsilon \in (0, p)$,*

$$x^* - \xi \in R((F'^*(x^*)F'(x^*))^{p-\varepsilon}).$$

Remark 4.3. In the case of $p = 1$, the following stronger variant of Theorem 4.4 is true: estimate (23) implies $x^* - \xi \in R(F'^*(x^*)F'(x^*))$ ([111]).

4.3 Generalized Newton–Kantorovich Type Methods for Nonlinear Irregular Equations in Banach Space

In this section we study the class of generalized Newton–Kantorovich type methods

$$x_0 \in \Omega_R(x^*), \quad x_{n+1} = \xi - \Theta(F'(x_n), \alpha_n)(F(x_n) - F'(x_n)(x_n - \xi)) \quad (1)$$

for solving an irregular operator equation

$$F(x) = 0, \quad x \in X$$

in a Banach space X. We obtain (1) by iterating the scheme (3.2.16), much as the process (1.3) was designed on the basis of the scheme (3.2.1). Suppose the operator $F : X \to X$ is Fréchet differentiable on $\Omega_R(x^*)$, and the derivative $F'(x)$ is Lipschitz continuous, i.e.,

$$\|F'(x) - F'(y)\|_{L(X)} \le L\|x - y\|_X \quad \forall x, y \in \Omega_R(x^*). \qquad (2)$$

Let the sequence of regularization parameters $\{\alpha_n\}$ satisfy Condition 4.1.

In this section and in Section 4.4 we shall extend the results of Sections 4.1 and 4.2 to the family of iterative processes (1) in a Banach space X. We shall assume that the operator $A = F'(x^*)$ satisfies Condition 1.1. In our context, the sourcewise representation (1.4) should be replaced by the equality

$$x^* - \xi = F'(x^*)^p v, \quad v \in X, \quad p \ge 1. \qquad (3)$$

Turning to the processes (1), let us first show that formula (1.1.6) is applicable to the operator $A = F'(x_n)$. Pick a number $R_0 > \|F'(x^*)\|_{L(X)}$ and suppose the generating function $\Theta(\lambda, \alpha)$ satisfies Condition 2.6.

Conditions 1.1 and 2.6 allow to present the operator function $\Theta(F'(x_n), \alpha_n)$ from (1) in the integral form

$$\Theta(F'(x_n), \alpha_n) = \frac{1}{2\pi i} \int_{\Gamma_n} \Theta(\lambda, \alpha_n) R(\lambda, F'(x_n)) d\lambda \tag{4}$$

with an appropriate contour Γ_n.

Lemma 4.1. *Suppose*

$$\|x_n - x^*\|_X \leq \min\left\{R, \frac{d_0 \nu_0 \alpha_n}{L r_0}\right\}, \tag{5}$$

where the constants r_0 and d_0 are specified in Conditions 1.1 and 2.6; $\nu_0 \in (0, 1)$. Then

$$\sigma(F'(x_n)) \subset \mathrm{int} K_{\alpha_n}(R_0, d_0, \varphi_0).$$

Proof. For each $\lambda \in \mathbf{C} \backslash \mathrm{int} K_{\alpha_n}(R_0, d_0, \varphi_0)$, by definition of $K_\alpha(R_0, d_0, \varphi_0)$ we have $|\lambda| \geq d_0 \alpha_n$. Note that by (5), $x_n \in \Omega_R(x^*)$. Setting in (1.1.11)

$$A = F'(x^*), \quad B = F'(x_n) - F'(x^*)$$

and using (2) and (1.1.8), we obtain

$$\|BR(\lambda, A)\|_{L(X)} \leq \|F'(x_n) - F'(x^*)\|_{L(X)} \|R(\lambda, F'(x^*))\|_{L(X)} \leq$$

$$\leq \frac{L r_0 \|x_n - x^*\|_X}{|\lambda|} \leq \nu_0 \quad \forall \lambda \in \mathbf{C} \backslash \mathrm{int} K_{\alpha_n}(R_0, d_0, \varphi_0).$$

Therefore $\lambda \in \rho(A + B) = \rho(F'(x_n))$. $\qquad\square$

As in Sections 2.2 and 2.3, suppose $\{\Gamma_\alpha\}_{\alpha \in (0,\infty)}$ is a family of contours on the complex plane such that $\Gamma_\alpha \subset D_\alpha$ and Γ_α surrounds $\mathrm{int} K_\alpha(R_0, d_0, \varphi_0)$ for each $\alpha > 0$. Also, assume that Γ_α doesn't contain the point $\lambda = -d_1 \alpha$ inside; $d_1 \in (d_0, 1)$. As an example we can set $\Gamma_\alpha = \gamma_\alpha$, where γ_α is the boundary of $K_\alpha(R_0, d_0, \varphi_0)$. Lemma 4.1 implies that representation (4) is valid with $\Gamma_n = \Gamma_{\alpha_n}$, provided condition (5) is fulfilled.

Let us now prove that the sourcewise representation (3) yields the power rate of convergence estimate for $\{x_n\}$ with the same exponent p, that is,

$$\|x_n - x^*\|_X \leq l \alpha_n^p, \quad n = 0, 1, \ldots. \tag{6}$$

The value l, as well as necessary restrictions on parameters of the procedure (1) will be specified below.

Suppose (6) is fulfilled for a number $n \in \mathbf{N}$. By (6) and Lemma 4.1 we conclude that conditions

$$\frac{\alpha_0^{p-1} L r_0 l}{d_0} \le \nu_0, \quad l\alpha_0^p \le R \tag{7}$$

are sufficient for representation (4) with $\Gamma_n = \Gamma_{\alpha_n}$. Throughout this section we suppose that conditions (7) are satisfied.

From (6) and (7) we get $x_n \in \Omega_R(x^*)$. Next, according to (1),

$$\begin{aligned}
x_{n+1} - x^* = &-\Theta(F'(x_n), \alpha_n)G(x_n, x^*)- \\
&- (E - \Theta(F'(x^*), \alpha_n)F'(x^*))(x^* - \xi)- \\
&- (\Theta(F'(x^*), \alpha_n)F'(x^*) - \Theta(F'(x_n), \alpha_n)F'(x_n))(x^* - \xi),
\end{aligned} \tag{8}$$

where

$$G(x_n, x^*) = F(x_n) - F(x^*) - F'(x_n)(x_n - x^*).$$

From (1.1.14) it follows that

$$\|G(x_n, x^*)\|_X \le \frac{1}{2}L\|x_n - x^*\|_X^2. \tag{9}$$

Combining (3) and (8), we obtain

$$\begin{aligned}
\|x_{n+1} - x^*\|_X \le &\|\Theta(F'(x_n), \alpha_n)\|_{L(X)}\|G(x_n, x^*)\|_X+ \\
&+ \|[E - \Theta(F'(x^*), \alpha_n)F'(x^*)]F'(x^*)^p v\|_X+ \\
&+ \|[\Theta(F'(x^*), \alpha_n)F'(x^*) - \Theta(F'(x_n), \alpha_n)F'(x_n)]F'(x^*)^p v\|_X.
\end{aligned} \tag{10}$$

In view of (4),

$$\begin{aligned}
\|\Theta(F'(x_n), \alpha_n)\|_{L(X)} &\le \\
&\le \frac{1}{2\pi} \int_{\Gamma_{\alpha_n}} |\Theta(\lambda, \alpha_n)|\|R(\lambda, F'(x_n))\|_{L(X)}|d\lambda|.
\end{aligned}$$

Further, from (1.1.8), (2), (6), and (7) we get

$$\|F'(x_n) - F'(x^*)\|_{L(X)}\|R(\lambda, F'(x^*))\|_{L(X)} \le \nu_0 < 1 \quad \forall \lambda \in \Gamma_{\alpha_n}. \tag{11}$$

Lemma 1.2 now implies that for each $\lambda \in \Gamma_{\alpha_n}$,

$$\begin{aligned}
\|R(\lambda, F'(x_n))\|_{L(X)} &\le \|R(\lambda, F'(x^*))\|_{L(X)} \cdot \\
&\cdot \sum_{k=0}^{\infty} \left(\|F'(x_n) - F'(x^*)\|_{L(X)}\|R(\lambda, F'(x^*))\|_{L(X)}\right)^k.
\end{aligned} \tag{12}$$

Using (1.1.8), (11), and (12), we conclude that

$$\|R(\lambda, F'(x_n))\|_{L(X)} \le \frac{r_0}{(1 - \nu_0)|\lambda|} \quad \forall \lambda \in \Gamma_{\alpha_n}.$$

Consequently,

$$\|\Theta(F'(x_n), \alpha_n)\|_{L(X)} \le \frac{r_0}{2\pi(1 - \nu_0)} \int_{\Gamma_{\alpha_n}} \frac{|\Theta(\lambda, \alpha_n)|}{|\lambda|}|d\lambda|. \tag{13}$$

From here on we suppose Condition 2.8 is satisfied, i.e.,

$$\int_{\Gamma_\alpha} \frac{|\Theta(\lambda, \alpha)|}{|\lambda|}|d\lambda| \le \frac{c_1}{\alpha} \quad \forall \alpha \in (0, \alpha_0]. \tag{14}$$

From (6), (9), (13), and (14) we obtain

$$\|\Theta(F'(x_n), \alpha_n)\|_{L(X)}\|G(x_n, x^*)\|_X \le c_2 l^2 \alpha_n^p. \tag{15}$$

Denote $m = [p]$ and $\mu = \{p\}$. Then the second summand in (10) can be estimated as follows:

$$\|[E - \Theta(F'(x^*), \alpha_n)F'(x^*)]F'(x^*)^p v\|_X \le$$
$$\le \|[E - \Theta(F'(x^*), \alpha_n)F'(x^*)]F'(x^*)^m(F'(x^*) + d_1\alpha_n E)^\mu v\|_X +$$
$$+ \|[E - \Theta(F'(x^*), \alpha_n)F'(x^*)]F'(x^*)^m \cdot$$
$$\cdot [(F'(x^*) + d_1\alpha_n E)^\mu - F'(x^*)^\mu]v\|_X. \tag{16}$$

In connection with (16), we now impose on $\Theta(\lambda, \alpha)$ Condition 2.7 and assume that $p \in [1, p_0]$ with p_0 specified in this condition.

Since the point $\lambda = -d_1\alpha_n$ lies outside the contour Γ_{α_n}, we have

$$[E - \Theta(F'(x^*), \alpha_n)F'(x^*)]F'(x^*)^m(F'(x^*) + d_1\alpha_n E)^\mu =$$
$$= \frac{1}{2\pi i} \int_{\Gamma_{\alpha_n}} (1 - \Theta(\lambda, \alpha_n)\lambda)\lambda^m(\lambda + d_1\alpha_n)^\mu R(\lambda, F'(x^*))d\lambda. \tag{17}$$

From (1.1.8), (17), and the inequality

$$|\lambda| \ge d_0\alpha_n \quad \forall \lambda \in \Gamma_{\alpha_n}$$

we obtain

$$\|[E - \Theta(F'(x^*), \alpha_n)F'(x^*)]F'(x^*)^m(F'(x^*) + d_1\alpha_n E)^\mu v\|_X \leq$$

$$\leq \frac{\|v\|_X}{2\pi} \int_{\Gamma_{\alpha_n}} |1 - \Theta(\lambda, \alpha_n)\lambda||\lambda|^m|\lambda + d_1\alpha_n|^\mu \|R(\lambda, F'(x^*))\|_{L(X)}|d\lambda| \leq$$

$$\leq \frac{r_0}{2\pi}\|v\|_X \int_{\Gamma_{\alpha_n}} |1 - \Theta(\lambda, \alpha_n)\lambda||\lambda|^{m-1}(|\lambda|^\mu + (d_1\alpha_n)^\mu)|d\lambda| \leq$$

$$\leq c_3\|v\|_X\alpha_n^p. \tag{18}$$

Since $m \in [1, p_0]$, by (1.1.10) and (2.2.9) we conclude that

$$\|[E - \Theta(F'(x^*), \alpha_n)F'(x^*)]F'(x^*)^m \cdot$$
$$\cdot[(F'(x^*) + d_1\alpha_n E)^\mu - F'(x^*)^\mu]v\|_X \leq$$
$$\leq c_4\alpha_n^\mu\|v\|_X\|[E - \Theta(F'(x^*), \alpha_n)F'(x^*)]F'(x^*)^m\|_{L(X)} \leq \tag{19}$$
$$\leq c_5\|v\|_X\alpha_n^p.$$

Now, by (16), (18), and (19),

$$\|[E - \Theta(F'(x^*), \alpha_n)F'(x^*)]F'(x^*)^p v\|_X \leq c_6\|v\|_X\alpha_n^p. \tag{20}$$

For the third summand in (10) we get

$$\|[\Theta(F'(x^*), \alpha_n)F'(x^*) - \Theta(F'(x_n), \alpha_n)F'(x_n)]F'(x^*)^p v\|_X \leq$$
$$\leq A_n + B_n; \tag{21}$$

$$A_n = \|[\Theta(F'(x^*), \alpha_n)F'(x^*) - \Theta(F'(x_n), \alpha_n)F'(x_n)] \cdot$$
$$\cdot F'(x^*)^m(F'(x^*) + d_1\alpha_n E)^\mu v\|_X,$$

$$B_n = \|[\Theta(F'(x^*), \alpha_n)F'(x^*) - \Theta(F'(x_n), \alpha_n)F'(x_n)]F'(x^*)^m \cdot$$
$$\cdot[(F'(x^*) + d_1\alpha_n E)^\mu - F'(x^*)^\mu]v\|_X.$$

Formula (4) then yields

$$A_n \leq \|[(E - \Theta(F'(x^*), \alpha_n)F'(x^*)) -$$
$$- (E - \Theta(F'(x_n), \alpha_n)F'(x_n))]F'(x^*)^m\|_{L(X)} \cdot$$
$$\cdot \|(F'(x^*) + d_1\alpha_n E)^\mu\|_{L(X)}\|v\|_X \leq \frac{1}{2\pi}\|(F'(x^*) + d_1\alpha_n E)^\mu\|_{L(X)}\|v\|_X \cdot$$
$$\cdot \int_{\Gamma_{\alpha_n}} |1 - \Theta(\lambda, \alpha_n)\lambda|\|[R(\lambda, F'(x^*)) - R(\lambda, F'(x_n))]F'(x^*)^m\|_{L(X)}|d\lambda|. \tag{22}$$

Next, by (1.1.10),

$$\|(F'(x^*) + d_1\alpha_n E)^\mu\|_{L(X)} \le \|F'(x^*)^\mu\|_{L(X)} + c_7\alpha_0^\mu. \tag{23}$$

Combining (1.1.12), (5), and (7), we conclude that for all $\lambda \in \Gamma_{\alpha_n}$,

$$
\begin{aligned}
\|[R(\lambda, F'(x^*)) &- R(\lambda, F'(x_n))]F'(x^*)^m\|_{L(X)} \le \\
&\le \frac{\|R(\lambda, F'(x^*))\|_{L(X)}\|F'(x_n) - F'(x^*)\|_{L(X)}}{1 - \|R(\lambda, F'(x^*))\|_{L(X)}\|F'(x_n) - F'(x^*)\|_{L(X)}} \cdot \\
&\quad \cdot \|R(\lambda, F'(x^*))F'(x^*)^m\|_{L(X)} \le \\
&\le \frac{Lr_0 l\alpha_n^p}{(1-\nu_0)|\lambda|}\|R(\lambda, F'(x^*))F'(x^*)^m\|_{L(X)}.
\end{aligned}
\tag{24}
$$

Since $m \ge 1$, the identity

$$R(\lambda, F'(x^*))F'(x^*) = -E + \lambda R(\lambda, F'(x^*))$$

and (1.1.8) imply

$$\|R(\lambda, F'(x^*))F'(x^*)^m\|_{L(X)} \le (1 + r_0)\|F'(x^*)\|_{L(X)}^{m-1}. \tag{25}$$

From (17) and (22)–(25) it follows that

$$
\begin{aligned}
B_n \le \frac{r_0(1 + r_0)L}{2\pi(1-\nu_0)}&\left(\|F'(x^*)^\mu\|_{L(X)} + c_7\alpha_n^\mu\right)\|F'(x^*)\|_{L(X)}^{m-1}l\|v\|_X\alpha_n^p \cdot \\
&\cdot \int_{\Gamma_{\alpha_n}} \frac{|1 - \Theta(\lambda, \alpha_n)\lambda|}{|\lambda|}|d\lambda| \le c_8 l\|v\|_X\alpha_n^p.
\end{aligned}
\tag{26}
$$

By the same argument, for the second summand in (21) we obtain

$$
\begin{aligned}
\|[\Theta(F'(x^*),\alpha_n)F'(x^*) &- \Theta(F'(x_n), \alpha_n)F'(x_n)]F'(x^*)^m \cdot \\
&\cdot [(F'(x^*) + d_1\alpha_n E)^\mu - F'(x^*)^\mu]v\|_X \le c_9 l\|v\|_X\alpha_n^p.
\end{aligned}
\tag{27}
$$

Combining (10), (15), (20), (21), (26), and (27), we come to the resulting estimate

$$\|x_{n+1} - x^*\|_X \le (c_2 l^2 + c_6\|v\|_X + c_{10}l\|v\|_X)\alpha_n^p. \tag{28}$$

Recall that (28) was derived in the assumption that $x_n \in \Omega_R(x^*)$.

The following theorem establishes conditions sufficient for the estimate (6).

Theorem 4.5. *Suppose (3), (7), and Conditions 2.6–2.8 and 4.1 are satisfied. Also, assume that*

$$\|x_0 - x^*\|_X \le l\alpha_0^p, \quad p \in [1, p_0]; \tag{29}$$

$$0 < l \leq \min\left\{ \frac{1}{2r^p c_2}, \frac{R}{\alpha_0^p} \right\}, \quad \|v\|_X \leq d = \min\left\{ \frac{1}{4r^p c_{10}}, \frac{l}{4r^p c_6} \right\}. \quad (30)$$

Then (6) is true.

Proof. The proof is by induction on n. For $n = 0$, there is nothing to prove. Observe that in view of (29) and (30), $x_0 \in \Omega_R(x^*)$. From (28) with the use of Condition 4.1 we obtain

$$\|x_{n+1} - x^*\|_X \leq r^p(c_2 l^2 + c_6\|v\|_X + c_{10}l\|v\|_X)\alpha_{n+1}^p. \quad (31)$$

Note that by (30),

$$c_2 l^2 + c_6\|v\|_X \leq \frac{3l}{4r^p}, \quad c_{10}l\|v\|_X \leq \frac{l}{4r^p}. \quad (32)$$

Combining (31) with (32), we conclude that

$$\|x_{n+1} - x^*\|_X \leq l\alpha_{n+1}^p \leq l\alpha_0^p \leq R,$$

i.e., $x_{n+1} \in \Omega_R(x^*)$. The assertion now follows by (31) and the induction hypothesis. This completes the proof. $\qquad\square$

Theorem 4.5 establishes that the process (1) locally converges to x^*, provided the controlling parameter ξ is taken from the ellipsoid

$$\mathbb{M}(x^*) = \{x^* + F'(x^*)^p v : \|v\|_X \leq d\}.$$

Observe that in the irregular case, where $0 \in \sigma(F'(x^*))$, the set $\mathbb{M}(x^*)$ has an empty interior, and hence the problem of finding an element of $\mathbb{M}(x^*)$ is not trivial.

Let us now address to the case where $F(x)$ is available with errors. Suppose instead of the exact operator $F(x)$ its approximation $\widetilde{F}(x)$ is given, where $\widetilde{F}(x)$ is Fréchet differentiable on $\Omega_R(x^*)$ and $\widetilde{F}'(x)$ satisfies (2). Also, assume that

$$\|\widetilde{F}(x^*)\|_X \leq \delta; \quad \|\widetilde{F}'(x) - F'(x)\|_{L(X)} \leq \delta \quad \forall x \in \Omega_R(x^*). \quad (33)$$

As a set of all feasible exact and approximate data we take the space

$$\mathfrak{F} = \{F : \|F'(x) - F'(y)\|_{L(X)} \leq L\|x - y\|_X \; \forall x, y \in \Omega_R(x^*)\}$$

supplied with the metric

$$\rho(F_1, F_2) =$$

$$= \max\left\{ \|F_1(x^*) - F_2(x^*)\|_X, \; \sup_{x \in \Omega_R(x^*)} \|F_1'(x) - F_2'(x)\|_{L(X)} \right\}.$$

Likewise Section 4.1, it is convenient to fix the element x^* and to associate with the original equation the mapping $G : \mathfrak{F} \to X$ as follows:

$$D(G) = \{F \in \mathfrak{F} : F(x^*) = 0\}, \quad G(F) = x^*.$$

Given a starting point $x_0 \in \Omega_R(x^*)$, define a sequence $\{x_n\}$:

$$x_{n+1} = \xi - \Theta(\widetilde{F}'(x_n), \alpha_n)(\widetilde{F}(x_n) - \widetilde{F}'(x_n)(x_n - \xi)). \tag{34}$$

Suppose the following approximate sourcewise representation is fulfilled:

$$x^* - \xi = F'(x^*)^p \widetilde{v} + \widetilde{w}, \quad p \geq 1, \\ \widetilde{v}, \widetilde{w} \in X, \quad \|\widetilde{w}\|_X \leq \Delta. \tag{35}$$

Here Δ is an error level in the exact representation (3).

By the same argument as in the proof of Theorem 4.5, using (33) and (35), in place of (28) we come to the following error estimate for the iteration (34):

$$\|x_{n+1} - x^*\|_X \leq (c_2 l^2 + c_{11}\|\widetilde{v}\|_X + c_{12}l\|\widetilde{v}\|_X)\alpha_n^p + c_{13}\left(\frac{\delta}{\alpha_n} + \Delta\right). \tag{36}$$

As an analog of (5), we now have the system of conditions

$$\frac{r_0 L\|x_n - x^*\|_X + r_0\delta}{d_0\alpha_n} \leq \nu_0 < 1, \quad \|x_n - x^*\|_X \leq R. \tag{37}$$

Conditions (37) imply the representation

$$\Theta(\widetilde{F}'(x_n), \alpha_n) = \frac{1}{2\pi i}\int_{\Gamma_{\alpha_n}} \Theta(\lambda, \alpha_n)R(\lambda, \widetilde{F}'(x_n))d\lambda.$$

The following theorem is an immediate consequence of (36) and (37).

Theorem 4.6. *Suppose Conditions 2.6–2.8 and 4.1 are satisfied. Assume that (35) is valid with $p \in [1, p_0]$, and*

$$\|x_0 - x^*\|_X \leq \widetilde{l}\alpha_0^p,$$

$$0 < \widetilde{l} \leq \min\left\{\frac{1}{4r^p c_2}, \frac{d_0\nu_0}{2r_0 L\alpha_0^{p-1}}, \frac{R}{\alpha_0^p}\right\},$$

$$\|\widetilde{v}\|_X \leq \widetilde{d} = \min\left\{\frac{\widetilde{l}}{4r^p c_{11}}, \frac{1}{4r^p c_{12}}\right\}. \tag{38}$$

Then for each $\overline{d}$ such that

$$0 < \overline{d} < \min\left\{\frac{d_0\nu_0}{2r_0\alpha_0^p}, \frac{\widetilde{l}}{4r^p c_{13}}\right\},$$

we have the estimate

$$\|x_n - x^*\|_X \leq \tilde{l}\alpha_n^p, \quad n = 0, 1, \ldots, N(\delta, \Delta),$$ (39)

where

$$N(\delta, \Delta) = \max\left\{ n \in \mathbf{N} \cup \{0\} : \frac{\delta}{\alpha_n^{p+1}} + \frac{\Delta}{\alpha_n^p} \leq \bar{d} \right\}.$$ (40)

The number $n = N(\delta, \Delta)$ specified in (40) indicates the stopping point of the process (34). We can now establish an estimate of the approximation error $\|x_{N(\delta,\Delta)} - x^*\|_X$ in terms of δ and Δ. From (39) and (40) we obtain

Corollary 4.2. *Suppose conditions of Theorem 4.6 are satisfied. Then*

$$\|x_{N(\delta,\Delta)} - x^*\|_X \leq r^p \tilde{l} \left(\frac{\delta + \alpha_{N(\delta,\Delta)}\Delta}{\bar{d}} \right)^{p/(p+1)}.$$ (41)

Inequality (41) means that in the case where $\Delta = 0$, the mapping $\mathfrak{R}_{1/N(\delta,0)}$ that takes each operator $\tilde{F} \in \mathfrak{F}$ to the element $x_{N(\delta,0)}$, defines a regularization algorithm for the original equation. The same is true if $\Delta = \Delta(\delta)$ and $\lim_{\delta \to 0} \Delta(\delta) = 0$. When ξ and $\{\alpha_n\}$ are fixed, conditions (35) (with $\tilde{w} = 0$) and (38) specify a subset of $D(G)$, on which the equality (1.2.5) is valid. In general situation, (41) provides a qualified estimate of the value

$$\|\mathfrak{R}_{1/N(\delta,\Delta)}(\tilde{F}) - G(F)\|_X$$

in (1.2.8) on this subset of $D(G)$.

According to (35) and (38), an admissible subset of parameters ξ coincides with the Δ–neighborhood $\tilde{\mathbb{M}}_\Delta(x^*)$ of the ellipsoid

$$\tilde{\mathbb{M}}(x^*) = \{x^* + F'(x^*)^p v : \|v\|_X \leq \tilde{d}\}.$$

Note that this neighborhood has a nonempty interior and hence the problem of finding an element of $\tilde{\mathbb{M}}_\Delta(x^*)$ is less complicated as compared to finding $\xi \in \mathbb{M}(x^*)$.

Let us remark that conditions of Theorems 4.5 and 4.6 are satisfied for all generating functions $\Theta(\lambda, \alpha)$ of Examples 4.1–4.5.

4.4 Necessity of Sourcewise Representation for Rate of Convergence Estimates of Newton–Kantorovich Type Methods

In this section, under appropriate conditions on the generating function $\Theta(\lambda, \alpha)$ we prove that representation (3.3) is close to a necessary condition for the estimate (3.6). In more exact terms, we claim that the estimate

$$\|x_n - x^*\|_X \le l\alpha_n^p, \quad n = 0, 1, \ldots \quad (p > 1) \tag{1}$$

implies the inclusion

$$x^* - \xi \in R(F'(x^*)^{p-\varepsilon}) \quad \forall \varepsilon \in (0, p)$$

(compare with Theorem 4.3).

Suppose $A = F'(x^*)$ satisfies Condition 1.1. Besides, let Conditions 2.6–2.8 and 4.1 be satisfied. Without loss of generality we can assume that inequality (3.5) is fulfilled for all numbers $n = 0, 1, \ldots$. Lemma 4.1 then yields the integral formula (3.4) with $\Gamma_n = \Gamma_{\alpha_n}, n = 0, 1, \ldots$. From (3.8) we obtain

$$\|[E - \Theta(F'(x^*), \alpha_n)F'(x^*)](x^* - \xi)\|_X \le$$
$$\le \|x_{n+1} - x^*\|_X + \|\Theta(F'(x_n), \alpha_n)\|_{L(X)}\|G(x_n, x^*)\|_X + \tag{2}$$
$$+ \|[\Theta(F'(x^*), \alpha_n)F'(x^*) - \Theta(F'(x_n), \alpha_n)F'(x_n)](x^* - \xi)\|_X.$$

Conditions 4.1, 2.8 and inequalities (3.9), (3.13), (1) now imply

$$\|x_{n+1} - x^*\|_X + \|\Theta(F'(x_n), \alpha_n)\|_{L(X)}\|G(x_n, x^*)\|_X \le c_1\alpha_n^p. \tag{3}$$

The third summand in (2) is estimated as follows:

$$\|[\Theta(F'(x^*), \alpha_n)F'(x^*) - \Theta(F'(x_n), \alpha_n)F'(x_n)](x^* - \xi)\|_X =$$
$$= \|[(E - \Theta(F'(x^*), \alpha_n)F'(x^*)) -$$
$$- (E - \Theta(F'(x_n), \alpha_n)F'(x_n))](x^* - \xi)\|_X \le$$
$$\le \frac{1}{2\pi}\|x^* - \xi\|_X \int_{\Gamma_{\alpha_n}} |1 - \Theta(\lambda, \alpha_n)\lambda| \cdot \tag{4}$$
$$\cdot \|R(\lambda, F'(x^*)) - R(\lambda, F'(x_n))\|_{L(X)} |d\lambda|.$$

Using (1) and the inequality

$$|\lambda| \ge d_0\alpha_n \quad \forall \lambda \in \Gamma_{\alpha_n}$$

with d_0 specified in Condition 2.6, as in (3.24) we get

$$\|R(\lambda, F'(x^*)) - R(\lambda, F'(x_n))\|_{L(X)} \le$$
$$\le \frac{c_2\|x_n - x^*\|_X}{|\lambda|^2} \le \frac{c_3\alpha_n^{p-1}}{|\lambda|} \quad \forall \lambda \in \Gamma_{\alpha_n}. \tag{5}$$

After setting into (2.2.9) $p = 0$, by (4) and (5) we obtain

$$\|[\Theta(F'(x^*), \alpha_n)F'(x^*) - \Theta(F'(x_n), \alpha_n)F'(x_n)](x^* - \xi)\|_X \le$$
$$\le c_4\alpha_n^{p-1}. \tag{6}$$

Given $\alpha > 0$, by analogy with (2.6) we denote

$$\Phi(\alpha) = \|[E - \Theta(F'(x^*), \alpha)F'(x^*)](x^* - \xi)\|_X. \qquad (7)$$

Combining (2), (3), (6), and (7), we come to the following auxiliary result.

Lemma 4.2. *Suppose inequality (1) and Conditions 2.6–2.8 and 4.1 are satisfied. Then*

$$\Phi(\alpha_n) \le c_5 \alpha_n^{p-1}, \quad n = 0, 1, \ldots. \qquad (8)$$

It will be more convenient to deal with a continuous regularization parameter $\alpha > 0$ rather than with the sequence $\{\alpha_n\}$. As in Section 2.2, we denote by γ_α the boundary of $K_\alpha(R_0, d_0, \varphi_0)$ (see (2.2.3) for definition). From Condition 2.6 it follows that the operator

$$\Theta(F'(x^*), \alpha) = \frac{1}{2\pi i} \int\limits_{\gamma_\alpha} \Theta(\lambda, \alpha) R(\lambda, F'(x^*)) d\lambda \qquad (9)$$

is well defined, so that $\Theta(F'(x^*), \alpha) \in L(X)$ for each $\alpha > 0$. Assume that $\Theta(\lambda, \alpha)$ satisfies Condition 2.13. Arguing as in the proof of Theorem 2.19, we denote

$$\chi(\lambda, \alpha, \beta) = \frac{1 - \Theta(\lambda, \alpha)\lambda}{1 - \Theta(\lambda, \beta)\lambda}; \quad \lambda \in \mathbf{C}, \quad \alpha, \beta > 0.$$

Suppose the following condition is satisfied.

Condition 4.8. *There exists a constant $t_0 \ge 0$ such that*

$$\int\limits_{\gamma_{\alpha_{n+1}}} \frac{|\chi(\lambda, \alpha, \alpha_n)|}{|\lambda|} |d\lambda| \le c_6(1 + |\ln \alpha_n|^{t_0}) \qquad (10)$$

$$\forall \alpha \in (\alpha_{n+1}, \alpha_n], \quad n = 0, 1, \ldots.$$

Due to Condition 2.13, the operator $\chi(F'(x^*), \alpha, \beta) \in L(X)$ is well defined for all $\alpha, \beta > 0$. Furthermore, by (8)–(10) we conclude that for all $\alpha \in$

$(\alpha_{n+1}, \alpha_n]$,

$$
\begin{aligned}
\Phi(\alpha) = \| & [E - \Theta(F'(x^*), \alpha)F'(x^*)][E - \Theta(F'(x^*), \alpha_n)F'(x^*)]^{-1} \cdot \\
& \cdot [E - \Theta(F'(x^*), \alpha_n)F'(x^*)](x^* - \xi)\|_X \leq \\
& \leq \|\chi(F'(x^*), \alpha, \alpha_n)\|_{L(X)} \Phi(\alpha_n) \leq \\
& \leq \frac{\Phi(\alpha_n)}{2\pi} \int\limits_{\gamma_{\alpha_{n+1}}} |\chi(\lambda, \alpha, \alpha_n)| \|R(\lambda, F'(x^*))\|_{L(X)} |d\lambda| \leq \\
& \leq c_7 (1 + |\ln \alpha_n|^{t_0}) \alpha_n^{p-1}.
\end{aligned}
\tag{11}
$$

Fix a number $\kappa \in (0, p-1)$. Condition 4.1 and (11) then imply that for all $\alpha \in (\alpha_{n+1}, \alpha_n]$ we have

$$
\frac{\Phi(\alpha)}{\alpha^\kappa} = \frac{\Phi(\alpha)\alpha^{p-1-\kappa}}{\alpha^{p-1}} \leq c_7(1 + |\ln \alpha_n|^{t_0})\alpha_n^{p-1-\kappa} \left(\frac{\alpha_n}{\alpha_{n+1}}\right)^{p-1} \leq c_8.
$$

Here the constant c_8 depends on the chosen κ but doesn't depend on n. Thus we obtain the following proposition.

Lemma 4.3. *Suppose inequality (1) and Conditions 2.6–2.8, 2.13, 4.1, and 4.8 are satisfied. Then for each $\kappa \in (0, p-1)$ there exists $c_8 = c_8(\kappa)$ such that*

$$
\Phi(\alpha) \leq c_8 \alpha^\kappa \quad \forall \alpha \in (0, \alpha_0].
\tag{12}
$$

The next statement is in some sense a converse one for Theorem 4.5.

Theorem 4.7. *Suppose Conditions 2.6–2.8, 2.11–2.15, 4.1, and 4.8 are satisfied. Let the process (3.1) generate a sequence $\{x_n\}$ for which estimate (1) is true. Then*

$$
x^* - \xi \in R(F'(x^*)^{p-\varepsilon}) \quad \forall \varepsilon \in (0, p).
$$

Proof. The proof is divided into several parts.
1) Let us apply Theorem 2.15 to the linear equation

$$
F'(x^*)(x - x^*) = 0, \quad x \in X.
$$

Since the value κ in (12) can be taken arbitrarily close to $p-1$, from (12) we deduce that for each $\varepsilon_1 \in (0, p-1)$ there exists an element $v^{(1)} \in X$ such that

$$
x^* - \xi = F'(x^*)^{p_1} v^{(1)}, \quad p_1 = p - 1 - \varepsilon_1.
\tag{13}
$$

Representations of type (13) can be used for improving estimates of the third summand in (2). Following the proof of Theorem 4.3, we consider two cases.

2) If $p_1 \geq 1$, then by (3.21) and (3.26) we have

$$\|[\Theta(F'(x^*), \alpha_n)F'(x^*) - \Theta(F'(x_n), \alpha_n)F'(x_n)](x^* - \xi)\|_X \leq$$
$$\leq \|[\Theta(F'(x^*), \alpha_n)F'(x^*) - \Theta(F'(x_n), \alpha_n)F'(x_n)]F'(x^*)\|_{L(X)} \cdot$$
$$\cdot \|F'(x^*)^{p_1-1}v^{(1)}\|_X \leq c_9 \alpha_n^p.$$

Using this estimate, in place of (8) we obtain the inequality

$$\Phi(\alpha_n) \leq c_{10}\alpha_n^p, \quad n = 0, 1, \ldots.$$

Then we conclude that for each $\kappa \in (0, p)$,

$$\Phi(\alpha) = O(\alpha^\kappa), \quad \alpha \in (0, \alpha_0],$$

much as (12) was obtained by (8). From Theorem 2.15 it now follows that for each $\varepsilon \in (0, p)$ there exists $\widetilde{v} = \widetilde{v}(\varepsilon) \in X$ such that

$$x^* - \xi = F'(x^*)^{p-\varepsilon}\widetilde{v}.$$

Hence in this case the assertion of the theorem is true.

3) Now assume that $p_1 \in (0, 1)$. Without loss of generality, in (3.4) we can set $\Gamma_n = \gamma_{\alpha_n}$. Analogously to (4), by (13) we get

$$\|[\Theta(F'(x^*), \alpha_n)F'(x^*) - \Theta(F'(x_n), \alpha_n)F'(x_n)](x^* - \xi)\|_X \leq$$
$$\leq c_{11} \int_{\gamma_{\alpha_n}} |1 - \Theta(\lambda, \alpha_n)\lambda| \cdot \tag{14}$$
$$\cdot \|[R(\lambda, F'(x^*)) - R(\lambda, F'(x_n))]F'(x^*)^{p_1}\|_{L(X)}|d\lambda|.$$

As in (3.24), for all $\lambda \in \gamma_{\alpha_n}$ we have

$$\|[R(\lambda, F'(x^*)) - R(\lambda, F'(x_n))]F'(x^*)^{p_1}\|_{L(X)} \leq$$
$$\leq c_{12}|\lambda|^{-1}\|x_n - x^*\|_X \|R(\lambda, F'(x^*))F'(x^*)^{p_1}\|_{L(X)}. \tag{15}$$

From (1.1.9) it now follows that

$$\|R(\lambda, F'(x^*))F'(x^*)^{p_1}\|_{L(X)} \leq$$
$$\leq c_{13}\left(\int_0^{2d_0\alpha_n} t^{p_1-1}\|R(\lambda, F'(x^*))R(-t, F'(x^*))F'(x^*)\|_{L(X)}dt + \right.$$
$$\left. + \int_{2d_0\alpha_n}^\infty t^{p_1-1}\|R(\lambda, F'(x^*))R(-t, F'(x^*))F'(x^*)\|_{L(X)}dt \right).$$

Taking into account the identity

$$R(\lambda, F'(x^*))R(-t, F'(x^*)) =$$
$$= (\lambda + t)^{-1}[R(-t, F'(x^*)) - R(\lambda, F'(x^*))],$$

we obtain

$$\|R(\lambda, F'(x^*))F'(x^*)^{p_1}\|_{L(X)} \le$$

$$\le c_{13}\left(\int_0^{2d_0\alpha_n} t^{p_1-1}\|R(\lambda, F'(x^*))\|_{L(X)}\|R(-t, F'(x^*))F'(x^*)\|_{L(X)}dt + \right.$$

$$\left. + \int_{2d_0\alpha_n}^{\infty} \frac{t^{p_1-1}}{|\lambda + t|}\|[R(-t, F'(x^*)) - R(\lambda, F'(x^*))]F'(x^*)\|_{L(X)}dt \right) \le$$

$$\le c_{14}\left(\alpha_n^{-1}\int_0^{2d_0\alpha_n} t^{p_1-1}dt + \int_{2d_0\alpha_n}^{\infty} \frac{t^{p_1-1}}{|\lambda + t|}dt \right) \le c_{15}\alpha_n^{p_1-1} \quad \forall\lambda \in \gamma_{\alpha_n}. \quad (16)$$

From (1), (14)–(16), and (2.2.9) we deduce

$$\|[\Theta(F'(x^*), \alpha_n)F'(x^*) - \\ - \Theta(F'(x_n), \alpha_n)F'(x_n)](x^* - \xi)\|_X \le c_{16}\alpha_n^{p+p_1-1}.$$

Consequently, by (2) and (3),

$$\Phi(\alpha_n) \le c_{17}\alpha_n^{p+p_1-1}. \quad (17)$$

As in (8)–(12), from (17) we conclude that for each $\kappa \in (0, p + p_1 - 1)$,

$$\Phi(\alpha) = O(\alpha^\kappa), \quad \alpha \in (0, \alpha_0].$$

Theorem 2.15 now implies that for each $\varepsilon_2 \in (0, p - (1 - p_1))$ there exists an element $v^{(2)} \in X$ such that

$$x^* - \xi = F'(x^*)^{p_2}v^{(2)}, \quad p_2 = p - (1 - p_1) - \varepsilon_2.$$

Let us iterate the process of improving estimate (8).

4) The estimates of $\Phi(\alpha_n)$ are improved as follows. Take a sequence $\{\varepsilon_k\}$ such that

$$\varepsilon_k \in (0, p - 1), \quad \lim_{k\to\infty} \varepsilon_k = 0.$$

At the k-th step we have the estimate

$$\Phi(\alpha_n) \le c_{18}\alpha_n^{p+p_k-1}. \quad (18)$$

According to (8)–(12) and (18), for each $\kappa \in (0, p + p_k - 1)$,

$$\Phi(\alpha) = O(\alpha^\kappa), \quad \alpha \in (0, \alpha_0].$$

Theorem 2.15 then yields

$$x^* - \xi = F'(x^*)^{p_{k+1}} v^{(k+1)}, \quad v^{(k+1)} \in X, \tag{19}$$

where

$$p_{k+1} = p - (1 - p_k) - \varepsilon_{k+1}, \quad \varepsilon_{k+1} \in (0, p - (1 - p_k)). \tag{20}$$

If $p_k \geq 1$, then by (19) and (20),

$$x^* - \xi = F'(x^*)^{p - \varepsilon_{k+1}} \widetilde{v}^{(k+1)}$$

with $\widetilde{v}^{(k+1)} = F'(x^*)^{p_k - 1} v^{(k+1)}$, and the iteration terminates. Since ε_{k+1} can be chosen arbitrarily small, the assertion of the theorem is proved.

5) To conclude the proof it only requires to show that there exists a number $k = k_0$ such that $p_{k_0} \geq 1$. Assume the converse, i.e., $p_k < 1$ for all $k \in \mathbf{N}$. From (20) it follows that $p_k < p_{k+1} < p$. Therefore the sequence $\{p_k\}$ has a limit: $\lim_{k \to \infty} p_k = \widetilde{p} \leq 1$. Passing to limits in both parts of equality (20), we get $p = 1$. This contradicts the original assumption $p > 1$. Hence the process of iterating estimates is finite. This completes the proof of the theorem. $\square$

Remark 4.4. It can be checked that all generating functions from Examples 4.1–4.3 satisfy conditions of Theorem 4.7. The reader will easily prove a result similar to Theorem 4.7 also for the functions from Examples 4.4 and 4.5. The proof uses the same arguing with referring to Theorem 2.19 instead of Theorem 2.15.

4.5 Continuous Methods for Irregular Operator Equations in Hilbert and Banach Spaces

In this section we propose and study two additional classes of approximating schemes for irregular nonlinear operator equations in Hilbert and Banach spaces. The design of these schemes involves operator differential equations arising as continuous analogs of Gauss–Newton type and Newton–Kantorovich type methods (1.35) and (3.1). Error estimates similar to those of Theorems 4.2 and 4.6 are derived.

We start with an equation

$$F(x) = 0, \quad x \in X_1, \tag{1}$$

where the operator $F : X_1 \to X_2$ is Fréchet differentiable on $\Omega_R(x^*)$; x^* is a solution of (1) and X_1, X_2 are Hilbert spaces. As in Section 4.1, suppose $F(x)$ has a Lipschitz continuous derivative, that is,

$$\|F'(x) - F'(y)\|_{L(X_1, X_2)} \leq L\|x - y\|_{X_1} \quad \forall x, y \in \Omega_R(x^*). \tag{2}$$

Further, let $\widetilde{F}(x)$ be an available approximation of $F(x)$. Suppose $\widetilde{F}(x)$ is Fréchet differentiable on $\Omega_R(x^*)$, and

$$\|\widetilde{F}(x^*)\|_{X_2} \leq \delta; \quad \|\widetilde{F}'(x) - F'(x)\|_{L(X_1, X_2)} \leq \delta \quad \forall x \in \Omega_R(x^*). \tag{3}$$

There exists N_1 such that

$$\|F'(x)\|_{L(X_1, X_2)}, \quad \|\widetilde{F}'(x)\|_{L(X_1, X_2)} \leq N_1 \quad \forall x \in \Omega_R(x^*).$$

Assume that $x^* \in Q$, where Q is a given closed convex subset of X_1. In the first part of this section, we study the following continuous analog of the iterative process (1.35):

$$\dot{x} = P_Q\{\xi - \Theta(\widetilde{F}'^*(x)\widetilde{F}'(x), \alpha(t))\widetilde{F}'^*(x)(\widetilde{F}(x) - \widetilde{F}'(x)(x - \xi))\} - x, \tag{4}$$

$$x(0) = x_0 \in \Omega_R^0(x^*). \tag{5}$$

Here

$$\Omega_R^0(x^*) = \{x \in X_1 : \|x - x^*\|_{X_1} < R\},$$

x_0 is an initial approximation to x^*, and a differentiable function $\alpha = \alpha(t), t \geq 0$ satisfies the conditions

$$\alpha(t) > 0, \quad \dot{\alpha}(t) \leq 0 \quad \forall t \geq 0; \quad \lim_{t \to +\infty} \alpha(t) = 0. \tag{6}$$

By a solution of the Cauchy problem (4)–(5) we mean an X_1–valued function $x = x(t)$ defined on a segment $[0, \widetilde{T}](\widetilde{T} > 0)$, continuously differentiable in the strong sense on this segment and satisfying the equation (4) and the initial condition (5).

Suppose the function $\Theta(\lambda, \alpha)$ is analytic in λ on a domain D_α, where

$$D_\alpha \supset [0, N_1^2] \quad \forall \alpha \in (0, \alpha_0].$$

Also, assume that $\Theta(\lambda, \alpha)$ is continuous in (λ, α) on the set

$$\{(\lambda, \alpha) : \lambda \in D_\alpha, \alpha \in (0, \alpha_0]\}.$$

Moreover, suppose Conditions 4.2–4.5 are satisfied and the approximate source-wise representation (1.29) is fulfilled.

When considering the Cauchy problem (4)–(5) instead of the iterative process (1.35), it is natural to take as an approximation of x^* the element $x(t(\delta, \Delta))$, where $t = t(\delta, \Delta)$ is an appropriate stopping point of the trajectory $x = x(t)$. This means that solvability of (4)–(5) at least for $0 \le t \le t(\delta, \Delta)$ should be guaranteed. Below we propose an explicit rule for determination of the stopping point $t(\delta, \Delta)$ and establish the following error estimate analogous to (1.36):

$$\|x(t(\delta, \Delta)) - x^*\|_{X_1} = O((\delta + \Delta)^{2p/(2p+1)}). \tag{7}$$

For brevity sake, we denote

$$\widetilde{X}(t, x) = \xi - \Theta(\widetilde{F}'^*(x)\widetilde{F}'(x), \alpha(t))\widetilde{F}'^*(x)(\widetilde{F}(x) - \widetilde{F}'(x)(x - \xi)). \tag{8}$$

Due to (8), the Cauchy problem (4)–(5) takes the form

$$\dot{x} = P_Q(\widetilde{X}(t, x)) - x, \quad x(0) = x_0. \tag{9}$$

Using the Riesz–Dunford formula (1.1.6), inequality (1.1.16) and conditions imposed above on $\widetilde{F}(x)$ and $\Theta(\lambda, \alpha)$, it is not hard to prove that the operator $\widetilde{X}(t, x)$ is continuous in t whatever $x \in \Omega_R(x^*)$. Moreover, for all $z_0 \in \Omega_R^0(x^*), t_0 > 0$ there exist constants

$$\varepsilon_0 > 0; \quad M_1 = M_1(z_0, t_0, \varepsilon_0), \quad M_2 = M_2(z_0, t_0, \varepsilon_0)$$

such that

$$\|\widetilde{X}(t, x)\|_{X_1} \le M_1,$$
$$\|\widetilde{X}(t, x_1) - \widetilde{X}(t, x_2)\|_{X_1} \le M_2\|x_1 - x_2\|_{X_1} \tag{10}$$
$$\forall t \in [0, t_0] \quad \forall x, x_1, x_2 \in \Omega_{\varepsilon_0}(z_0).$$

From (10) it follows that the problem (4)–(5) has a unique solution $x = x(t)$ for $t \in [0, \widetilde{T}]$ with an appropriate $\widetilde{T} > 0$ (see, e.g., [31]). Since $x_0 \in \Omega_R^0(x^*)$ and the function $x = x(t)$ is continuous on $[0, \widetilde{T}]$, without loss of generality we can assume that $x(t) \in \Omega_R^0(x^*)$ for all $t \in [0, \widetilde{T}]$.

From (9) we deduce that for all $t \in [0, \widetilde{T}]$,

$$x(t) = \left(x_0 + \int_0^t P_Q(\widetilde{X}(\tau, x(\tau)))e^\tau d\tau\right)e^{-t} = \tag{11}$$

$$= e^{-t}x_0 + (1 - e^{-t})z_0(t),$$

where

$$z_0(t) = \int_0^t P_Q(\widetilde{X}(\tau, x(\tau)))\gamma(t, \tau)d\tau, \quad \gamma(t, \tau) = (e^t - 1)^{-1}e^\tau.$$

Note that

$$\int\limits_0^t \gamma(t,\tau)d\tau = 1, \quad \gamma(t,\tau) > 0;$$

$$P_Q(\widetilde{X}(\tau, x(\tau))) \in Q$$

$$\forall \tau \in [0, t] \quad \forall t \in (0, \widetilde{T}].$$

By [148, Theorem 1.6.13], it follows that $z_0(t) \in Q, t \in (0, \widetilde{T}]$. Therefore if the starting point $x_0 \in Q$, then (11) in view of the convexity of Q implies $x(t) \in Q, t \in [0, \widetilde{T}]$. Thus the trajectory $x(t), t \in [0, \widetilde{T}]$ defined by (4)–(5) lies in the subset Q, which contains the unknown solution x^*. Recall that iterative points $\{x_n\}$ generated by (1.35) possess an analogous property.

Using (9), (1.1.15), and (1.1.16), we obtain

$$
\begin{aligned}
\frac{d}{dt}\left(\frac{1}{2}\|x - x^*\|_{X_1}^2\right) &= (\dot{x}, x - x^*)_{X_1} = (\widetilde{X}(t,x) - x, x - x^*)_{X_1} + \\
&+ \left(P_Q(\widetilde{X}(t,x)) - \widetilde{X}(t,x), x - \widetilde{X}(t,x)\right)_{X_1} + \\
&+ \left(P_Q(\widetilde{X}(t,x)) - \widetilde{X}(t,x), \widetilde{X}(t,x) - P_Q(\widetilde{X}(t,x))\right)_{X_1} + \\
&+ \left(P_Q(\widetilde{X}(t,x)) - \widetilde{X}(t,x), P_Q(\widetilde{X}(t,x)) - x^*\right)_{X_1} \leq \qquad (12) \\
&\leq \left(\widetilde{X}(t,x) - x, x - x^*\right)_{X_1} - \|P_Q(\widetilde{X}(t,x)) - \widetilde{X}(t,x)\|_{X_1}^2 + \\
&+ \left(P_Q(\widetilde{X}(t,x)) - \widetilde{X}(t,x), x - \widetilde{X}(t,x)\right)_{X_1} \leq \\
&\leq (\widetilde{X}(t,x) - x, x - x^*)_{X_1} + \frac{1}{4}\|\widetilde{X}(t,x) - x\|_{X_1}^2,
\end{aligned}
$$

$$0 \leq t \leq \widetilde{T}.$$

Let us now estimate individual summands in the last inequality in (12). First note that by (2) and (1.1.14),

$$\widetilde{F}(x) = \widetilde{F}(x^*) + \widetilde{F}'(x)(x - x^*) + \widetilde{G}(x, x^*),$$

where

$$\|\widetilde{G}(x, x^*)\|_{X_2} \leq \frac{1}{2}L\|x - x^*\|_{X_1}^2 \quad \forall x \in \Omega_R(x^*).$$

According to (8),

$$
\begin{aligned}
(\widetilde{X}(t,x) - x, x - x^*)_{X_1} = \\
- \left(\Theta(\widetilde{F}'^*(x)\widetilde{F}'(x), \alpha(t))\widetilde{F}'^*(x)\widetilde{F}(x^*), x - x^*\right)_{X_1} - \\
- \left(\Theta(\widetilde{F}'^*(x)\widetilde{F}'(x), \alpha(t))\widetilde{F}'^*(x)\widetilde{G}(x), x - x^*\right)_{X_1} - \\
- \left([E - \Theta(F'^*(x^*)F'(x^*), \alpha(t))F'^*(x^*)F'(x^*)] \cdot \right. \\
\left. \cdot (x^* - \xi), x - x^*\right)_{X_1} - \|x - x^*\|^2_{X_1} + \\
+ \left([\Theta(\widetilde{F}'^*(x)\widetilde{F}'(x), \alpha(t))\widetilde{F}'^*(x)\widetilde{F}'(x) - \right. \\
\left. - \Theta(F'^*(x^*)F'(x^*), \alpha(t))F'^*(x^*)F'(x^*)](x^* - \xi), x - x^*\right)_{X_1}.
\end{aligned}
\tag{13}
$$

Furthermore with the use of (3) and Remark 4.1 we get

$$
\begin{aligned}
\|\Theta(\widetilde{F}'^*(x)\widetilde{F}'(x), \alpha(t))\widetilde{F}'^*(x)\widetilde{F}(x^*)\|_{X_1} \leq \\
\leq \|\Theta(\widetilde{F}'^*(x)\widetilde{F}'(x), \alpha(t))\widetilde{F}'^*(x)\|_{L(X_2, X_1)} \|\widetilde{F}(x^*) - F(x^*)\|_{X_2} \leq \\
\leq \frac{c_1 \delta}{\sqrt{\alpha(t)}},
\end{aligned}
\tag{14}
$$

$$
\|\Theta(\widetilde{F}'^*(x)\widetilde{F}'(x), \alpha(t))\widetilde{F}'^*(x)\widetilde{G}(x, x^*)\|_{X_1} \leq \frac{c_1 L \|x - x^*\|^2_{X_1}}{2\sqrt{\alpha(t)}}.
\tag{15}
$$

Condition 4.3 and representation (1.29) then yield

$$
\begin{aligned}
\|[E - \Theta(F'^*(x^*)F'(x^*), \alpha(t))F'^*(x^*)F'(x^*)](x^* - \xi)\|_{X_1} \leq \\
\leq \|[E - \Theta(F'^*(x^*)F'(x^*), \alpha(t))F'^*(x^*)F'(x^*)](F'^*(x^*)F'(x^*))^p v\|_{X_1} + \\
+ \|[E - \Theta(F'^*(x^*)F'(x^*), \alpha(t))F'^*(x^*)F'(x^*)]w\|_{X_1} \leq \\
\leq c_2(\|v\|_{X_1}\alpha^p(t) + \Delta).
\end{aligned}
\tag{16}
$$

The last summand in (13) can be estimated similarly to (1.11)–(1.20). Using (1.1.6) and (1.29), we obtain

$$
\begin{aligned}
[\Theta(\widetilde{F}'^*(x)\widetilde{F}'(x), \alpha)\widetilde{F}'^*(x)\widetilde{F}'(x) - \\
- \Theta(F'^*(x^*)F'(x^*), \alpha(t))F'^*(x^*)F'(x^*)](x^* - \xi) = \\
= \frac{1}{2\pi i} \int_{\Gamma_\alpha} (1 - \Theta(\lambda, \alpha)\lambda)[R(\lambda, \widetilde{F}'^*(x)\widetilde{F}'(x)) - \\
- R(\lambda, F'^*(x^*)F'(x^*))](F'^*(x^*)F'(x^*))^p v d\lambda + \\
+ \frac{1}{2\pi i} \int_{\Gamma_\alpha} (1 - \Theta(\lambda, \alpha)\lambda)[R(\lambda, \widetilde{F}'^*(x)\widetilde{F}'(x)) - \\
- R(\lambda, F'^*(x^*)F'(x^*))]w d\lambda.
\end{aligned}
\tag{17}
$$

For the first integrand in (17) we get the estimate

$$\|[R(\lambda, \widetilde{F}'^*(x)\widetilde{F}'(x)) -$$
$$- R(\lambda, F'^*(x^*)F'(x^*))](F'^*(x^*)F'(x^*))^p v\|_{X_1} \le$$
$$\le \frac{c_3\|v\|_{X_1}(L\|x - x^*\|_{X_1} + \delta)}{|\lambda|} \tag{18}$$
$$\forall \lambda \in \Gamma_\alpha \quad \forall \alpha \in (0, \alpha_0].$$

Since by Condition 4.4,

$$\|R(\lambda, \widetilde{F}'^*(x)\widetilde{F}'(x))\|_{L(X_1)} \le \sup_{\mu \in [0, N_1^2]} \frac{1}{|\lambda - \mu|} \le \frac{c_4}{|\lambda|}$$
$$\forall \lambda \in \Gamma_\alpha \quad \forall \alpha \in (0, \alpha_0],$$

we have

$$\|R(\lambda, \widetilde{F}'^*(x)\widetilde{F}'(x)) - R(\lambda, F'^*(x^*)F'(x^*))\|_{L(X_1)} \le \frac{2c_4}{|\lambda|} \tag{19}$$
$$\forall \lambda \in \Gamma_\alpha \quad \forall \alpha \in (0, \alpha_0].$$

Estimates (17)–(19) and Condition 4.5 yield

$$\|[\Theta(\widetilde{F}'^*(x)\widetilde{F}'(x), \alpha)\widetilde{F}'^*(x)\widetilde{F}'(x) -$$
$$- \Theta(F'^*(x^*)F'(x^*), \alpha)F'^*(x^*)F'(x^*)](x^* - \xi)\|_{X_1} \le$$
$$\le \frac{1}{2\pi} \int_{\Gamma_\alpha} \frac{|1 - \Theta(\lambda, \alpha)\lambda|}{|\lambda|} \cdot \tag{20}$$
$$\cdot \left(c_3\|v\|_{X_1}(L\|x - x^*\|_{X_1} + \delta) + 2c_4\Delta\right)|d\lambda| \le$$
$$\le c_5\left(L\|v\|_{X_1}\|x - x^*\|_{X_1} + \|v\|_{X_1}\delta + \Delta\right).$$

From (13)–(16) and (20) it now follows that

$$(\widetilde{X}(t,x) - x, x - x^*)_{X_1} \le$$
$$\le \left(\frac{c_1\delta}{\sqrt{\alpha(t)}} + c_2\|v\|_{X_1}\alpha^p(t) + c_5\|v\|_{X_1}\delta + c_6\Delta\right)\|x - x^*\|_{X_1} + \tag{21}$$
$$+ c_5 L\|v\|_{X_1}\|x - x^*\|_{X_1}^2 + \frac{c_1 L\|x - x^*\|_{X_1}^3}{2\sqrt{\alpha(t)}} - \|x - x^*\|_{X_1}^2.$$

On the other hand, by (13)–(16) and (20),

$$\|\widetilde{X}(t, x) - x\|_{X_1} \le \frac{c_1\delta}{\sqrt{\alpha(t)}} + c_2\|v\|_{X_1}\alpha^p(t) + c_5\|v\|_{X_1}\delta + c_6\Delta +$$
$$+ \|x - x^*\|_{X_1} + c_5 L\|v\|_{X_1}\|x - x^*\|_{X_1} + \frac{c_1 L\|x - x^*\|_{X_1}^2}{2\sqrt{\alpha(t)}}. \tag{22}$$

From (21) and (22) we derive

$$\frac{d}{dt}\left(\frac{1}{2}\|x - x^*\|_{X_1}^2\right) \le -\|x - x^*\|_{X_1}^2 +$$

$$+ \left(\frac{c_1\delta}{\sqrt{\alpha(t)}} + c_2\|v\|_{X_1}\alpha^p(t) + c_5\|v\|_{X_1}\delta + c_6\Delta\right)\|x - x^*\|_{X_1} +$$

$$+ c_5 L\|v\|_{X_1}\|x - x^*\|_{X_1}^2 + \frac{c_1 L\|x - x^*\|_{X_1}^3}{2\sqrt{\alpha(t)}} +$$

$$+ \frac{1}{4}\left(\frac{c_1\delta}{\sqrt{\alpha(t)}} + c_2\|v\|_{X_1}\alpha^p(t) + c_5\|v\|_{X_1}\delta + c_6\Delta +\right.$$

$$\left. + \|x - x^*\|_{X_1} + c_5 L\|v\|_{X_1}\|x - x^*\|_{X_1} + \frac{c_1 L\|x - x^*\|_{X_1}^2}{2\sqrt{\alpha(t)}}\right)^2.$$

The application of elementary inequality

$$(a_1 + \cdots + a_l)^2 \le l(a_1^2 + \cdots + a_l^2) \tag{23}$$

now yields

$$\frac{d}{dt}\left(\frac{1}{2}\|x - x^*\|_{X_1}^2\right) \le -\frac{1}{2}\|x - x^*\|_{X_1}^2 +$$

$$+ \left(\frac{c_1\delta}{\sqrt{\alpha(t)}} + c_2\|v\|_{X_1}\alpha^p(t) + c_5\|v\|_{X_1}\delta + c_6\Delta\right)\|x - x^*\|_{X_1} +$$

$$+ c_5 L\|v\|_{X_1}\|x - x^*\|_{X_1}^2 + \frac{c_1 L\|x - x^*\|_{X_1}^3}{2\sqrt{\alpha(t)}} +$$

$$+ \frac{3}{2}\left(\frac{c_1\delta}{\sqrt{\alpha(t)}} + c_2\|v\|_{X_1}\alpha^p(t) + c_5\|v\|_{X_1}\delta + c_6\Delta\right)^2 +$$

$$+ \frac{3}{2}(c_5 L\|v\|_{X_1})^2\|x - x^*\|_{X_1}^2 + \frac{3(c_1 L)^2\|x - x^*\|_{X_1}^4}{8\alpha(t)},$$

$$0 \le t \le \tilde{T}.$$

Using the inequality

$$|ab| \le \frac{\varepsilon}{2}a^2 + \frac{1}{2\varepsilon}b^2, \quad \varepsilon > 0 \tag{24}$$

with $\varepsilon = 1/4$ and $\varepsilon = (c_1 L)^{-1}\alpha(t)$, we obtain

$$\left(\frac{c_1\delta}{\sqrt{\alpha(t)}} + c_2\|v\|_{X_1}\alpha^p(t) + c_5\|v\|_{X_1}\delta + c_6\Delta\right)\|x - x^*\|_{X_1} \le$$

$$\le 2\left(\frac{c_1\delta}{\sqrt{\alpha(t)}} + c_2\|v\|_{X_1}\alpha^p(t) + c_5\|v\|_{X_1}\delta + c_6\Delta\right)^2 + \qquad (25)$$

$$+ \frac{1}{8}\|x - x^*\|_{X_1}^2,$$

$$\frac{c_1 L\|x - x^*\|_{X_1}^3}{2\sqrt{\alpha(t)}} = c_1 L\|x - x^*\|_{X_1}^2\left(\frac{1}{2\sqrt{\alpha(t)}}\|x - x^*\|_{X_1}\right) \le$$

$$\le \frac{1}{8}\|x - x^*\|_{X_1}^2 + \frac{(c_1 L)^2}{2\alpha(t)}\|x - x^*\|_{X_1}^4. \qquad (26)$$

Combining the estimates (25) and (26), we finally get

$$\frac{d}{dt}\left(\frac{1}{2}\|x - x^*\|_{X_1}^2\right) \le \frac{7(c_1 L)^2}{8\alpha(t)}\|x - x^*\|_{X_1}^4 -$$

$$- \left(\frac{1}{4} - c_5 L\|v\|_{X_1} - \frac{3}{2}(c_5 L\|v\|_{X_1})^2\right)\|x - x^*\|_{X_1}^2 + \qquad (27)$$

$$+ \frac{7}{2}\left(\frac{c_1\delta}{\sqrt{\alpha(t)}} + c_2\|v\|_{X_1}\alpha^p(t) + c_5\|v\|_{X_1}\delta + c_6\Delta\right)^2,$$

$$0 \le t \le \tilde{T}.$$

Let us denote

$$u(t) = \frac{1}{2}\|x(t) - x^*\|_{X_1}^2.$$

From (27) we deduce the following auxiliary result.

Theorem 4.8. *Suppose Conditions 4.2–4.5 and (6) are satisfied. Also, assume that (1.29) is fulfilled with*

$$\|v\|_{X_1} \le \frac{\sqrt{7} - 2}{6c_5 L}. \qquad (28)$$

Then for all $t \in [0, \tilde{T}]$,

$$\dot{u}(t) \le -\frac{1}{4}u(t) + \frac{c_7}{\alpha(t)}u^2(t) +$$

$$+ c_8\left(\frac{\delta^2}{\alpha(t)} + \|v\|_{X_1}^2\alpha^{2p}(t) + (\|v\|_{X_1}\delta + \Delta)^2\right). \qquad (29)$$

Note that in the case $\delta + \Delta > 0$, the last summand in (29) doesn't tend to zero as $t \to +\infty$. Therefore in conditions of Theorem 4.8, it is in general impossible to guarantee the convergence $x(t) \to x^*$ as $t \to +\infty$. Hence in order to get an acceptable approximation to x^*, the trajectory $x = x(t)$ should be stopped at an appropriate moment $t = t(\delta, \Delta) < \infty$.

Let us give a stopping rule for the process (4)–(5). Choose $m > 0$ such that

$$\alpha^{p+1/2}(0) > m(\delta + \Delta). \tag{30}$$

We determine the stopping point $t = t(\delta, \Delta) > 0$ as a root of the equation

$$\alpha^{p+1/2}(t(\delta, \Delta)) = m(\delta + \Delta). \tag{31}$$

Since the function $\alpha = \alpha(t)$, $t \geq 0$ is continuous and nonincreasing, and $\lim_{t \to +\infty} \alpha(t) = 0$, condition (30) implies that equation (31) has at least one solution $t = t(\delta, \Delta) > 0$. Moreover,

$$\lim_{\delta, \Delta \to 0} t(\delta, \Delta) = \infty,$$
$$\alpha^{p+1/2}(t) \geq m(\delta + \Delta), \quad 0 \leq t \leq t(\delta, \Delta). \tag{32}$$

We now need to establish the solvability of (4)–(5) for $0 \leq t \leq t(\delta, \Delta)$.

According to (32), for each $0 \leq t \leq t(\delta, \Delta)$ we have

$$\frac{\delta^2}{\alpha(t)} \leq \frac{\alpha^{2p}(t)}{m^2},$$

$$(\|v\|_{X_1}\delta + \Delta)^2 \leq \max\{1, \|v\|^2_{X_1}\}(\delta + \Delta)^2 \leq$$
$$\leq \frac{\alpha(0)}{m^2} \max\{1, \|v\|^2_{X_1}\}\alpha^{2p}(t).$$

Let us denote

$$\beta(t) = \alpha^{2p}(t).$$

From (29) and the last inequalities, for all $0 \leq t \leq \min\{t(\delta, \Delta), \tilde{T}\}$ we obtain

$$\dot{u}(t) \leq -\frac{1}{4}u(t) + \frac{c_7\alpha^{2p-1}(0)}{\beta(t)}u^2(t) + c_8 D(v, m)\beta(t), \tag{33}$$

where, by definition,

$$D(v, m) = \|v\|^2_{X_1} + \frac{1}{m^2}\left(1 + \alpha(0)\max\{1, \|v\|^2_{X_1}\}\right).$$

Suppose $\alpha(t)$ satisfies the following additional condition

$$\frac{\dot{\alpha}(t)}{\alpha(t)} \geq -\frac{1}{16p}, \quad t \geq 0. \tag{34}$$

Assumption (34) can be considered as a continuous analog of the last inequality in Condition 4.1. By (34),

$$\dot{\beta}(t) \geq -\frac{1}{8}\beta(t), \quad t \geq 0.$$

Therefore (33) implies

$$\frac{d}{dt}\left(\frac{u(t)}{\beta(t)}\right) \leq -\frac{1}{8}\left(\frac{u(t)}{\beta(t)}\right) + c_7\alpha^{2p-1}(0)\left(\frac{u(t)}{\beta(t)}\right)^2 + c_8 D(v, m), \tag{35}$$
$$0 \leq t \leq \min\{t(\delta, \Delta), \widetilde{T}\}.$$

Along with inequality (35), we consider the majorizing differential equation

$$\dot{y}(t) = -\frac{1}{8}y(t) + c_7\alpha^{2p-1}(0)y^2(t) + c_8 D(v, m), \quad t \geq 0 \tag{36}$$

with the initial condition

$$y(0) = \frac{u(0)}{\beta(0)}.$$

By the well-known lemma on differential inequalities (see, e.g., [93]),

$$\frac{u(t)}{\beta(t)} \leq y(t), \quad 0 \leq t \leq \min\{t(\delta, \Delta), \widetilde{T}\}. \tag{37}$$

Let $\alpha(0)$ satisfy the condition

$$D(v, m)\alpha^{2p-1}(0) < \frac{1}{256c_7c_8}. \tag{38}$$

It follows easily that the quadratic equation

$$-\frac{1}{8}y + c_7\alpha^{2p-1}(0)y^2 + c_8 D(v, m) = 0$$

has two positive roots $y_1 < y_2$. Elementary analysis of equation (36) shows that $0 \leq y(0) \leq y_2$ implies $y(t) \leq y_2$ for all $t \geq 0$. From (37) we now deduce that the inequality $u(0) \leq y_2\beta(0)$ yields

$$u(t) \leq y_2\beta(t) \leq y_2\beta(0), \quad 0 \leq t \leq \min\{t(\delta, \Delta), \widetilde{T}\}. \tag{39}$$

Denote by $\widetilde{T}^* \geq \widetilde{T}$ the full time of existence of the solution $x = x(t)$ to (4)–(5). In other words, suppose that the function $x(t)$ is defined for all $t \in [0, \widetilde{T}^*)$ but $x(t)$ can't be uniquely continued forward by time through $t = \widetilde{T}^*$.

Further, suppose

$$\alpha(0) < 4c_7 R^2, \tag{40}$$

$$\|x_0 - x^*\|_{X_1} \leq \frac{1}{2}\sqrt{\frac{\alpha(0)}{2c_7}}. \tag{41}$$

Since

$$\frac{1}{16c_7\alpha^{2p-1}(0)} \leq y_2 \leq \frac{1}{8c_7\alpha^{2p-1}(0)},$$

by (39)–(41) it follows that the trajectory $x = x(t)$ lies in the ball

$$\left\{ x \in X_1 : \|x - x^*\|_{X_1} \leq \frac{1}{2}\sqrt{\frac{\alpha(0)}{c_7}} \right\} \subset \Omega_R^0(x^*)$$

for all t from the time interval $0 \leq t < \min\{t(\delta, \Delta), \tilde{T}^*\}$.

We now claim that $\tilde{T}^* > t(\delta, \Delta)$. Indeed, assume the converse, i.e., $\tilde{T}^* \leq t(\delta, \Delta)$. Then the above considerations yield

$$\sup\{\|x(t) - x^*\|_{X_1} : t \in [0, \tilde{T}^*)\} < R.$$

By (2) and (4) it now follows that

$$\sup\{\|\dot{x}(t)\|_{X_1} : t \in [0, \tilde{T}^*)\} < \infty.$$

Using the representation

$$x(t) = x_0 + \int_0^t \dot{x}(\tau)d\tau, \quad t \in [0, \tilde{T}^*),$$

we conclude that for each sequence $\{t_n\}$ such that $t_n < \tilde{T}^*$ and $\lim_{n \to \infty} t_n = \tilde{T}^*$, the elements $x(t_n)$ form a Cauchy sequence in X_1. Therefore,

$$\lim_{t \to \tilde{T}^* - 0} \|x(t) - \tilde{x}\|_{X_1} = 0,$$

where $\tilde{x} \in \Omega_R^0(x^*)$. Consequently inequalities (10) are valid with $z_0 = \tilde{x}$ and an appropriate $\varepsilon_0 > 0$ and hence the solution $x(t)$ can be continued forward by time through $t = \tilde{T}^*$ ([55, Theorem 3.3.3]), contrary to the definition of $\tilde{T}^*$. This contradiction proves that $\tilde{T}^* > t(\delta, \Delta)$. The following theorem summarizes the above discussion.

Theorem 4.9. *Suppose Conditions 4.2–4.5 are satisfied. Assume that (1.29), (6), (28), (30), (34), (38), (40), and (41) are valid. Then a solution to the Cauchy* *problem (4)–(5) is uniquely determined for $0 \leq t \leq t(\delta, \Delta)$, where $t(\delta, \Delta)$ is specified by (31).*

In the conditions of Theorem 4.9, inequality (39) is fulfilled for all $0 \leq t \leq t(\delta, \Delta)$. Therefore,

$$\|x(t) - x^*\|_{X_1} \leq \sqrt{2y_2\beta(t)} \leq \frac{\alpha^p(t)}{2\sqrt{c_7}\alpha^{p-1/2}(0)}, \quad 0 \leq t \leq t(\delta, \Delta).$$

Setting here $t = t(\delta, \Delta)$, with the use of (31) we obtain the estimate (7):

$$\|x(t(\delta, \Delta)) - x^*\|_{X_1} \leq \frac{m^{2p/(2p+1)}}{2\sqrt{c_7}\alpha^{p-1/2}(0)}(\delta + \Delta)^{2p/(2p+1)}. \tag{42}$$

Thus we proved the following statement.

Theorem 4.10. *Suppose conditions of Theorem 4.9 are satisfied. Let $x(t(\delta, \Delta))$ be defined by (4)–(5) and (31). Then (42) is true.*

In the case $\delta = \Delta = 0$, when the exact operator $F(x)$ is available and the exact sourcewise representation

$$x^* - \xi = (F'^*(x^*)F'(x^*))^p v, \quad v \in X_1, \quad p \geq \frac{1}{2} \tag{43}$$

is fulfilled, the preceding conditions and estimates take the following form.

Theorem 4.11. *Suppose Conditions 4.2–4.5 are satisfied. Assume that (6), (34), (40), (41), and (43) are valid and*

$$\|v\|_{X_1} < \min\left\{\frac{\sqrt{7} - 2}{6c_5L}, \frac{1}{16\sqrt{c_7c_8}\alpha^{p-1/2}(0)}\right\}.$$

Then a solution $x(t)$ to the problem (4)–(5) with $\widetilde{F}(x) = F(x)$ is uniquely determined for all $t \geq 0$. Moreover,

$$\|x(t) - x^*\|_{X_1} \leq \frac{\alpha^p(t)}{2\sqrt{c_7}\alpha^{p-1/2}(0)}, \quad t \geq 0. \tag{44}$$

Let us analyze the conditions of Theorems 4.9–4.11 and discuss, how these conditions can be ensured in practice. The function $\alpha(t)$ can be chosen subject to (34) and (40) as follows:

$$\alpha(t) = \frac{a}{(t+1)^s}, \quad 0 < s \leq \frac{1}{16p}, \quad 0 < a < 4c_7R^2.$$

Condition (38) is fulfilled if m and $\|v\|_{X_1}$ are taken subject to the conditions

$$m > 16\sqrt{2c_7c_8(1+a)}a^{p-1/2}, \qquad (45)$$

$$\|v\|_{X_1} \le \min\left\{1, \frac{1}{16\sqrt{2c_7c_8}a^{p-1/2}}\right\}. \qquad (46)$$

Inequalities (40) and (45) put restrictions on error levels in the operator $F(x)$ and in the sourcewise representation (43); (1.29), (28), and (46) imply that the controlling parameter ξ should be chosen from the Δ–neighborhood of the ellipsoid

$$\mathbb{M}(x^*) = \{x^* + (F'^*(x^*)F'(x^*))^p v : \|v\|_{X_1} \le d_0\},$$

where

$$d_0 = \min\left\{1, \frac{\sqrt{7}-2}{6c_5L}, \frac{1}{16\sqrt{2c_7c_8}a^{p-1/2}}\right\}.$$

If $\Delta > 0$, then this neighborhood has a nonempty interior, whereas the interior of $\mathbb{M}(x^*)$ is empty when $0 \in \sigma(F'^*(x^*)F'(x^*))$. At last, (41) determines the degree of closeness of x_0 to x^* sufficient for the estimates (42) and (44).

The most restrictive condition on ξ we get when $\Delta = 0$, that is, when the element ξ should be chosen from the ellipsoid $\mathbb{M}(x^*)$. Denote by $\mathfrak{R}_{1/t(\delta,0)}$ the mapping that takes each approximate operator $\widetilde{F}(x)$ to the element $x(t(\delta,0))$. From (42) it follows that $\mathfrak{R}_{1/t(\delta,0)}$ defines a regularization algorithm for the original problem (1). When $\Delta > 0$, estimate (42) indicates that this regularization algorithm is stable with respect to small perturbations in the sourcewise representation (43).

Practical implementation of the presented approximating scheme may base upon various approaches to finite difference discretization of the trajectory $x = x(t)$ defined by (4)–(5). The original iterative process (1.35) indicates a possible way of such discretization.

Let us now address to an equation

$$F(x) = 0, \quad x \in X \qquad (47)$$

with a Fréchet differentiable operator $F : X \to X$ in a Banach space X. Suppose X has a Gâteaux differentiable norm. The most familiar spaces satisfying this condition are Hilbert spaces and the spaces l_p, L_p, W_p^m with $1 < p < \infty$ [142]. Let x^* be a solution to (47). Suppose $F(x)$ has on $\Omega_R(x^*)$ a Lipschitz continuous Fréchet derivative, i.e.,

$$\|F'(x) - F'(y)\|_{L(X)} \le L\|x - y\|_X \quad \forall x, y \in \Omega_R(x^*). \qquad (48)$$

As in Section 4.3, we assume that operator $A = F'(x^*)$ satisfies Condition 1.1.

Let $\widetilde{F}(x)$ be an available approximation to $F(x)$. Suppose $\widetilde{F}(x)$ is Fréchet differentiable on $\Omega_R(x^*)$, and $\widetilde{F}'(x)$ satisfies (48). Also, assume that

$$\|\widetilde{F}(x^*)\|_X \le \delta; \quad \|\widetilde{F}'(x) - F'(x)\|_{L(X)} \le \delta \quad \forall x \in \Omega_R(x^*). \tag{49}$$

Now consider the following continuous analog of the iterative scheme (3.1):

$$\dot{x} = \xi - \Theta(\widetilde{F}'(x), \alpha(t))(\widetilde{F}(x) - \widetilde{F}'(x)(x - \xi)) - x, \tag{50}$$

$$x(0) = x_0 \in \Omega_R^0(x^*). \tag{51}$$

A differentiable function $\alpha = \alpha(t), t \ge 0$ is chosen subject to conditions (6). The definition of a solution to the Cauchy problem (50)–(51) doesn't differ from that of the problem (3)–(4).

Below we give a stopping rule $t = t(\delta, \Delta)$ ensuring for approximations $x(t(\delta, \Delta))$ the error estimate similar to (3.41):

$$\|x(t(\delta, \Delta)) - x^*\|_X = O((\delta + \Delta)^{p/(p+1)}).$$

Suppose the approximate sourcewise representation (3.35) is fulfilled, and the generating function $\Theta(\lambda, \alpha)$ satisfies Conditions 2.6–2.8.

Let $R_0 > \|F'(x^*)\|_{L(X)}$. By the scheme of the proof of Lemma 4.1 we get the following result (compare with (3.37)).

Lemma 4.4. *Suppose*

$$\|x - x^*\|_X < \frac{d_0\nu_0\alpha(0)}{2r_0L}, \quad \delta < \frac{d_0\nu_0\alpha(0)}{2r_0}, \quad \nu_0 \in (0, 1), \tag{52}$$

where r_0 and d_0 are specified in Condition 1.1. Then

$$\sigma(\widetilde{F}'(x)) \subset \operatorname{int} K_{\alpha(t)}(R_0, \varphi_0, d_0)$$

for all sufficiently small $t > 0$.

Lemma 4.4 with an appropriate choice of contours $\{\Gamma_\alpha\}_{\alpha \in (0, \alpha_0]}$ $(\alpha_0 \ge \alpha(0))$ yields

$$\Theta(\widetilde{F}'(x), \alpha(t)) = \frac{1}{2\pi i} \int_{\Gamma_{\alpha(t)}} \Theta(\lambda, \alpha(t)) R(\lambda, \widetilde{F}'(x)) d\lambda. \tag{53}$$

Using (53) and Conditions 2.6–2.8, we obtain the following existence result for the problem (50)–(51).

Theorem 4.12. *Suppose Conditions 1.1 and 2.6–2.8 are satisfied. Assume that inequalities (52) are valid. Then there exists $\widetilde{T} > 0$ such that a solution*

$x = x(t)$ *to the Cauchy problem (50)–(51) is uniquely determined for* $t \in [0, \widetilde{T}]$. *Moreover,* $x(t) \in \Omega_R^0(x^*)$ *for all* $t \in [0, \widetilde{T}]$.

Let X^* be a dual space for X. By $\langle f, z \rangle$ we denote the value of a functional $f \in X^*$ at an element $z \in X$.

According to our conditions, the functional $x \to \|x\|_X$, $x \in X$ is Gâteaux differentiable when $x \neq 0$. Let $U : X \to X^*$ be the Gâteaux derivative of the functional

$$g(x) = \frac{1}{2}\|x\|_X^2,$$

then

$$U(x) = \begin{cases} \|x\|_X \operatorname{grad}\|x\|_X, & x \neq 0, \\ 0, & x = 0. \end{cases} \tag{54}$$

The operator U defined by (54) is called the normalized dual mapping from X into X^*. The following proposition is well known ([142]).

Theorem 4.13. *For all* $x \in X$,

$$\langle U(x), x \rangle = \|x\|_X^2, \quad \|U(x)\|_{X^*} = \|x\|_X. \tag{55}$$

Using (50) and (54), we obtain

$$\frac{d}{dt}\left(\frac{1}{2}\|x - x^*\|_X^2\right) = \langle U(x - x^*), \dot{x} \rangle =$$

$$= \langle U(x - x^*), \xi - x - \Theta(\widetilde{F}'(x), \alpha(t))\widetilde{F}(x) +$$
$$+ \Theta(\widetilde{F}'(x), \alpha(t))\widetilde{F}'(x)(x - \xi)\rangle, \quad 0 \le t < \widetilde{T}.$$

Since

$$\widetilde{F}(x) = \widetilde{F}(x^*) + \widetilde{F}'(x)(x - x^*) + \widetilde{G}(x, x^*),$$

$$\|\widetilde{G}(x, x^*)\|_X \le \frac{1}{2}L\|x - x^*\|_X^2 \quad \forall x \in \Omega_R(x^*),$$

this yields

$$\frac{d}{dt}\left(\frac{1}{2}\|x - x^*\|_X^2\right) = -\langle U(x - x^*), x - x^* \rangle -$$

$$- \langle U(x - x^*), \Theta(\widetilde{F}'(x), \alpha(t))\widetilde{F}(x^*) + \Theta(\widetilde{F}'(x), \alpha(t))\widetilde{G}(x, x^*) +$$
$$+ [E - \Theta(\widetilde{F}'(x), \alpha(t))\widetilde{F}'(x)](x^* - \xi)\rangle.$$

Taking into account (55) and the estimate

$$|\langle f, z \rangle| \le \|f\|_{X^*} \|z\|_X,$$

we get

$$\frac{d}{dt}\left(\frac{1}{2}\|x - x^*\|_X^2\right) \le -\|x - x^*\|_X^2 +$$

$$+ \Big(\|\Theta(\tilde{F}'(x), \alpha(t))\tilde{F}(x^*)\|_X + \|\Theta(\tilde{F}'(x), \alpha(t))\tilde{G}(x, x^*)\|_X +$$

$$+ \|[E - \Theta(\tilde{F}'(x), \alpha(t))\tilde{F}'(x)](x^* - \xi)\|_X \Big) \|x - x^*\|_X,$$

$$(56)$$

$$0 \le t \le \tilde{T}.$$

Arguing as in Section 4.3, it is not difficult to deduce from (49) and (56) that there exists a constant $c_9 > 0$ such that if

$$\|v\|_X \le \frac{1}{c_9 L}, \tag{57}$$

then

$$\frac{d}{dt}\left(\frac{1}{2}\|x - x^*\|_X^2\right) \le$$

$$\le -\frac{1}{4}\|x - x^*\|_X^2 + \frac{c_{10} L^2}{\alpha^2(t)}\|x - x^*\|_X^4 +$$

$$+ c_{11}\left(\frac{\delta^2}{\alpha^2(t)} + \|v\|_X^2 \alpha^{2p}(t) + (\|v\|_X \delta + \Delta)^2 \right),$$

$$0 \le t \le \tilde{T}.$$

Thus with the notation

$$u(t) = \frac{1}{2}\|x(t) - x^*\|_X^2$$

we have the following auxiliary result.

Theorem 4.14. *Suppose Conditions 1.1 and 2.6–2.8 are satisfied. Assume that (6), (52), and (57) are valid. Then*

$$\dot{u}(t) \le -\frac{1}{2}u(t) + \frac{c_{12}}{\alpha^2(t)}u^2(t) +$$

$$+ c_{11}\left(\frac{\delta^2}{\alpha^2(t)} + \|v\|_X^2 \alpha^{2p}(t) + (\|v\|_X \delta + \Delta)^2 \right),$$

$$0 \leq t \leq \tilde{T}.$$

For the process (50)–(51) we propose a stopping rule similar to that of (4)–(5). Pick a number $m > 0$ such that

$$\alpha^{p+1}(0) > m(\delta + \Delta) \tag{58}$$

and define the stopping point $t = t(\delta, \Delta) > 0$ as a root of the equation

$$\alpha^{p+1}(t(\delta, \Delta)) = m(\delta + \Delta). \tag{59}$$

Arguing as in (30)–(40), we obtain the following results analogous to Theorems 4.9–4.11.

Theorem 4.15. *Suppose Conditions 1.1 and 2.6–2.8 are satisfied. Assume that (6), (57), and (58) are valid. Also, let*

$$\alpha^p(0) < \frac{md_0\nu_0}{2r_0},$$

$$\frac{\dot{\alpha}(t)}{\alpha(t)} \geq -\frac{1}{8p}, \quad t \geq 0, \tag{60}$$

$$\left(\frac{1}{m^2}\left(1 + \max\{1, \|v\|_X^2\}\alpha^2(0)\right) + \|v\|_X^2\right)\alpha^{2p-2}(0) <$$

$$< \frac{1}{64c_{11}c_{12}},$$

$$\|x_0 - x^*\|_X \leq \frac{\alpha(0)}{2\sqrt{c_{12}}}, \quad \alpha(0) < \sqrt{2c_{12}}R. \tag{61}$$

Then a solution to the Cauchy problem (50)–(51) is uniquely determined for $0 \leq t \leq t(\delta, \Delta)$, where $t(\delta, \Delta)$ is specified by (59).

Theorem 4.16. *In the conditions of Theorem 4.15, we have*

$$\|x(t(\delta, \Delta)) - x^*\|_X \leq \frac{m^{p/(p+1)}}{\sqrt{2c_{12}}\alpha^{p-1}(0)}(\delta + \Delta)^{p/(p+1)}.$$

Theorem 4.17. *Suppose Conditions 1.1 and 2.6–2.8 are satisfied. Assume that (57), (60), and (61) are valid, and*

$$x^* - \xi = F'(x^*)^p v, \quad v \in X, \quad p \geq 1,$$

$$\|v\|_X \alpha^{2p-2}(0) \leq \frac{1}{64 c_{11} c_{12}}.$$

Then a solution $x(t)$ to the problem (50)–(51) with $\widetilde{F}(x) = F(x)$ is uniquely determined for all $t \geq 0$, and the estimate is true:

$$\|x(t) - x^*\|_X \leq \frac{\alpha^p(t)}{\sqrt{2 c_{12}} \alpha^{p-1}(0)}, \quad t \geq 0.$$

Chapter 5

STABLE ITERATIVE PROCESSES
FOR IRREGULAR EQUATIONS

This chapter is devoted to construction and investigation of stable iterative processes for irregular nonlinear equations in a Hilbert space. Here the stability is understood with respect both to variations of a starting point in a neighborhood of a solution and to errors in input data and in sourcewise–like representations of the solution. The stability of the processes means that iterative points generated by them are attracted to a neighborhood of the solution, as an iteration number increases. Diameters of the attracting neighborhoods arise to be proportional to levels of the mentioned errors. Therefore there is no a necessity to equip such iterative processes with stopping criterions. Let us remember that in Sections 4.1, 4.3, and 4.5, the construction of approximations adequate to error levels was ensured just by the stopping of iterative or continuous processes at an appropriate point. In Sections 5.1 and 5.2 we develop a technique for constructing stable iterative methods on the basis of the gradient iteration combined with a projection onto specially chosen finite–dimensional subspaces. Section 5.3 is devoted to an iterative method for finding quasisolutions to irregular equations. In Section 5.4 we present a unified approach to constructing stable iterative methods of parametric approximations. Section 5.5 is devoted to continuous analogs of the iterative processes based on the parametric approximations. In Section 5.6 using the ideas of iterative regularization, we study a regularized modification of the gradient process. A continuous version of the method is also analyzed. In Section 5.7 we develop an approach to constructing stable iterative processes applicable to discontinuous operator equations. The technique of approximation of an irregular or even discontinuous equation by a smooth strongly convex optimization problem plays a key role here.

5.1   Stable Gradient Projection Methods with Adaptive Choice of Projectors

In this section we construct and study several stable iterative gradient type methods for approximate solution of nonlinear irregular equations given with errors. We prove that iterative points generated by the methods are attracted to a small neighborhood of the solution. This is ensured by a combination of the standard gradient process

$$x_{n+1} = x_n - \gamma F'^*(x_n)F(x_n), \quad \gamma > 0$$

for the problem

$$\min_{x \in X_1} \frac{1}{2}\|F(x)\|_{X_2}^2$$

and projecting of x_{n+1} onto appropriate finite–dimensional subspaces.

We consider an equation

$$F(x) = 0, \quad x \in X_1, \tag{1}$$

where $F : X_1 \to X_2$ is a nonlinear operator and X_1, X_2 are Hilbert spaces. Let $F(x)$ be a Fréchet differentiable operator, and the linear operator $F'(x) :$ $X_1 \to X_2$ be compact for all $x \in \Omega_R(x^*)$. Also, suppose the Lipschitz condition (4.1.2) is satisfied. As above, x^* denotes a solution to (1). Assume that instead of $F(x)$ an approximation $\widetilde{F}(x)$ is accessible. Let the operator $\widetilde{F} : X_1 \to X_2$ be Fréchet differentiable on $\Omega_R(x^*)$, and $\widetilde{F}(x)$, $\widetilde{F}'(x)$ satisfy (4.1.2) and the conditions

$$\|\widetilde{F}(x^*)\|_{X_2} \le \delta, \tag{2}$$

$$\|\widetilde{F}'^*(x_0)\widetilde{F}'(x_0) - F'^*(x_0)F'(x_0)\|_{L(X_1)} \le \delta, \tag{3}$$

where $x_0 \in \Omega_R(x^*)$ is an initial approximation to the solution x^*. There exists N_1 such that

$$\|F'(x)\|_{L(X_1,X_2)}, \quad \|\widetilde{F}'(x)\|_{L(X_1,X_2)} \le N_1 \quad \forall x \in \Omega_R(x^*).$$

Assume that the operator $\widetilde{F}'(x)$ is compact along with $F'(x)$, $x \in \Omega_R(x^*)$. Since neither operators $F'(x)$, $\widetilde{F}'(x)$ nor $F'^*(x)F'(x)$, $\widetilde{F}'^*(x)\widetilde{F}'(x)$ are continuously invertible in a neighborhood of x^*, the equation (1) is irregular.

In Chapter 4 we have proposed and studied a wide class of iterative methods for solving equation (1). Recall that in the case of exact operators $F(x)$, sequences $\{x_n\}$ generated by the methods converge to x^* (Theorems 4.1 and 4.5). On the other hand, when $F(x)$ is given approximately, the methods generate iterative points that usually don't converge as $n \to \infty$. For this reason, in order to obtain an acceptable approximation to x^* we must terminate these iterations at some finite step $n = N(\delta, \Delta)$ such that

$$\lim_{\delta, \Delta \to 0} N(\delta, \Delta) = \infty, \quad \lim_{\delta, \Delta \to 0} x_{N(\delta, \Delta)} = x^*.$$

Effective numerical implementation of related computational schemes is hampered by the fact that the stopping point $N(\delta, \Delta)$ depends on several constants, which characterize both the method and the equation under consideration (Theorems 4.2 and 4.6). Therefore approximating properties of $x_{N(\delta, \Delta)}$ essentially depend on a proper choice of these constants.

An alternative approach to constructing iterative processes for irregular problems was proposed in [14, 16; 17, Section 7.2]. For each $x \in \Omega_R(x^*)$ and $N \in (0, N_1^2)$, we denote by $\Pi_N(x)$ an orthoprojector from X_1 onto a finite-dimensional subspace that spans all eigenvectors of $F'^*(x)F'(x)$ corresponding to eigenvalues $\lambda \geq N$. In the case where $\sigma(F'^*(x)F'(x)) \cap [N, +\infty) = \emptyset$, we put $\Pi_N(x) = O$. In the similar way, let $\widetilde{\Pi}_N(x)$ be an orthoprojector from X_1 onto a subspace that spans all eigenvectors of $\widetilde{F}'^*(x)\widetilde{F}'(x)$ corresponding to eigenvalues $\lambda \geq N$. Assume that $N \notin \sigma(F'^*(x_0)F'(x_0))$ and denote

$$\rho = \inf\{|\lambda - N| : \lambda \in \sigma(F'^*(x_0)F'(x_0)) \cup \{N_1^2\}\}. \tag{4}$$

Let us define the iterative sequence $\{x_n\}$ ([14, 16, 17]):

$$x_0 \in \Omega_R(x^*), \quad x_{n+1} = \widetilde{\Pi}_N(z_n)(x_n - \xi - \gamma \widetilde{F}'^*(x_n)\widetilde{F}(x_n)) + \xi. \tag{5}$$

Here $\gamma > 0$ is a constant stepsize, $\xi \in X_1$ a controlling parameter, and $z_n = x_n$ ([14]) or $z_n = x_0$ ([16]). In [14, 16], the method (5) was studied in the assumption that ξ satisfies the condition

$$\|(\Pi_N(x^*) - E)(x^* - \xi)\|_{X_1} \leq \Delta. \tag{6}$$

It was shown that if the starting point x_0 is sufficiently close to the solution x^*, then there exists a constant $c_1 > 0$ independent of δ and Δ such that

$$\limsup_{n \to \infty} \|x_n - x^*\|_{X_1} \leq c_1(\delta + \Delta). \tag{7}$$

According to (7), the iterative points x_n are attracted to a neighborhood of x^* as $n \to \infty$. The attracting neighborhood has the radius $O(\delta + \Delta)$ and thus the problem of stopping criterions for (5) does not arise.

We stress that exact determination of the projector $\widetilde{\Pi}_N(z_n)$ is a nontrivial computational problem even if $\{z_n\}$ is stationary, i.e., $z_n = x_0$. Below we study a family of iterative processes of type (5) with $z_n = x_0$, and with the use of suitable approximations $\widetilde{\Pi}_N^{(m)}(x_0)$ instead of the true projector $\widetilde{\Pi}_N(x_0)$. We assume that operators $\widetilde{\Pi}_N^{(m)}(x_0)$ satisfy the following condition.

Condition 5.1.

$$\|\widetilde{\Pi}_N^{(m)}(x_0) - \widetilde{\Pi}_N(x_0)\|_{L(X_1)} \leq \kappa_m, \qquad \lim_{m\to\infty} \kappa_m = 0. \qquad (8)$$

Thus the process we intend to study in this section has the form

$$x_0 \in \Omega_R(x^*), \quad x_{n+1} = \widetilde{\Pi}_N^{(m)}(x_0)(x_n - \xi - \gamma \widetilde{F}'^*(x_n)\widetilde{F}(x_n)) + \xi. \qquad (9)$$

According to (8), the operator $\widetilde{\Pi}_N^{(m)}(x_0)$ should be constructed in dependence of the initial point x_0. Below we present two implementable schemes of such adaptive choice of $\widetilde{\Pi}_N^{(m)}(x_0)$. In Section 5.2 we shall consider a similar process with an a priori choice

$$\widetilde{\Pi}_N^{(m)}(x_0) = P_{\mathcal{M}},$$

where $\mathcal{M}$ is a fixed finite–dimensional subspace of X_1.

Fix a number $\Delta \geq 0$. Throughout this section we assume that the following condition is satisfied.

Condition 5.2.

$$\|(\Pi_N(x_0) - E)(x^* - \xi)\|_{X_1} \leq \Delta. \qquad (10)$$

Let us remark that in (10), contrary to (6), the projector $\Pi_N(x_0)$ is determined by the operator $F'(x)$ at a given point x_0 not at the unknown solution x^*. Observe that inequality (6) can be considered as a weakened sourcewise–like representation

$$x^* - \xi \in R(\Pi_N(x^*)),$$

which plays here the same role as the classical source condition

$$x^* - \xi \in R((F'^*(x^*)F'(x^*))^p)$$

played in Chapter 4. The condition (10) allows for a similar interpretation. Hence the value Δ in (6) and (10) has the meaning of an error in corresponding sourcewise–like representations.

Let us turn to asymptotic properties of the process (9). Assuming that $x_n \in \Omega_R(x^*)$, from (9) we get

$$x_{n+1} - x^* = \widetilde{\Pi}_N^{(m)}(x_0)(E - \gamma \widetilde{F}'^*(x_n)\widetilde{F}'(x_n))(x_n - x^*) -$$
$$- \gamma \widetilde{\Pi}_N^{(m)}(x_0)\widetilde{F}'^*(x_n)\widetilde{G}(x_n, x^*) + (\widetilde{\Pi}_N^{(m)}(x_0) - \Pi_N(x_0))(x^* - \xi) + \qquad (11)$$
$$+ (\Pi_N(x_0) - E)(x^* - \xi) - \gamma \widetilde{\Pi}_N^{(m)}(x_0)\widetilde{F}'^*(x_n)\widetilde{F}(x^*),$$

where

$$\widetilde{G}(x_n, x^*) = \widetilde{F}(x_n) - \widetilde{F}(x^*) - \widetilde{F}'(x_n)(x_n - x^*).$$

We already know that

$$\|\widetilde{G}(x_n, x^*)\|_{X_2} \le \frac{1}{2} L \|x_n - x^*\|_{X_1}^2. \tag{12}$$

Using (11) and the equality

$$\widetilde{\Pi}_N^{(m)}(x_0) - \Pi_N(\dot{x}_0) =$$
$$= (\widetilde{\Pi}_N^{(m)}(x_0) - \widetilde{\Pi}_N(x_0)) + (\widetilde{\Pi}_N(x_0) - \Pi_N(x_0)),$$

we obtain

$$\|x_{n+1} - x^*\|_{X_1} \le$$
$$\le \|\widetilde{\Pi}_N^{(m)}(x_0)(E - \gamma \widetilde{F}'^*(x_n)\widetilde{F}'(x_n))\|_{L(X_1)}\|x_n - x^*\|_{X_1} +$$
$$+ \gamma \|\widetilde{\Pi}_N^{(m)}(x_0)\|_{L(X_1)}\|\widetilde{F}'^*(x_n)\widetilde{G}(x_n, x^*)\|_{X_1} +$$
$$+ \|\widetilde{\Pi}_N^{(m)}(x_0) - \widetilde{\Pi}_N(x_0)\|_{L(X_1)}\|x^* - \xi\|_{X_1} + \tag{13}$$
$$+ \|\widetilde{\Pi}_N(x_0) - \Pi_N(x_0)\|_{L(X_1)}\|x^* - \xi\|_{X_1} +$$
$$+ \|(\Pi_N(x_0) - E)(x^* - \xi)\|_{X_1} +$$
$$+ \gamma \|\widetilde{\Pi}_N^{(m)}(x_0)\widetilde{F}'^*(x_n)\widetilde{F}(x^*)\|_{X_1}.$$

From (8) it follows that

$$\|\widetilde{\Pi}_N^{(m)}(x_0)\|_{L(X_1)} \le$$
$$\le \|\widetilde{\Pi}_N(x_0)\|_{L(X_1)} + \|\widetilde{\Pi}_N^{(m)}(x_0) - \widetilde{\Pi}_N(x_0)\|_{L(X_1)} \le 1 + \kappa_m. \tag{14}$$

Let us define the operator

$$T(x) = \widetilde{\Pi}_N^{(m)}(x_0)(E - \gamma \widetilde{F}'^*(x)\widetilde{F}'(x)), \quad x \in \Omega_R(x^*).$$

We obviously have

$$\|T(x)\|_{L(X_1)} \le$$
$$\le \|\widetilde{\Pi}_N^{(m)}(x_0) - \widetilde{\Pi}_N(x)\|_{L(X_1)}\|E - \gamma \widetilde{F}'^*(x)\widetilde{F}'(x)\|_{L(X_1)} +$$
$$+ \|\widetilde{\Pi}_N(x)(E - \gamma \widetilde{F}'^*(x)\widetilde{F}'(x))\|_{L(X_1)} \le \tag{15}$$
$$\le \overline{q}\|\widetilde{\Pi}_N^{(m)}(x_0) - \widetilde{\Pi}_N(x)\|_{L(X_1)} + q(N),$$

where

$$\overline{q} = \sup_{0 \le \lambda \le N_1^2} |1 - \gamma\lambda|, \quad q(N) = \sup_{N \le \lambda \le N_1^2} |1 - \gamma\lambda|.$$

It is readily seen that the inequality

$$0 < \gamma < \frac{2}{N_1^2} \tag{16}$$

implies $\overline{q} = 1, q(N) < 1$. Let condition (16) be satisfied. Then by (8) and (15) we obtain the estimate

$$\|T(x)\|_{L(X_1)} \leq \kappa_m + \|\widetilde{\Pi}_N(x_0) - \Pi_N(x_0)\|_{L(X_1)} + $$
$$+ \|\Pi_N(x_0) - \widetilde{\Pi}_N(x)\|_{L(X_1)} + q(N). \tag{17}$$

According to (4.1.2), (2), and (14), for the last norm in (13) we have

$$\|\widetilde{\Pi}_N^{(m)}(x_0)\widetilde{F}'^*(x_n)\widetilde{F}(x^*)\|_{X_1} \leq (1 + \kappa_m) N_1 \delta. \tag{18}$$

Combining (13) and inequalities (10), (12), (14), (17), and (18), we get

$$\|x_{n+1} - x^*\|_{X_1} \leq$$
$$\leq \Big(\kappa_m + \|\widetilde{\Pi}_N(x_0) - \Pi_N(x_0)\|_{L(X_1)} + $$
$$+ \|\widetilde{\Pi}_N(x_n) - \Pi_N(x_0)\|_{L(X_1)} + q(N)\Big)\|x_n - x^*\|_{X_1} + $$
$$+ \frac{1}{2}\gamma(1 + \kappa_m) N_1 L \|x_n - x^*\|_{X_1}^2 + \kappa_m \|x^* - \xi\|_{X_1} + $$
$$+ \|\widetilde{\Pi}_N(x_0) - \Pi_N(x_0)\|_{L(X_1)}\|x^* - \xi\|_{X_1} + $$
$$+ \Delta + \gamma(1 + \kappa_m) N_1 \delta. \tag{19}$$

Consider a circle $\mathcal{J}$ with the diameter $[N, N_1^2 + a]$, where $a > \rho$. Obviously, $\sigma(F'^*(x_0)F'(x_0)) \cap (N, N_1^2]$ lies inside $\mathcal{J}$. To estimate the norm

$$\|\widetilde{\Pi}_N(x_n) - \Pi_N(x_0)\|_{L(X_1)},$$

we need the following lemma.

Lemma 5.1. *Assume that*

$$\delta \leq \frac{\rho}{4(1 + N_1^2)}, \tag{20}$$

$$\|x_n - x_0\|_{X_1} \leq \frac{\rho}{8N_1 L(1 + N_1^2)}. \tag{21}$$

Then

$$\|\widetilde{\Pi}_N(x_n) - \Pi_N(x_0)\|_{L(X_1)} \leq M(N)(2N_1 L \|x_n - x_0\|_{X_1} + \delta), \tag{22}$$

where

$$M(N) = \frac{(N_1^2 + a - N)(1 + N_1^2)}{\rho^2}.$$

In particular, (22) with $n = 0$ yields

$$\|\widetilde{\Pi}_N(x_0) - \Pi_N(x_0)\|_{L(X_1)} \le M(N)\delta. \tag{23}$$

Proof. Using representation (1.1.6) for $\widetilde{\Pi}_N(x_0)$ and $\Pi_N(x_0)$, we obtain (see [123, Ch.9] for details)

$$\|\widetilde{\Pi}_N(x_n) - \Pi_N(x_0)\|_{L(X_1)} \le$$

$$\le \frac{N_1^2 + a - N}{2\rho} \left(1 - \left(\|R(\lambda, F'^*(x_0)F'(x_0))\|_{L(X_1)} + \right. \right.$$

$$+ \|F'^*(x_0)F'(x_0)R(\lambda, F'^*(x_0)F'(x_0))\|_{L(X_1)} \right) \cdot$$

$$\cdot \|\widetilde{F}'^*(x_n)\widetilde{F}'(x_n) - F'^*(x_0)F'(x_0)\|_{L(X_1)} \right)^{-1} \cdot \tag{24}$$

$$\cdot \left(\|R(\lambda, F'^*(x_0)F'(x_0))\|_{L(X_1)} + \right.$$

$$+ \|F'^*(x_0)F'(x_0)R(\lambda, F'^*(x_0)F'(x_0))\|_{L(X_1)} \right) \cdot$$

$$\cdot \|\widetilde{F}'^*(x_n)\widetilde{F}'(x_n) - F'^*(x_0)F'(x_0)\|_{L(X_1)}$$

$$\forall \lambda \in \mathcal{J}.$$

Furthermore,

$$\|R(\lambda, F'^*(x_0)F'(x_0))\|_{L(X_1)} +$$

$$+ \|F'^*(x_0)F'(x_0)R(\lambda, F'^*(x_0)F'(x_0))\|_{L(X_1)} \le \tag{25}$$

$$\le \frac{1 + N_1^2}{\rho} \quad \forall \lambda \in \mathcal{J}.$$

Next, from (4.1.2) and (3) it follows that

$$\|\widetilde{F}'^*(x_n)\widetilde{F}'(x_n) - F'^*(x_0)F'(x_0)\|_{L(X_1)} \le$$

$$\le \|\widetilde{F}'^*(x_n)\widetilde{F}'(x_n) - \widetilde{F}'^*(x_0)\widetilde{F}'(x_0)\|_{L(X_1)} +$$

$$+ \|\widetilde{F}'^*(x_0)\widetilde{F}'(x_0) - F'^*(x_0)F'(x_0)\|_{L(X_1)} \le$$

$$\le 2N_1 L \|x_n - x_0\|_{X_1} + \delta.$$

Combining this inequality with (24) and (25), we deduce

$$\|\tilde{\Pi}_N(x_n) - \Pi_N(x_0)\|_{L(X_1)} \le$$
$$\le \frac{(N_1^2 + a - N)(1 + N_1^2)(2N_1L\|x_n - x_0\|_{X_1} + \delta)}{2\rho[\rho - (1 + N_1^2)(2N_1L\|x_n - x_0\|_{X_1} + \delta)]}.$$

Taking into account (20) and (21), we obtain the required estimate (22).   □

By (19), (22), and (23), it follows that if (20) and (21) are satisfied, then

$$\|x_{n+1} - x^*\|_{X_1} \le \Big(\kappa_m + q(N) + 2M(N)\delta +$$
$$+ 2M(N)N_1L\|x_0 - x^*\|_{X_1}\Big)\|x_n - x^*\|_{X_1} +$$
$$+ \Big(2M(N)N_1L + \frac{1}{2}\gamma(1 + \kappa_m)N_1L\Big)\|x_n - x^*\|_{X_1}^2 + \qquad (26)$$
$$+ \kappa_m\|x^* - \xi\|_{X_1} + M(N)\|x^* - \xi\|_{X_1}\delta +$$
$$+ \Delta + \gamma(1 + \kappa_m)N_1\delta.$$

We claim that there exist constants $l, C > 0$, and $q \in (0, 1)$ such that

$$\|x_n - x^*\|_{X_1} \le lq^n + C(\delta + \Delta + \kappa_m). \qquad (27)$$

Indeed, suppose $x_n \in \Omega_R(x^*)$ and (27) is fulfilled for $n = 0$ and for some number $n \in \mathbf{N}$. Also, assume that

$$\kappa_m + q(N) +$$
$$+ 2M(N)(\delta + \Delta + \kappa_m) + 2M(N)N_1L(l + C(\delta + \Delta + \kappa_m)) +$$
$$+ \Big(2M(N)N_1L + \frac{1}{2}\gamma(1 + \kappa_m)N_1L\Big)(lq + 2C(\delta + \Delta + \kappa_m)) \le q, \qquad (28)$$

$$\Big(\kappa_m + q(N) + 2M(N)(\delta + \Delta + \kappa_m) +$$
$$+ 2M(N)N_1L(l + C(\delta + \Delta + \kappa_m))\Big)C +$$
$$+ \Big(2M(N)N_1L + \frac{1}{2}\gamma(1 + \kappa_m)N_1L\Big)C^2(\delta + \Delta + \kappa_m) + \qquad (29)$$
$$+ \max\{1, \gamma(1 + \kappa_m)N_1 + M(N)\|x^* - \xi\|_{X_1}, \|x^* - \xi\|_{X_1}\} \le C,$$
$$l + C(\delta + \Delta + \kappa_m) \le R. \qquad (30)$$

Then from (26), (28), and (29) we obtain

$$\|x_{n+1} - x^*\|_{X_1} \le lq^{n+1} + C(\delta + \Delta + \kappa_m).$$

Since

$$\|x_{n+1} - x^*\|_{X_1} \le lq^{n+1} + C(\delta + \Delta + \kappa_m) \le$$
$$\le l + C(\delta + \Delta + \kappa_m) \le R,$$

we have $x_{n+1} \in \Omega_R(x^*)$. Notice that the point x_{n+1} satisfies condition (21) if

$$l(1 + q) + 2C(\delta + \Delta + \kappa_m) \le \frac{\rho}{8N_1 L(1 + N_1^2)}. \tag{31}$$

In fact, from (31) it follows that

$$\|x_{n+1} - x_0\|_{X_1} \le \|x_{n+1} - x^*\|_{X_1} + \|x_0 - x^*\|_{X_1} \le$$
$$\le l(1 + q^{n+1}) + 2C(\delta + \Delta + \kappa_m) \le \frac{\rho}{8N_1 L(1 + N_1^2)}.$$

Thus we proved the following statement.

Theorem 5.1. *Suppose Conditions 5.1 and 5.2 are satisfied, and the initial point x_0 is chosen such that*

$$\|x_0 - x^*\|_{X_1} \le l + C(\delta + \Delta + \kappa_m).$$

Let conditions (16), (20), and (28)–(31) be fulfilled. Then (27) is true.

Corollary 5.1. *Under the conditions of Theorem 5.1, we have*

$$\limsup_{n \to \infty} \|x_n - x^*\|_{X_1} \le C(\delta + \Delta + \kappa_m).$$

According to (28)–(31), the constants $l = l(\delta, \Delta, \kappa_m)$ and $C = C(\delta, \Delta, \kappa_m)$ can be chosen such that

$$\lim_{\substack{\delta, \Delta \to 0 \\ m \to \infty}} l(\delta, \Delta, \kappa_m) > 0, \qquad \lim_{\substack{\delta, \Delta \to 0 \\ m \to \infty}} C(\delta, \Delta, \kappa_m) < \infty.$$

Corollary 5.1 then implies that the iterative points $x_n, n \to \infty$ stabilize in a neighborhood of x^* with diameter of order $O(\delta + \Delta + \kappa_m)$, while the starting point x_0 can be chosen from a neighborhood of diameter $O(1)$. Therefore the process (9) doesn't require a stopping criterion to generate an approximation adequate to the error levels δ, Δ, and κ_m.

Let us turn to examples of operator families $\{\widetilde{\Pi}_N^{(m)}(x_0)\}$ and corresponding specifications of estimate (8). We shall need the following auxiliary result.

Lemma 5.2. *Suppose*

$$\delta \leq \frac{\rho}{2}, \tag{32}$$

where ρ is defined by (4); then

$$\left(N - \frac{\rho}{2}, N + \frac{\rho}{2}\right) \cap \sigma(\tilde{F}'^*(x_0)\tilde{F}'(x_0)) = \emptyset.$$

Proof. By (4) we conclude that the operator

$$((N + \varepsilon)E - F'^*(x_0)F'(x_0))^{-1} = R\left(N + \varepsilon, F'^*(x_0)F'(x_0)\right), \quad |\varepsilon| < \frac{\rho}{2}$$

is bounded, and

$$\|R\left(N + \varepsilon, F'^*(x_0)F'(x_0)\right)\|_{L(X_1)} < \frac{2}{\rho}.$$

Then from (3) and (32) it follows that

$$\|[((N + \varepsilon)E - \tilde{F}'^*(x_0)\tilde{F}'(x_0)) - ((N + \varepsilon)E - F'^*(x_0)F'(x_0))] \cdot$$
$$\cdot ((N + \varepsilon)E - F'^*(x_0)F'(x_0))^{-1}\|_{L(X_1)} \leq$$
$$\leq \|\tilde{F}'^*(x_0)\tilde{F}'(x_0) - F'^*(x_0)F'(x_0)\|_{L(X_1)} \cdot$$
$$\cdot \|R(N + \varepsilon, F'^*(x_0)F'(x_0))\|_{L(X_1)} < \frac{2\delta}{\rho} \leq 1.$$

The application of Lemma 1.2 yields the required assertion.   □

Suppose (32) is valid.

Example 5.1. We have the representation

$$\tilde{\Pi}_N(x_0) = \int\limits_{0-0}^{N_1^2} \chi(\lambda)d\tilde{E}_\lambda(x_0),$$

where

$$\chi(\lambda) = \begin{cases} 0, & \lambda < N, \\ 1, & \lambda \geq N. \end{cases}$$

and $\{\widetilde{E}_\lambda(x_0)\}_{\lambda \in [0, N_1^2]}$ is the family of spectral projectors of $\widetilde{F}'^*(x_0)\widetilde{F}'(x_0)$. Let us define the function $g(\lambda)$ as follows:

$$g(\lambda) = \begin{cases} 0, & \lambda \leq N - \rho/4, \\ \rho^{-1}(2\lambda - 2N + \rho/2), & N - \rho/4 < \lambda < N + \rho/4, \\ 1, & \lambda \geq N + \rho/4. \end{cases}$$

By Lemma 5.2,

$$\left(N - \frac{\rho}{2}, N + \frac{\rho}{2}\right) \subset \rho(\widetilde{F}'^*(x_0)\widetilde{F}'(x_0)).$$

Therefore,

$$\widetilde{\Pi}_N(x_0) = \int\limits_{0-0}^{N_1^2} \chi(\lambda)d\widetilde{E}_\lambda(x_0) = \int\limits_{0-0}^{N_1^2} g(\lambda)d\widetilde{E}_\lambda(x_0). \tag{33}$$

Observe that the function $g(\lambda)$, unlike $\chi(\lambda)$, is continuous on the segment $[0, N_1^2]$. Suppose polynomials $g_m(\lambda)$ approximate $g(\lambda)$ as $m \to \infty$ such that

$$\lim_{m \to \infty} \|g_m - g\|_{C[0, N_1^2]} = 0.$$

We put

$$\widetilde{\Pi}_N^{(m)}(x_0) = \int\limits_{0-0}^{N_1^2} g_m(\lambda)d\widetilde{E}_\lambda(x_0). \tag{34}$$

In particular, as $g_m(\lambda)$ we can take a segment of the expansion of $g(\lambda)$ in a series of Legendre polynomials on $[0, N_1^2]$:

$$g_m(\lambda) = \sum_{k=0}^{m} a_k \widehat{P}_k(\lambda).$$

Here

$$a_k = \int\limits_{0}^{N_1^2} g(\lambda)\widehat{P}_k(\lambda)d\lambda, \quad \widehat{P}_k(\lambda) = \frac{\sqrt{2n+1}}{N_1}P_k\left(\frac{2\lambda}{N_1^2} - 1\right), \quad k = 0, \ldots, m;$$

$P_k(\lambda)$ are the classical Legendre polynomials orthogonal on $[-1, 1]$ ([132]):

$$P_0(\lambda) = 1, \quad P_1(\lambda) = \lambda;$$
$$P_{k+1}(\lambda) = \frac{2k+1}{k+1}\lambda P_k(\lambda) - \frac{k}{k+1}P_{k-1}(\lambda), \quad k \in \mathbf{N}. \tag{35}$$

Since $g(\lambda)$ satisfies

$$|g(\lambda_1) - g(\lambda_2)| \leq \frac{2}{\rho}|\lambda_1 - \lambda_2| \quad \forall \lambda_1, \lambda_2 \in [0, N_1^2],$$

from (1.1.3), (33), and (34) we get (see, e.g., [131])

$$\|\widetilde{\Pi}_N^{(m)}(x_0) - \widetilde{\Pi}_N(x_0)\|_{L(X_1)} = \operatorname*{ess\,sup}_{\{\widetilde{E}_\lambda(x_0)\}} |g_m(\lambda) - g(\lambda)| \leq$$

$$\leq \max_{\lambda \in [0, N_1^2]} \left| \sum_{k=0}^{m} a_k \widehat{P}_k(\lambda) - g(\lambda) \right| \leq \frac{c_2}{\rho\sqrt{m}},$$

where c_2 is independent of ρ and m. Therefore in (8) we can set

$$\kappa_m = \frac{c_2}{\rho\sqrt{m}}.$$

Moreover, from (34) we obtain an explicit expression for the operator $\widetilde{\Pi}_N^{(m)}(x_0)$:

$$\widetilde{\Pi}_N^{(m)}(x_0) = \int_{0-0}^{N_1^2} \left(\sum_{k=0}^{m} a_k \widehat{P}_k(\lambda) \right) d\widetilde{E}_\lambda(x_0) = \sum_{k=0}^{m} a_k \widehat{P}_k(\widetilde{F}'^*(x_0)\widetilde{F}'(x_0)).$$

Now, let $s > 0$ be arbitrarily large. Then, taking as $g(\lambda)$ a sufficiently smooth function with

$$g(\lambda) = 0, \lambda \leq N - \frac{\rho}{4}; \quad g(\lambda) = 1, \lambda \geq N + \frac{\rho}{4},$$

by the same scheme we get approximations $\widetilde{\Pi}_N^{(m)}(x_0)$ such that

$$\|\widetilde{\Pi}_N^{(m)}(x_0) - \widetilde{\Pi}_N(x_0)\|_{L(X_1)} = O(m^{-s}).$$

Example 5.2. Without loss of generality we can assume that

$$\|\widetilde{F}'(x_0)\|_{L(X_1, X_2)} \leq N_1 \leq 1.$$

Consider an auxiliary operator $B = \widetilde{F}'^*(x_0)\widetilde{F}'(x_0) - NE$. Let

$$\{E_\lambda(x_0)\}_{\lambda \in [-N, N_1^2 - N]}$$

be the family of spectral projectors of B. We denote

$$P_I = \int_{-N-0}^{0} dE_\lambda(x_0), \quad P_{II} = \int_{0}^{N_1^2 - N} dE_\lambda(x_0).$$

By Lemma 5.2,

$$\left(-\frac{\rho}{2}, \frac{\rho}{2}\right) \cap \sigma(B) = \emptyset.$$

Next, we set

$$\varphi(\lambda) = \frac{1}{2}(3\lambda - \lambda^3)$$

and define the operator iteration

$$B_0 = B; \quad B_{m+1} = \varphi(B_m), \quad m = 0, 1, \dots . \tag{36}$$

From (36) it follows that

$$B_{m+1} = \psi_{m+1}(B) = \int_{-N-0}^{N_1^2-N} \psi_{m+1}(\lambda)dE_\lambda(x_0), \tag{37}$$

$$\psi_{m+1}(\lambda) = \varphi(\psi_m(\lambda)), \quad m = 0, 1, \dots,$$

where $\psi_0(\lambda) = \lambda$. Also, define the function

$$\psi(\lambda) = \begin{cases} -1, & \lambda \le 0, \\ 1, & \lambda > 0. \end{cases}$$

Then we have

$$P_I = \frac{1}{2}(E - \psi(B)), \quad P_{II} = \frac{1}{2}(E + \psi(B)).$$

Let us prove that the sequence $\{B_m\}$ converges to $\psi(B)$ as $m \to \infty$. From (37) we obtain

$$\|B_m - \psi(B)\|_{L(X_1)} = \operatorname*{ess\,sup}_{\{E_\lambda(x_0)\}} |\psi_m(\lambda) - \psi(\lambda)| \le$$

$$\le \sup\left\{ |\psi_m(\lambda) - \psi(\lambda)| : \lambda \in \left[-N, -\frac{\rho}{2}\right] \cup \left[\frac{\rho}{2}, N_1^2 - N\right] \right\}. \tag{38}$$

Pick an arbitrary $\varepsilon \in (1/\sqrt{3}, 1)$. It is readily seen that for all $\lambda_1, \lambda_2 \in [-1, -\varepsilon]$ and all $\lambda_1, \lambda_2 \in [\varepsilon, 1]$,

$$|\varphi(\lambda_1) - \varphi(\lambda_2)| \le q_0(\varepsilon)|\lambda_1 - \lambda_2|, \quad q_0(\varepsilon) = \frac{3}{2}(1 - \varepsilon^2) \in (0, 1).$$

Therefore for each $\lambda \in [-1, -\varepsilon] \cup [\varepsilon, 1]$, the simple iteration process

$$\psi_0(\lambda) = \lambda, \quad \psi_{m+1}(\lambda) = \varphi(\psi_m(\lambda))$$

converges to $\psi(\lambda)$, and the estimate is valid:

$$|\psi_m(\lambda) - \psi(\lambda)| \leq$$

$$\leq \frac{q_0(\varepsilon)^m}{1 - q_0(\varepsilon)} \sup\{|\lambda - \varphi(\lambda)| : \lambda \in [-1, -\varepsilon] \cup [\varepsilon, 1]\} = \qquad (39)$$

$$= \frac{\varepsilon(1 - \varepsilon)}{3\varepsilon^2 - 1} q_0(\varepsilon)^m.$$

Now let us consider the behavior of iterations $\{\psi_m(\lambda)\}$ for $\lambda \in [-\varepsilon, -\rho/2] \cup [\rho/2, \varepsilon]$. Assume that $\rho/2 < \varepsilon$. Since $\varphi(-\lambda) = -\varphi(\lambda)$, it is sufficient to examine $\lambda \in [\rho/2, \varepsilon]$. Direct calculations prove that

$$\varphi(\lambda) \geq k_0(\varepsilon)\lambda \quad \forall \lambda \in [0, \varepsilon],$$

where $k_0(\varepsilon) = 1/2(3 - \varepsilon^2)$. Hence,

$$\psi_m(\lambda) \geq k_0(\varepsilon)^m \lambda \quad \forall \lambda \in \left[\frac{\rho}{2}, \varepsilon\right].$$

For the numbers

$$m \leq K(\lambda, \rho, \varepsilon) = \max\left\{k : \psi_k(\lambda) \in \left[\frac{\rho}{2}, \varepsilon\right]\right\}$$

we get

$$|\psi_m(\lambda) - \psi(\lambda)| = 1 - \psi_m(\lambda) \leq 1 - \frac{\rho}{2} \leq$$

$$\leq \left(1 - \frac{\rho}{2}\right) q_0(\varepsilon)^{m - K(\lambda, \rho, \varepsilon)} \quad \forall \lambda \in \left[\frac{\rho}{2}, \varepsilon\right].$$

Let us remark that

$$K(\lambda, \rho, \varepsilon) \leq K_0(\rho, \varepsilon) \equiv \left[\frac{\ln(2\varepsilon/\rho)}{\ln k_0(\varepsilon)}\right] + 1.$$

As before, $[x]$ stands for the integer part of x. Thus for the numbers $m \leq K(\lambda, \rho, \varepsilon)$ we have

$$|\psi_m(\lambda) - \psi(\lambda)| \leq \left(1 - \frac{\rho}{2}\right) q_0(\varepsilon)^{m - K_0(\rho, \varepsilon)}. \qquad (40)$$

Since $N \leq 1$, $N_1^2 - N \leq 1$, combining inequalities (39) and (40), we conclude that for all $\lambda \in [-N, -\rho/2] \cup [\rho/2, N_1^2 - N]$,

$$|\psi_m(\lambda) - \psi(\lambda)| \leq c_3(\rho, \varepsilon) q_0(\varepsilon)^m,$$

where

$$c_3(\rho, \varepsilon) = \max\left\{\frac{\varepsilon(1 - \varepsilon)}{3\varepsilon^2 - 1}, \left(1 - \frac{\rho}{2}\right)\left(\frac{3}{2}(1 - \varepsilon^2)\right)^{-K_0(\rho, \varepsilon)}\right\}.$$

By (38) we get

$$\|B_m - \psi(B)\|_{L(X_1)} \leq c_3(\rho, \varepsilon)q_0(\varepsilon)^m. \tag{41}$$

Observe that proper subspaces of eigenvalues $\lambda \in \sigma(\tilde{F}'^*(x_0)\tilde{F}'(x_0))$ coincide with those of $\lambda - N \in \sigma(B)$. Therefore,

$$P_{II} = \tilde{\Pi}_N(x_0).$$

Now let us set

$$\tilde{\Pi}_N^{(m)}(x_0) = \frac{1}{2}(E + B_m);$$

then (41) yields

$$\|\tilde{\Pi}_N^{(m)}(x_0) - \tilde{\Pi}_N(x_0)\|_{L(X_1)} =$$

$$= \left\|\frac{1}{2}(E + B_m) - \frac{1}{2}(E + \psi(B))\right\|_{L(X_1)} =$$

$$= \frac{1}{2}\|B_m - \psi(B)\|_{L(X_1)} \leq \frac{1}{2}c_3(\rho, \varepsilon)q_0(\varepsilon)^m. \tag{42}$$

From (42) it follows that in (8) we can set

$$\kappa_m = \frac{1}{2}c_3(\rho, \varepsilon)q_0(\varepsilon)^m.$$

Since $q_0(\varepsilon) \in (0, 1)$, we have $\lim_{m\to\infty} \kappa_m = 0$.

In a conclusion let us show how conditions of Theorem 5.1 can be ensured in practice. First, pick q such that

$$q(N) < q < 1. \tag{43}$$

Then the following inequalities put restrictions on the constants l and C:

$$0 < l \leq \min\left\{\frac{R}{2}, \frac{q - q(N)}{(8M(N)(1 + q) + 3\gamma q)N_1 L}, \right.$$

$$\left. \frac{\rho}{16(1 + q)(1 + N_1^2)N_1 L}\right\}, \tag{44}$$

$$C \geq \frac{2}{1 - q}\max\left\{1, M(N)\|x^* - \xi\|_{X_1} + \frac{3}{2}\gamma N_1, \|x^* - \xi\|_{X_1}\right\}. \tag{45}$$

The error levels δ, Δ, and κ_m must satisfy (20) and the conditions

$$\kappa_m \leq \frac{q - q(N)}{2}, \tag{46}$$

$$\delta + \Delta + \kappa_m \leq \min\left\{ \frac{R}{2C}, \frac{\rho}{32(1 + N_1^2)N_1 LC}, \right.$$
$$\left. \frac{1 - q}{(4M(N) + 3\gamma/2)N_1 LC}, \frac{q - q(N)}{8M(N) + 6(4M(N) + \gamma)N_1 LC} \right\}. \tag{47}$$

The starting point x_0 and the parameter ξ must be chosen such that

$$\|x_0 - x^*\|_{X_1} \leq l + C(\delta + \Delta + \kappa_m), \tag{48}$$

$$\|(\Pi_N(x_0) - E)(x^* - \xi)\|_{X_1} \leq \Delta. \tag{49}$$

Inequalities (43)–(49) guarantee the fulfilment of conditions (28), (29), and (31). On the other hand, (30) is a direct consequence of (44) and (47).

Remark 5.1. The reader will easily prove an analog of Theorem 5.1 with (3) replaced by the conditions

$$N \in \rho(\widetilde{F}'^*(x_0)\widetilde{F}'(x_0)), \quad \|(\widetilde{\Pi}_N(x_0) - E)(x^* - \xi)\|_{X_1} \leq \Delta.$$

Remark 5.2. Using estimate (27), it is not difficult to indicate a number $N = N(\delta, \Delta, \kappa_m)$ such that

$$\|x_{N(\delta,\Delta,\kappa_m)} - x^*\|_{X_1} \leq 2C(\delta + \Delta + \kappa_m). \tag{50}$$

From (50) we deduce that if $\Delta = \kappa_m = 0$, then the mapping

$$\mathfrak{R}_{1/N(\delta,0,0)}(\widetilde{F}) = x_{N(\delta,0,0)}$$

defines a regularization algorithm for the original problem (1). From (10) it follows however that in order to get $\Delta = 0$, we must choose ξ from a finite–dimensional affine subspace $\mathbb{M}(x^*) = x^* + R(\Pi_N(x_0))$. It is unlikely that any algorithmic ways of finding such ξ exist. In this situation, inequality (50) establishes the stability of $\mathfrak{R}_{1/N(\delta,0,0)}$ with respect to small perturbations of the exact sourcewise–like inclusion $\xi \in x^* + R(\Pi_N(x_0))$, and to perturbations of $\Pi_N(x_0)$. Here the class of approximate data $\mathfrak{F}$ and the mapping $G : \mathfrak{F} \to X_1$ can be defined analogously to Section 4.1. Inequality (50) can be considered as an estimate for the norm from (1.2.8).

5.2   Projection Method with a Priori Choice of Projectors

In this section we address to the case where z_n in (1.5) do not necessarily coincide with the starting point x_0. In such an event, it is convenient to assume that the orthoprojector $\widetilde{\Pi}_N(z_n)$ is constant throughout the iterative process. So let $\widetilde{\Pi}_N(z_n) = P_{\mathcal{M}}$ be a projector onto a fixed finite–dimensional subspace $\mathcal{M} \subset X_1$. From the formal point of view, $\mathcal{M}$ is not connected with the original operator $F(x)$. Suppose $F(x)$ and $\widetilde{F}(x)$ satisfy all the conditions introduced in Section 5.1. We consider the iterative process

$$x_0 \in \Omega_R(x^*), \quad x_{n+1} = P_{\mathcal{M}}(x_n - \xi - \gamma \widetilde{F}'^*(x_n)\widetilde{F}(x_n)) + \xi, \quad (1)$$

where γ satisfies (1.16). Observe that (1) implies $x_n - \xi \in \mathcal{M} \quad \forall n \in \mathbf{N}$ and hence the process (1) can be also presented in the form

$$x_1 = P_{\mathcal{M}}(x_0 - \gamma \widetilde{F}'^*(x_0)\widetilde{F}(x_0)) + \xi - P_{\mathcal{M}}\xi;$$

$$x_{n+1} = x_n - \gamma P_{\mathcal{M}}\widetilde{F}'^*(x_n)\widetilde{F}(x_n), \quad n = 1, 2, \ldots.$$

Let the subspace $\mathcal{M}$ satisfy the following condition.

Condition 5.3.
$$N(F'(x_0)) \cap \mathcal{M} = \{0\}. \tag{2}$$

Fix a number $\Delta \geq 0$. The next condition poses a restriction on the parameter $\xi \in X_1$ in (1).

Condition 5.4.
$$\|(P_{\mathcal{M}} - E)(x^* - \xi)\|_{X_1} \leq \Delta. \tag{3}$$

Here Δ can be considered as an error level in the exact sourcewise–like representation

$$x^* - \xi \in \mathcal{M}.$$

Assume that $\Delta \leq R$ and $x_n \in \Omega_R(x^*)$. Using (3), we obtain

$$\|x_{n+1} - x^*\|_{X_1} \leq \|x_{n+1} - P_{\mathcal{M}}(x^* - \xi) - \xi\|_{X_1} + \Delta. \tag{4}$$

Next, similarly to (1.11),

$$
\begin{aligned}
x_{n+1} - P_{\mathcal{M}}(x^* - \xi) - \xi &= \\
= (x_n - P_{\mathcal{M}}(x^* - \xi) - \xi) &- \gamma P_{\mathcal{M}}\widetilde{F}'^*(x_n)\widetilde{F}(x_n) = \\
= [E - \gamma P_{\mathcal{M}}\widetilde{F}'^*(x_n)\widetilde{F}'(x_n)P_{\mathcal{M}}](x_n &- P_{\mathcal{M}}(x^* - \xi) - \xi) - \\
- \gamma P_{\mathcal{M}}\widetilde{F}'^*(x_n)\overline{G}(x_n, x^*) &- \gamma P_{\mathcal{M}}\widetilde{F}'^*(x_n)\widetilde{F}(P_{\mathcal{M}}(x^* - \xi) + \xi),
\end{aligned} \tag{5}
$$

where
$$\overline{G}(x_n, x^*) =$$
$$= \widetilde{F}(x_n) - \widetilde{F}(P_{\mathcal{M}}(x^* - \xi) + \xi) - \widetilde{F}'(x_n)(x_n - P_{\mathcal{M}}(x^* - \xi) - \xi).$$

Since
$$\|x^* - (P_{\mathcal{M}}(x^* - \xi) + \xi)\|_{X_1} \le \Delta \le R,$$
we have $P_{\mathcal{M}}(x^* - \xi) + \xi \in \Omega_R(x^*)$. Therefore by (1.1.14) it follows that

$$\|\overline{G}(x_n, x^*)\|_{X_2} \le \frac{1}{2}L\|x_n - P_{\mathcal{M}}(x^* - \xi) - \xi\|_{X_1}^2. \tag{6}$$

Combining (5) and (6), we get the estimate

$$\|x_{n+1} - P_{\mathcal{M}}(x^* - \xi) - \xi\|_{X_1} \le$$
$$\le \|[E - \gamma P_{\mathcal{M}}\widetilde{F}'^*(x_n)\widetilde{F}'(x_n)P_{\mathcal{M}}](x_n - P_{\mathcal{M}}(x^* - \xi) - \xi)\|_{X_1} +$$
$$+ \frac{1}{2}\gamma N_1 L\|x_n - P_{\mathcal{M}}(x^* - \xi) - \xi\|_{X_1}^2 +$$
$$+ \gamma\|\widetilde{F}'^*(x_n)\widetilde{F}(P_{\mathcal{M}}(x^* - \xi) + \xi)\|_{X_1}. \tag{7}$$

Recall that
$$\|\widetilde{F}'(x)\|_{L(X_1, X_2)} \le N_1 \quad \forall x \in \Omega_R(x^*).$$

We shall consider $E - \gamma P_{\mathcal{M}}\widetilde{F}'^*(x_n)\widetilde{F}'(x_n)P_{\mathcal{M}}$ as an operator acting from $\mathcal{M}$ into $\mathcal{M}$. It is obvious that the spectrum of this selfadjoint operator consists of a finite number of real eigenvalues; the spectrum of $P_{\mathcal{M}}F'^*(x_0)F'(x_0)P_{\mathcal{M}}$ has the same structure. According to (2),

$$0 \in \rho(P_{\mathcal{M}}F'^*(x_0)F'(x_0)P_{\mathcal{M}}). \tag{8}$$

Therefore the spectrum $\sigma(P_{\mathcal{M}}F'^*(x_0)F'(x_0)P_{\mathcal{M}})$ lies on the positive semiaxis of $\mathbf{R}$. Let us denote

$$\widetilde{\rho} = \inf\{\lambda : \lambda \in \sigma(P_{\mathcal{M}}F'^*(x_0)F'(x_0)P_{\mathcal{M}})\}.$$

Notice that (8) yields $\widetilde{\rho} > 0$.

Repeating with evident changes the proof of Lemma 5.2, we get the following proposition.

Lemma 5.3. *Let Condition 5.3 be satisfied and*

$$\delta < \frac{\widetilde{\rho}}{2}, \quad \|x_n - x_0\|_{X_1} \le \frac{\widetilde{\rho} - 2\delta}{4N_1 L}.$$

Then

$$\left(-\infty, \frac{\widetilde{\rho}}{2}\right) \cap \sigma(P_{\mathcal{M}}\widetilde{F}'^*(x_n)\widetilde{F}'(x_n)P_{\mathcal{M}}) = \emptyset.$$

For the first summand in (7) we have the estimate

$$\|[E - \gamma P_{\mathcal{M}}\widetilde{F}'^*(x_n)\widetilde{F}'(x_n)P_{\mathcal{M}}](x_n - P_{\mathcal{M}}(x^* - \xi) - \xi)\|_{X_1} \le$$
$$\le \|E - \gamma P_{\mathcal{M}}\widetilde{F}'^*(x_n)\widetilde{F}'(x_n)P_{\mathcal{M}}\|_{L(\mathcal{M})}\|x_n - P_{\mathcal{M}}(x^* - \xi) - \xi\|_{X_1} \le$$
$$\le \widetilde{q}\|x_n - P_{\mathcal{M}}(x^* - \xi) - \xi\|_{X_1},$$

where in the conditions of Lemma 5.3,

$$\widetilde{q} = \sup_{\lambda \in [\widetilde{\rho}/2, N_1^2]} |1 - \gamma\lambda| < 1.$$

Combining (3) and (7), with the use of (4.1.2) and (1.2), we finally obtain

$$\|x_{n+1} - P_{\mathcal{M}}(x^* - \xi) - \xi\|_{X_1} \le$$
$$\le \frac{1}{2}\gamma N_1 L\|x_n - P_{\mathcal{M}}(x^* - \xi) - \xi\|_{X_1}^2 + \tag{9}$$
$$+ \widetilde{q}\|x_n - P_{\mathcal{M}}(x^* - \xi) - \xi\|_{X_1} + \gamma N_1(\delta + N_1\Delta).$$

From (4) and (9), by the scheme of Section 5.1 we get the following statement on asymptotic properties of the iteration (1).

Theorem 5.2. *Suppose Conditions 5.3 and 5.4 and inequality (1.16) are satisfied. Assume that*

$$\widetilde{q} < q < 1, \quad 0 < l \le \min\left\{\frac{R}{2}, \frac{q - \widetilde{q}}{\gamma q N_1 L}, \frac{\widetilde{\rho} - 2\delta}{8(1 + q)N_1 L}\right\},$$

$$C \ge \max\left\{1, \frac{\gamma N_1 \max\{1, N_1\}}{1 - q}\right\}.$$

Let the error levels δ and Δ satisfy the conditions

$$\delta < \frac{\widetilde{\rho}}{2}, \quad C(\delta + \Delta) + \Delta \le \min\left\{\frac{R}{2}, \frac{\widetilde{\rho} - 2\delta}{16 N_1 L}\right\},$$

$$\delta + \Delta \le \frac{q - \widetilde{q}}{2\gamma C N_1 L}.$$

Suppose the initial point x_0 is chosen such that

$$\|x_0 - P_{\mathcal{M}}(x^* - \xi) - \xi\|_{X_1} \le l + C(\delta + \Delta). \tag{10}$$

Then

$$\|x_n - P_{\mathcal{M}}(x^* - \xi) - \xi\|_{X_1} \le lq^n + C(\delta + \Delta). \tag{11}$$

Corollary 5.2. *In the conditions of Theorem 5.2, we have*

$$\limsup_{n\to\infty} \|x_n - P_{\mathcal{M}}(x^* - \xi) - \xi\|_{X_1} \le C(\delta + \Delta),$$

$$\limsup_{n\to\infty} \|x_n - x^*\|_{X_1} \le C(\delta + \Delta) + \Delta.$$

Remark 5.3. Condition (10) is fulfilled if

$$\|x_0 - x^*\|_{X_1} \le l + C\delta.$$

Remark 5.4. Using estimate (11), it is not hard to indicate a number $N = N(\delta, \Delta)$ such that

$$\|x_{N(\delta,\Delta)} - x^*\|_{X_1} \le 2C(\delta + \Delta) + \Delta. \tag{12}$$

From (12) it follows that if $\Delta = 0$, that is, $\xi \in \mathbb{M}(x^*) \equiv x^* + \mathcal{M}$, then the operator

$$\mathfrak{R}_{1/N(\delta,0)}(\widetilde{F}) = x_{N(\delta,0)}$$

defines a regularization algorithm for the problem (1.1). According to (12), this algorithm is stable with respect to small deviations of ξ from the affine subspace $x^* + \mathcal{M}$. Inequality (12) establishes an estimate for

$$\|\mathfrak{R}_{1/N(\delta,\Delta)}(\tilde{F}) - G(F)\|_{X_1}$$

in the limit relation (1.2.8).

5.3   Projection Method for Finding Quasisolutions

In this section we deal with the problem of finding a quasisolution to a nonlinear equation

$$F(x) = 0, \quad x \in X_1, \tag{1}$$

where $F : X_1 \to X_2$ is a smooth operator and X_1, X_2 are Hilbert spaces. A point $x^* \in X_1$ is said to be a quasisolution of equation (1) if x^* is a solution to the variational problem

$$\min_{x \in X_1} \Psi(x), \quad \Psi(x) = \frac{1}{2}\|F(x)\|_{X_2}^2. \tag{2}$$

Suppose x^* is a quasisolution of (1); at the same time we don't assume that the original equation (1) is solvable. Since x^* is a stationary point of the problem (2), we have

$$\Psi'(x^*) = F'^*(x^*)F(x^*) = 0.$$

Assume that $F(x)$ is twice Fréchet differentiable and

$$\|F'(x)\|_{L(X_1, X_2)} \le N_1,$$

$$\|F''(x) - F''(y)\|_{L(X_1, L(X_1, X_2))} \le \Lambda \|x - y\|_{X_1} \quad \forall x, y \in \Omega_R(x^*). \quad (3)$$

We don't impose on $F(x)$ any regularity conditions, so the equation (1) is in general irregular. Suppose instead of the true operator $F(x)$ an approximation $\widetilde{F}(x) \colon X_1 \to X_2$ is accessible. We assume that $\widetilde{F}(x)$ is twice Fréchet differentiable and the derivatives of $\widetilde{F}(x)$ satisfy conditions (3) with the same constants N_1 and Λ. Moreover, let the following error estimates be satisfied:

$$\|\widetilde{F}(x) - F(x)\|_{X_2} \le \delta, \quad \|\widetilde{F}'(x) - F'(x)\|_{L(X_1, X_2)} \le \delta$$
$$\forall x \in \Omega_R(x^*). \quad (4)$$

By (3) it follows that there exists N_2 such that

$$\|F''(x)\|_{L(X_1, L(X_1, X_2))}, \quad \|\widetilde{F}''(x)\|_{L(X_1, L(X_1, X_2))} \le N_2$$

$$\forall x \in \Omega_R(x^*).$$

Let us apply the iteration (2.1) to the problem of finding a quasisolution of (1). Thus we have the iterative process

$$x_0 \in \Omega_R(x^*), \quad x_{n+1} = P_{\mathcal{M}}(x_n - \xi - \gamma \widetilde{F}'^*(x_n)\widetilde{F}(x_n)) + \xi, \quad (5)$$

where

$$0 < \gamma < \frac{2}{N_1^2}.$$

Suppose $\mathcal{M} \subset X_1$ is a finite-dimensional subspace satisfying Condition 5.3, and the parameter $\xi \in X_1$ is chosen subject to Condition 5.4. Below we shall prove that the iterative points x_n, $n \to \infty$ are attracted to a small neighborhood of the quasisolution x^*.

From (5) it follows that

$$x_{n+1} - P_{\mathcal{M}}(x^* - \xi) - \xi =$$
$$= (x_n - P_{\mathcal{M}}(x^* - \xi) - \xi) - \gamma P_{\mathcal{M}}\widetilde{F}'^*(x_n)\widetilde{F}(x_n). \quad (6)$$

Suppose $x_n \in \Omega_R(x^*)$ and $\Delta \le R$; then by (6) we obtain

$$x_{n+1} - P_{\mathcal{M}}(x^* - \xi) - \xi =$$
$$= [E - \gamma P_{\mathcal{M}}\widetilde{F}'^*(x_n)\widetilde{F}'(x_n)P_{\mathcal{M}}](x_n - P_{\mathcal{M}}(x^* - \xi) - \xi) - \quad (7)$$
$$- \gamma P_{\mathcal{M}}\widetilde{F}'^*(x_n)\overline{G}(x_n, x^*) - \gamma P_{\mathcal{M}}\widetilde{F}'^*(x_n)\widetilde{F}(P_{\mathcal{M}}(x^* - \xi) + \xi),$$

where $\overline{G}(x_n, x^*)$ has the same form as in (2.5). Furthermore,

$$\|[E - \gamma P_{\mathcal{M}}\widetilde{F}'^*(x_n)\widetilde{F}'(x_n)P_{\mathcal{M}}](x_n - P_{\mathcal{M}}(x^* - \xi) - \xi)\|_{X_1} \le$$
$$\le \|E - \gamma P_{\mathcal{M}}\widetilde{F}'^*(x_n)\widetilde{F}'(x_n)P_{\mathcal{M}}\|_{L(\mathcal{M})}\|x_n - P_{\mathcal{M}}(x^* - \xi) - \xi\|_{X_1}.$$
(8)

As in Section 5.2, we have

$$\widetilde{\rho} \equiv \inf\{\lambda : \lambda \in \sigma(P_{\mathcal{M}}F'^*(x_0)F'(x_0)P_{\mathcal{M}})\} > 0.$$

Here we need the following analog of Lemma 5.3.

Lemma 5.4. *Suppose Condition 5.3 is satisfied and*

$$\delta < \frac{\widetilde{\rho}}{4N_1}; \quad \|x - x_0\|_{X_1} \le \frac{\widetilde{\rho} - 4N_1\delta}{4N_1N_2}, \quad x \in \Omega_R(x^*).$$
(9)

Then

$$\left(-\infty, \frac{\widetilde{\rho}}{2}\right) \cap \sigma(P_{\mathcal{M}}\widetilde{F}'^*(x)\widetilde{F}'(x)P_{\mathcal{M}}) = \emptyset.$$

From Lemma 5.4 we deduce that for all points $x \in \Omega_R(x^*)$ satisfying conditions (9),

$$\|E - \gamma P_{\mathcal{M}}\widetilde{F}'^*(x)\widetilde{F}'(x)P_{\mathcal{M}}\|_{L(\mathcal{M})} \le \widetilde{q} < 1,$$
(10)

where

$$\widetilde{q} = \sup_{\lambda \in [\widetilde{\rho}/2, N_1^2]} |1 - \gamma\lambda|.$$

Let us denote

$$\mathcal{S}(x)y = (F''(x)y)^*F(x^*), \quad y \in X_1; \quad x \in \Omega_R(x^*).$$
(11)

Suppose $F(x)$ satisfies the following condition.

Condition 5.5. *There exists $\overline{q} \in (\widetilde{q}, 1)$ such that*

$$\sup_{x \in \Omega_R(x^*)} \|\mathcal{S}(x)\|_{L(X_1, X_2)} < \frac{\overline{q} - \widetilde{q}}{\gamma}.$$
(12)

According to (11),

$$\|\mathcal{S}(x)\|_{L(X_1, X_2)} \le \|F''(x)\|_{L(X_1, L(X_1, X_2))}\|F(x^*)\|_{X_2}.$$

Therefore inequality (12) is fulfilled, e.g., in each of the following cases.

1) The discrepancy of equation (1) at the quasisolution is sufficiently small:

$$\|F(x^*)\|_{X_2} < \frac{\overline{q} - \widetilde{q}}{\gamma N_2}. \tag{13}$$

2) The second derivative $F''(x)$ is bounded from above:

$$\sup_{x \in \Omega_R(x^*)} \|F''(x)\|_{L(X_1, L(X_1, X_2))} < \frac{\overline{q} - \widetilde{q}}{\gamma \|F(x^*)\|_{X_2}}.$$

Using (1.1.14), (2.3), (3), and (4), we can estimate the last summand in (7) as follows:

$$\|P_{\mathcal{M}} \widetilde{F}'^*(x_n) \widetilde{F}(P_{\mathcal{M}}(x^* - \xi) + \xi)\|_{X_1} \leq$$
$$\leq \|P_{\mathcal{M}} \widetilde{F}'^*(x_n)[\widetilde{F}(P_{\mathcal{M}}(x^* - \xi) + \xi) -$$
$$- \widetilde{F}(x^*) - \widetilde{F}'(P_{\mathcal{M}}(x^* - \xi) + \xi)(P_{\mathcal{M}}(x^* - \xi) + \xi - x^*)]\|_{X_1} +$$
$$+ \|P_{\mathcal{M}} \widetilde{F}'^*(x_n) \widetilde{F}'(P_{\mathcal{M}}(x^* - \xi) + \xi)(P_{\mathcal{M}}(x^* - \xi) + \xi - x^*)\|_{X_1} +$$
$$+ \|P_{\mathcal{M}} \widetilde{F}'^*(x_n) \widetilde{F}(x^*)\|_{X_1} \leq$$
$$\leq \frac{1}{2} N_1 N_2 \Delta^2 + N_1^2 \Delta +$$
$$+ \|P_{\mathcal{M}} \widetilde{F}'^*(x_n)(\widetilde{F}(x^*) - F(x^*))\|_{X_1} +$$
$$+ \|P_{\mathcal{M}}(\widetilde{F}'(x_n) - F'(x_n))^* F(x^*)\|_{X_1} +$$
$$+ \|P_{\mathcal{M}}[F'(x_n) - F'(P_{\mathcal{M}}(x^* - \xi) + \xi) -$$
$$- F''(x_n)(x_n - P_{\mathcal{M}}(x^* - \xi) - \xi)]^* F(x^*)\|_{X_1} +$$
$$+ \|P_{\mathcal{M}} F'^*(P_{\mathcal{M}}(x^* - \xi) + \xi) F(x^*)\|_{X_1} +$$
$$+ \|P_{\mathcal{M}}[F''(x_n)(x_n - P_{\mathcal{M}}(x^* - \xi) - \xi)]^* F(x^*)\|_{X_1}.$$

Hence,

$$\|P_{\mathcal{M}} \widetilde{F}'^*(x_n) \widetilde{F}(P_{\mathcal{M}}(x^* - \xi) + \xi)\|_{X_1} \leq$$
$$\leq \frac{1}{2} N_1 N_2 \Delta^2 + N_1^2 \Delta + (N_1 + \|F(x^*)\|_{X_2})\delta +$$
$$+ \frac{1}{2} \Lambda \|x_n - P_{\mathcal{M}}(x^* - \xi) - \xi\|_{X_1}^2 \|F(x^*)\|_{X_2} + N_2 \|F(x^*)\|_{X_2} \Delta + \tag{14}$$
$$+ \|[F''(x_n)(x_n - P_{\mathcal{M}}(x^* - \xi) - \xi)]^* F(x^*)\|_{X_1}.$$

Combining (7), (8), (10)–(12), and (14), we finally obtain an inequality analogous to (2.9):

$$\|x_{n+1} - P_{\mathcal{M}}(x^* - \xi) - \xi\|_{X_1} \leq \overline{q}\|x_n - P_{\mathcal{M}}(x^* - \xi) - \xi\|_{X_1} +$$
$$+ \gamma M_1 \|x_n - P_{\mathcal{M}}(x^* - \xi) - \xi\|_{X_1}^2 + \gamma \max\{M_2(\Delta), M_3\}(\delta + \Delta). \tag{15}$$

Here, we have introduced the notation

$$M_1 = N_1 N_2 + \frac{1}{2}\Lambda\|F(x^*)\|_{X_2}, \quad M_2(\Delta) = N_1^2 + \frac{1}{2}N_1 N_2 \Delta + N_2\|F(x^*)\|_{X_2},$$

$$M_3 = N_1 + \|F(x^*)\|_{X_2}.$$

As in Section 5.1, from (15) we get the following theorem concerning asymptotic properties of the iteration (5).

Theorem 5.3. *Assume that Conditions 5.3–5.5 and inequality (1.16) are satisfied. Suppose $q, l, C > 0$ are chosen such that*

$$\overline{q} < q < 1, \quad 0 < l \le \min\left\{\frac{R}{2}, \frac{q-\overline{q}}{2\gamma q M_1}, \frac{\widetilde{\rho} - 4N_1 \delta}{8(1+q)N_1 N_2}\right\},$$

$$C \ge \max\left\{1, \frac{\gamma}{1-q}\max\{M_2(\Delta), M_3\}\right\},$$

and the error levels δ and Δ satisfy

$$\delta < \frac{\widetilde{\rho}}{4N_1}, \quad C(\delta + \Delta) + \Delta \le \min\left\{\frac{R}{2}, \frac{\widetilde{\rho} - 4N_1 \delta}{16 N_1 N_2}\right\},$$

$$\delta + \Delta \le \frac{q-\overline{q}}{4\gamma C M_1}.$$

Also, assume that

$$\|x_0 - x^*\|_{X_1} \le l + C(\delta + \Delta) - \Delta. \tag{16}$$

Then

$$\|x_n - x^*\|_{X_1} \le lq^n + C(\delta + \Delta) + \Delta,$$

$$\limsup_{n\to\infty}\|x_n - x^*\|_{X_1} \le C(\delta + \Delta) + \Delta.$$

Remark 5.5. Inequality (16) is true if

$$\|x_0 - x^*\|_{X_1} \le l + C\delta.$$

Remark 5.6. Suppose (13) is fulfilled and $F(x)$ instead of (3) satisfies the condition

$$\|F''(x)\|_{L(X_1, L(X_1, X_2))} \le N_2 \quad \forall x \in \Omega_R(x^*);$$

then estimate (15) and Theorem 5.3 remain valid. This follows by the chain of inequalities

$$\|P_{\mathcal{M}}\widetilde{F}'^*(x_n)\widetilde{F}(x^*)\|_{X_1} \leq$$
$$\leq \|(\widetilde{F}'^*(x_n) - F'^*(x_n))F(x^*)\|_{X_1} + \|F'^*(x_n)F(x^*)\|_{X_1} +$$
$$+ \|\widetilde{F}'^*(x_n)(\widetilde{F}(x^*) - F(x^*))\|_{X_1} \leq (N_1 + \|F(x^*)\|_{X_2})\delta +$$
$$+ \|(F'^*(x_n) - F'^*(P_{\mathcal{M}}(x^* - \xi) + \xi))F(x^*)\|_{X_1} +$$
$$+ \|(F'^*(P_{\mathcal{M}}(x^* - \xi) + \xi) - F'^*(x^*))F(x^*)\|_{X_1} \leq$$
$$\leq (N_1 + \|F(x^*)\|_{X_2})\delta + N_2\|F(x^*)\|_{X_2}\Delta +$$
$$+ N_2\|x_n - P_{\mathcal{M}}(x^* - \xi) - \xi\|_{X_1}\|F(x^*)\|_{X_2}.$$

5.4 Stable Methods on the Basis of Parametric Approximations

This section is devoted to application of the technique developed in Chapters 3 and 4 to construction of stable iterative processes for irregular equations. Consider an equation

$$F(x) = 0, \quad x \in X_1, \tag{1}$$

where $F : X_1 \rightarrow X_2$ is a nonlinear operator and X_1, X_2 are Hilbert spaces. Let x^* be a solution of (1). Suppose $F(x)$ is a Fréchet differentiable operator and the derivative $F'(x)$ satisfies the Lipschitz condition (4.1.2). Assume that instead of the true operator $F(x)$ an approximation $\widetilde{F} : X_1 \rightarrow X_2$ is given. Suppose $\widetilde{F}(x)$ satisfies (4.1.2) and the inequalities

$$\|\widetilde{F}(x^*)\|_{X_2} \leq \delta, \tag{2}$$

$$\|\widetilde{F}'(x) - F'(x)\|_{L(X_1,X_2)} \leq \delta \quad \forall x \in \Omega_R(x^*). \tag{3}$$

We know that there exists N_1 such that

$$\|F'(x)\|_{L(X_1,X_2)}, \quad \|\widetilde{F}'(x)\|_{L(X_1,X_2)} \leq N_1 \quad \forall x \in \Omega_R(x^*).$$

Let $Q \subset X_1$ be a closed convex subset containing the solution x^*. If no a priori information concerning x^* is available, we can simply set $Q = X_1$. Below we study the following family of iterative processes for equation (1) (compare with (4.1.35)):

$$x_0 \in \Omega_R(x^*), \quad x_{n+1} = P_Q\{\xi - \Theta(\widetilde{F}'^*(x_n)\widetilde{F}'(x_n), \alpha_0)\widetilde{F}'^*(x_n) \cdot$$
$$\cdot (\widetilde{F}(x_n) - \widetilde{F}'(x_n)(x_n - \xi))\}. \tag{4}$$

Here $\xi \in X_1$; $\alpha_0 > 0$ is the regularization parameter. Unlike the processes of Chapter 4, the regularization parameter α_0 in (4) is fixed and doesn't vanish as $n \to \infty$. The generating function $\Theta(\lambda, \alpha)$ is assumed to be analytic in λ on an open neighborhood of the segment $[0, N_1^2]$. Also, suppose $\Theta(\lambda, \alpha)$ satisfies Conditions 4.2, 4.3, and 4.5. Let a family of contours $\{\Gamma_\alpha\}$ satisfy Condition 4.4.

Below, under appropriate conditions on the parameter $\xi \in X_1$ and the initial approximation x_0 we shall prove that iterative points x_n are attracted to a neighborhood of the solution of (1).

Let ξ satisfy the perturbed sourcewise representation condition (4.1.29). Assume that the current point $x_n \in \Omega_R(x^*)$. Using (1.1.16), from (4) we obtain

$$
\begin{aligned}
\|x_{n+1} - x^*\|_{X_1} &\leq \|P_Q\{\xi - \Theta(\widetilde{F}'^*(x_n)\widetilde{F}'(x_n), \alpha_0)\widetilde{F}'^*(x_n)\cdot \\
&\quad \cdot (\widetilde{F}(x_n) - \widetilde{F}'(x_n)(x_n - \xi))\} - P_Q(x^*)\|_{X_1} \leq \\
&\leq \|\xi - \Theta(\widetilde{F}'^*(x_n)\widetilde{F}'(x_n), \alpha_0)\widetilde{F}'^*(x_n)\cdot \\
&\quad \cdot (\widetilde{F}(x_n) - \widetilde{F}'(x_n)(x_n - \xi)) - x^*\|_{X_1}.
\end{aligned} \tag{5}
$$

Observe that

$$
\widetilde{F}(x_n) = \widetilde{F}'(x_n)(x_n - x^*) + \widetilde{G}(x_n, x^*), \tag{6}
$$

where in view of (2),

$$
\begin{aligned}
\|\widetilde{G}(x_n, x^*)\|_{X_2} &\leq \|\widetilde{F}(x_n) - \widetilde{F}(x^*) - \widetilde{F}'(x_n)(x_n - x^*)\|_{X_2} + \\
&\quad + \|\widetilde{F}(x^*)\|_{X_2} \leq \frac{1}{2}L\|x_n - x^*\|_{X_1}^2 + \delta.
\end{aligned} \tag{7}
$$

From (6), (7), and (4.1.29) we deduce

$$
\begin{aligned}
\|\xi - \Theta(\widetilde{F}'^*(x_n)\widetilde{F}'(x_n), \alpha_0)&\widetilde{F}'^*(x_n)(\widetilde{F}(x_n) - \widetilde{F}'(x_n)(x_n - \xi)) - x^*\|_{X_1} \leq \\
&\leq \|\Theta(\widetilde{F}'^*(x_n)\widetilde{F}'(x_n), \alpha_0)\widetilde{F}'^*(x_n)\widetilde{G}(x_n, x^*)\|_{X_1} + \\
&+ \|[E - \Theta(\widetilde{F}'^*(x_n)\widetilde{F}'(x_n), \alpha_0)\widetilde{F}'^*(x_n)\widetilde{F}'(x_n)](x^* - \xi)\|_{X_1} \leq \\
&\leq \|\Theta(\widetilde{F}'^*(x_n)\widetilde{F}'(x_n), \alpha_0)\widetilde{F}'^*(x_n)\|_{L(X_2, X_1)}\left(\frac{1}{2}L\|x_n - x^*\|_{X_1}^2 + \delta\right) + \\
&\quad + \|[E - \Theta(\widetilde{F}'^*(x_n)\widetilde{F}'(x_n), \alpha_0)\widetilde{F}'^*(x_n)\widetilde{F}'(x_n)]\cdot \\
&\qquad\qquad\qquad \cdot ((F'^*(x^*)F'(x^*))^p\widetilde{v} + \widetilde{w})\|_{X_1}.
\end{aligned} \tag{8}
$$

Combining (5) and (8), with the use of Conditions 4.2, 4.3 and Remark 4.1 we get

$$\|x_{n+1} - x^*\|_{X_1} \leq \frac{c_1}{\sqrt{\alpha_0}}\left(\frac{1}{2}L\|x_n - x^*\|_{X_1}^2 + \delta\right) + c_2\Delta +$$
$$+ \|[E - \Theta(\widetilde{F}'^*(x_n)\widetilde{F}'(x_n), \alpha_0)\widetilde{F}'^*(x_n)\widetilde{F}'(x_n)](F'^*(x^*)F'(x^*))^p\widetilde{v}\|_{X_1}.$$
(9)

Condition 4.3 allows to estimate the last norm in (9) as follows:

$$\|[E - \Theta(\widetilde{F}'^*(x_n)\widetilde{F}'(x_n), \alpha_0)\widetilde{F}'^*(x_n)\widetilde{F}'(x_n)](F'^*(x^*)F'(x^*))^p\widetilde{v}\|_{X_1} =$$
$$= \|[E - \Theta(F'^*(x^*)F'(x^*), \alpha_0)F'^*(x^*)F'(x^*)](F'^*(x^*)F'(x^*))^p\widetilde{v}\|_{X_1} +$$
$$+ \|[\Theta(F'^*(x^*)F'(x^*), \alpha_0)F'^*(x^*)F'(x^*) -$$
$$- \Theta(\widetilde{F}'^*(x_n)\widetilde{F}'(x_n), \alpha_0)\widetilde{F}'^*(x_n)\widetilde{F}'(x_n)](F'^*(x^*)F'(x^*))^p\widetilde{v}\|_{X_1} \leq$$
$$\leq c_3\alpha_0^p\|\widetilde{v}\|_{X_1} + \|[\Theta(F'^*(x^*)F'(x^*), \alpha_0)F'^*(x^*)F'(x^*) -$$
$$- \Theta(\widetilde{F}'^*(x_n)\widetilde{F}'(x_n), \alpha_0)\widetilde{F}'^*(x_n)\widetilde{F}'(x_n)](F'^*(x^*)F'(x^*))^p\widetilde{v}\|_{X_1}.$$
(10)

Application of (1.1.6) yields

$$[\Theta(F'^*(x^*)F'(x^*), \alpha_0)F'^*(x^*)F'(x^*) -$$
$$- \Theta(\widetilde{F}'^*(x_n)\widetilde{F}'(x_n), \alpha_0)\widetilde{F}'^*(x_n)\widetilde{F}'(x_n)](F'^*(x^*)F'(x^*))^p\widetilde{v} =$$
$$= \Big([\Theta(F'^*(x^*)F'(x^*), \alpha_0)F'^*(x^*)F'(x^*) - E] -$$
$$- [\Theta(\widetilde{F}'^*(x_n)\widetilde{F}'(x_n), \alpha_0)\widetilde{F}'^*(x_n)\widetilde{F}'(x_n) - E]\Big)(F'^*(x^*)F'(x^*))^p\widetilde{v} =$$
$$= \frac{1}{2\pi i}\int_{\Gamma_{\alpha_0}}(\Theta(\lambda, \alpha_0)\lambda - 1)[R(\lambda, F'^*(x^*)F'(x^*)) -$$
$$- R(\lambda, \widetilde{F}'^*(x_n)\widetilde{F}'(x_n))](F'^*(x^*)F'(x^*))^p\widetilde{v}d\lambda.$$
(11)

From (11) it follows that

$$\|[\Theta(F'^*(x^*)F'(x^*), \alpha_0)F'^*(x^*)F'(x^*) -$$
$$- \Theta(\widetilde{F}'^*(x_n)\widetilde{F}'(x_n), \alpha_0)\widetilde{F}'^*(x_n)\widetilde{F}'(x_n)](F'^*(x^*)F'(x^*))^p\widetilde{v}\|_{X_1} \leq$$
$$\leq \frac{1}{2\pi}\int_{\Gamma_{\alpha_0}}|1 - \Theta(\lambda, \alpha_0)\lambda| \|[R(\lambda, F'^*(x^*)F'(x^*)) - R(\lambda, \widetilde{F}'^*(x_n)\widetilde{F}'(x_n))] \cdot$$
$$\cdot (F'^*(x^*)F'(x^*))^p\|_{L(X_1)}\|\widetilde{v}\|_{X_1}|d\lambda|.$$
(12)

Taking into account (3), we get the estimate

$$\|[R(\lambda, F'^*(x^*)F'(x^*)) -$$
$$- R(\lambda, \widetilde{F}'^*(x_n)\widetilde{F}'(x_n))](F'^*(x^*)F'(x^*))^p\|_{L(X_1)} \le$$
$$\le \frac{c_4(L\|x_n - x^*\|_{X_1} + \delta)}{|\lambda|} \quad \forall \lambda \in \Gamma_{\alpha_0}.$$

Then from (12) it follows that

$$\|[\Theta(F'^*(x^*)F'(x^*), \alpha_0)F'^*(x^*)F'(x^*) -$$
$$- \Theta(\widetilde{F}'^*(x_n)\widetilde{F}'(x_n), \alpha_0)\widetilde{F}'^*(x_n)\widetilde{F}'(x_n)](F'^*(x^*)F'(x^*))^p\widetilde{v}\|_{X_1} \le$$
$$\le c_5(L\|x_n - x^*\|_{X_1} + \delta)\|\widetilde{v}\|_{X_1} \int_{\Gamma_{\alpha_0}} \frac{|1 - \Theta(\lambda, \alpha_0)\lambda|}{|\lambda|}|d\lambda|. \tag{13}$$

By (13) and Condition 4.5 we deduce

$$\|[\Theta(F'^*(x^*)F'(x^*), \alpha_0)F'^*(x^*)F'(x^*) -$$
$$- \Theta(\widetilde{F}'^*(x_n)\widetilde{F}'(x_n), \alpha_0)\widetilde{F}'^*(x_n)\widetilde{F}'(x_n)](F'^*(x^*)F'(x^*))^p\widetilde{v}\|_{X_1} \le \tag{14}$$
$$\le c_6(L\|x_n - x^*\|_{X_1} + \delta)\|\widetilde{v}\|_{X_1}.$$

Combining estimates (10) and (14), from (9) we obtain

$$\|x_{n+1} - x^*\|_{X_1} \le \frac{c_1}{\sqrt{\alpha_0}}\left(\frac{1}{2}L\|x_n - x^*\|_{X_1}^2 + \delta\right) + c_3\|\widetilde{v}\|_{X_1}\alpha_0^p +$$
$$+ c_6(L\|x_n - x^*\|_{X_1} + \delta)\|\widetilde{v}\|_{X_1} + c_2\Delta. \tag{15}$$

We now intend to establish the following rate of convergence estimate for the iteration (4):

$$\|x_n - x^*\|_{X_1} \le lq^n\sqrt{\alpha_0} + m\left(\frac{c_1\delta}{\sqrt{\alpha_0}} + c_3\|\widetilde{v}\|_{X_1}\alpha_0^p + c_6\|\widetilde{v}\|_{X_1}\delta + c_2\Delta\right); \tag{16}$$

$$m, l > 0, \quad 0 < q < 1.$$

Suppose (16) is valid for $n = 0$ and for a number $n \in \mathbf{N}$. To prove (16) by induction it suffices to show that

$$\frac{c_1 L}{2\sqrt{\alpha_0}} \left[lq^n \sqrt{\alpha_0} + m\left(\frac{c_1\delta}{\sqrt{\alpha_0}} + c_3\|\tilde{v}\|_{X_1}\alpha_0^p + c_6\|\tilde{v}\|_{X_1}\delta + c_2\Delta \right) \right]^2 +$$

$$+ c_6 L\|\tilde{v}\|_{X_1} \left[lq^n \sqrt{\alpha_0} + m\left(\frac{c_1\delta}{\sqrt{\alpha_0}} + c_3\|\tilde{v}\|_{X_1}\alpha_0^p + c_6\|\tilde{v}\|_{X_1}\delta + c_2\Delta \right) \right] +$$

$$+ \left(\frac{c_1\delta}{\sqrt{\alpha_0}} + c_3\|\tilde{v}\|_{X_1}\alpha_0^p + c_6\|\tilde{v}\|_{X_1}\delta + c_2\Delta \right) \le$$

$$\le lq^{n+1}\sqrt{\alpha_0} + m\left(\frac{c_1\delta}{\sqrt{\alpha_0}} + c_3\|\tilde{v}\|_{X_1}\alpha_0^p + c_6\|\tilde{v}\|_{X_1}\delta + c_2\Delta \right). \tag{17}$$

Indeed, (15) and (17) imply

$$\|x_{n+1} - x^*\|_{X_1} \le$$

$$\le lq^{n+1}\sqrt{\alpha_0} + m\left(\frac{c_1\delta}{\sqrt{\alpha_0}} + c_3\|\tilde{v}\|_{X_1}\alpha_0^p + c_6\|\tilde{v}\|_{X_1}\delta + c_2\Delta \right). \tag{18}$$

Inequality (17) is satisfied if

$$\frac{1}{2}c_1 Llq + c_6 L\|\tilde{v}\|_{X_1} \le q, \tag{19}$$

$$c_1 Llmq + \frac{c_1 Lm^2}{2\sqrt{\alpha_0}}\left(\frac{c_1\delta}{\sqrt{\alpha_0}} + c_3\|\tilde{v}\|_{X_1}\alpha_0^p + c_6\|\tilde{v}\|_{X_1}\delta + c_2\Delta \right) +$$

$$+ c_6 Lm\|\tilde{v}\|_{X_1} + 1 \le m. \tag{20}$$

Moreover, from (18) we conclude that the conditions

$$0 < l \le \frac{R}{2\sqrt{\alpha_0}}, \quad \|\tilde{v}\|_{X_1} \le \frac{R}{4c_3 m\alpha_0^p},$$

$$\frac{c_1\delta}{2\sqrt{\alpha_0}} + c_6\|\tilde{v}\|_{X_1}\delta + c_2\Delta \le \frac{R}{4m} \tag{21}$$

imply $x_{n+1} \in \Omega_R(x^*)$.

Let us now fix $q \in (0, 1/3)$ and derive conditions on l, m, $\|\tilde{v}\|_{X_1}$, δ, and Δ under which (19)–(21) are fulfilled. Condition (19) is fulfilled if

$$c_1 Ll \le 1, \quad \|\tilde{v}\|_{X_1} \le \frac{q}{2c_6 L}. \tag{22}$$

Next, from (22) we deduce that (20) is valid if

$$m = \frac{2}{1 - 3q},$$

$$c_1 Lm\left(\frac{c_1\delta}{\sqrt{\alpha_0}} + c_3\|\tilde{v}\|_{X_1}\alpha_0^p + c_6\|\tilde{v}\|_{X_1}\delta + c_2\Delta \right) \le \sqrt{\alpha_0}. \tag{23}$$

At last, (23) follows by the inequalities

$$\|\tilde{v}\|_{X_1} \le \frac{1}{2c_1c_3Lma_0^{p-1/2}}, \quad \frac{c_1\delta}{\sqrt{\alpha_0}} + c_6\|\tilde{v}\|_{X_1}\delta + c_2\Delta \le \frac{\sqrt{\alpha_0}}{2c_1Lm}. \quad (24)$$

Taking into account the first conditions in (21) and (22), we set

$$l = \min\left\{\frac{1}{c_1L}, \frac{R}{2\sqrt{\alpha_0}}\right\}. \quad (25)$$

According to (21), (22), and (24), we impose on $\tilde{v}$ the condition

$$\|\tilde{v}\|_{X_1} \le d = \min\left\{\frac{R}{4c_3ma_0^p}, \frac{q}{2c_6L}, \frac{1}{2c_1c_3Lma_0^{p-1/2}}\right\}. \quad (26)$$

From (21) and (24) it follows that δ and Δ must satisfy the inequality

$$\frac{c_1\delta}{\sqrt{\alpha_0}} + c_6\|\tilde{v}\|_{X_1}\delta + c_2\Delta \le \min\left\{\frac{R}{4m}, \frac{\sqrt{\alpha_0}}{2c_1Lm}\right\}. \quad (27)$$

Thus we proved that if

$$\|x_0 - x^*\|_{X_1} \le l\sqrt{\alpha_0} + m\left(\frac{c_1\delta}{\sqrt{\alpha_0}} + c_3\|\tilde{v}\|_{X_1}a_0^p + c_6\|\tilde{v}\|_{X_1}\delta + c_2\Delta\right) \quad (28)$$

and conditions (26) and (27) are satisfied, then (17) is valid. The next theorem summarizes the obtained results.

Theorem 5.4. *Suppose Conditions 4.2–4.5 and (26)–(28) are satisfied. Then estimate (16) is fulfilled. Moreover,*

$$\limsup_{n\to\infty} \|x_n - x^*\|_{X_1} \le m\left(\frac{c_1\delta}{\sqrt{\alpha_0}} + c_3\|\tilde{v}\|_{X_1}a_0^p + c_6\|\tilde{v}\|_{X_1}\delta + c_2\Delta\right). \quad (29)$$

From (25) and (26) we conclude that the values l and d are separated from zero and m is bounded from above for all sufficiently small $\alpha_0 > 0$. According to (28), the initial point x_0 can be chosen subject to the condition

$$\|x_0 - x^*\|_{X_1} \le l\sqrt{\alpha_0}.$$

Suppose $\delta + \Delta = O(\alpha_0^p)$; then from (29) we obtain

$$\limsup_{n\to\infty} \|x_n - x^*\|_{X_1} = O(\alpha_0^{p-1/2}),$$

where $\alpha_0^{p-1/2}$ is much less than $\sqrt{\alpha_0}$ if $p > 1$ and $\alpha_0 \to 0$. Therefore the iterative points x_n are attracted to a neighborhood of diameter $O(\alpha_0^{p-1/2})$, while the initial point x_0 can be taken from a larger neighborhood of diameter $O(\sqrt{\alpha_0})$ centered at x^*. Consequently the iterative process (4) is stable with respect to variations of x_0 and to errors in $F(x)$ and in the sourcewise representation of $x^* - \xi$. For this reason, the scheme (4) is free of the problem of stopping criterions.

Remark 5.7. The preceding arguing can be extended to the case where $x^* \notin Q$. In this case, instead of (5) we obtain

$$\|x_{n+1} - x^*\|_{X_1} \leq \|\xi - \Theta(\widetilde{F}'^*(x_n)\widetilde{F}'(x_n), \alpha_0)\widetilde{F}'^*(x_n) \cdot$$
$$\cdot (\widetilde{F}(x_n) - \widetilde{F}'(x_n)(x_n - \xi)) - x^*\|_{X_1} + \operatorname{dist}(x^*, Q).$$

Therefore (15) takes the form

$$\|x_{n+1} - x^*\|_{X_1} \leq \frac{c_1}{\sqrt{\alpha_0}}\left(\frac{1}{2}L\|x_n - x^*\|_{X_1}^2 + \delta\right) + c_3\|\widetilde{v}\|_{X_1}\alpha_0^p +$$
$$+ c_6(L\|x_n - x^*\|_{X_1} + \delta)\|\widetilde{v}\|_{X_1} + c_2\Delta + \operatorname{dist}(x^*, Q).$$

Arguing as above, we obtain a result analogous to Theorem 5.4. It is sufficient to replace the value Δ in (16), (27), and (29) by the sum

$$\Delta + c_2^{-1}\operatorname{dist}(x^*, Q).$$

5.5 Stable Continuous Approximations and Attractors of Dynamical Systems in Hilbert Space

In this section we propose and study a class of continuous methods for approximation of solutions to smooth irregular operator equations in a Hilbert space. The methods under consideration are continuous analogs of the iterative processes of Section 5.4. Here, a solution of the equation is approximated by a trajectory of an appropriate dynamical system. We prove that the trajectory is attracted to a small ball centered at the solution.

We consider a nonlinear equation

$$F(x) = 0, \quad x \in X_1. \tag{1}$$

Suppose $F(x)$ and an approximation $\widetilde{F}(x)$ satisfy conditions of Section 5.4, namely, the Lipschitz condition (4.1.2), the inequalities (4.2) and (4.3) and the sourcewise representation (4.1.29). Let x^* be a solution of (1).

A popular and widely accepted approach to solution methods for equations (1) consists in constructing on a phase space X_1 a dynamical system for which

x^* is a stable stationary point. In more exact terms, it is required to find a dynamical system

$$\dot{x} = \Phi(x), \quad x = x(t) \in X_1, \quad t \geq 0 \quad (\Phi : X_1 \to X_1) \qquad (2)$$

such that $\Phi(x^*) = 0$, and $\lim\limits_{t \to +\infty} \|x(t) - x^*\|_{X_1} = 0$ at least for initial states $x(0) = x_0$ from a neighborhood of x^*. Then, the element $x(t), t \to \infty$ can be considered as an approximation of x^*. In the regular case, where $F'(x)$ or $F'^*(x)F'(x)$ is continuously invertible, as a dynamical system (2) we can take, e.g., continuous analogs of the classical Newton–Kantorovich and Gauss–Newton iterative schemes (3.1.9) and (3.1.10):

$$\dot{x} = -(F'(x))^{-1}F(x), \quad \dot{x} = -(F'^*(x)F'(x))^{-1}F'^*(x)F(x). \qquad (3)$$

However when equation (1) is irregular, the schemes (3) are not even implementable for the reasons discussed in Section 3.1. In this situation, the iteration (4.4) can serve as a model for implementable continuous approximation schemes applicable to arbitrary smooth nonlinear equations.

Below we study the following continuous analog of the process (4.4), in which for simplicity we set $Q = X_1$:

$$\dot{x} = \xi - \Theta(\widetilde{F}'^*(x)\widetilde{F}'(x), \alpha_0)\widetilde{F}'^*(x)(\widetilde{F}(x) - \widetilde{F}'(x)(x - \xi)) - x, \qquad (4)$$

$$x(0) = x_0 \in \Omega_R^0(x^*). \qquad (5)$$

Here x_0 is an initial approximation to x^*; $\alpha_0 > 0$. Suppose $\Theta(\lambda, \alpha)$ satisfies all the conditions of Section 5.4. Below it will be shown that a trajectory $x = x(t)$ of the dynamical system (4) is defined for all $t \geq 0$. Also, we shall establish the following continuous analog of estimate (4.29):

$$\limsup_{t \to +\infty} \|x(t) - x^*\|_{X_1} < c_1\left(\frac{\delta}{\sqrt{\alpha_0}} + \|\widetilde{v}\|_{X_1}\alpha_0^p + \|\widetilde{v}\|_{X_1}\delta + \Delta\right). \qquad (6)$$

Arguing as in Section 4.5, we conclude that there exists a constant $\widetilde{T} = \widetilde{T}(x_0) > 0$ such that the Cauchy problem (4)–(5) has a unique solution $x = x(t)$ defined on the segment $[0, \widetilde{T}]$. Since $x_0 \in \Omega_R^0(x^*)$, we can assume without loss of generality that $x(t) \in \Omega_R(x^*)$ for all $t \in [0, \widetilde{T}]$.

Let us denote

$$u(t) = \frac{1}{2}\|x(t) - x^*\|_{X_1}^2.$$

If we repeat with $\alpha(t) \equiv \alpha_0$ the arguing of Section 4.5, then we obtain the following proposition.

Theorem 5.5. *Suppose Conditions 4.2–4.5 are satisfied. Let the sourcewise representation (4.1.29) be fulfilled. Then for all $t \in [0, \widetilde{T}]$,*

$$\dot{u}(t) \leq -u(t) + \frac{c_2}{\alpha_0} u^2(t) + c_3 \left(\frac{\delta^2}{\alpha_0} + \|\tilde{v}\|_{X_1}^2 \alpha_0^{2p} + (\|\tilde{v}\|_{X_1} \delta + \Delta)^2 \right). \quad (7)$$

Let us turn to asymptotic properties of trajectories $x = x(t)$ of the system (4). To establish existence and uniqueness of a solution $x(t)$ to (4)–(5) for all $t \geq 0$, consider for (7) the majorizing equation

$$\dot{y}(t) = -y(t) + \frac{c_2}{\alpha_0} y^2(t) +$$
$$+ c_3 \left(\frac{\delta^2}{\alpha_0} + \|\tilde{v}\|_{X_1}^2 \alpha_0^{2p} + (\|\tilde{v}\|_{X_1} \delta + \Delta)^2 \right), \quad t \geq 0 \quad (8)$$

with the initial condition

$$y(0) = u(0). \quad (9)$$

By the lemma on differential inequalities ([93]),

$$u(t) \leq y(t), \quad 0 \leq t \leq \widetilde{T}. \quad (10)$$

Suppose

$$\frac{\delta^2}{\alpha_0^2} + \frac{(\|\tilde{v}\|_{X_1} \delta + \Delta)^2}{\alpha_0} + \|\tilde{v}\|_{X_1}^2 \alpha_0^{2p-1} < \frac{1}{4c_2 c_3}; \quad (11)$$

then as in Section 4.5 we conclude that the equation

$$-y + \frac{c_2}{\alpha_0} y^2 + c_3 \left(\frac{\delta^2}{\alpha_0} + \|\tilde{v}\|_{X_1}^2 \alpha_0^{2p} + (\|\tilde{v}\|_{X_1} \delta + \Delta)^2 \right) = 0$$

has two positive roots $y_1 < y_2$, where

$$y_1 < 2c_3 \left(\frac{\delta^2}{\alpha_0} + \|\tilde{v}\|_{X_1}^2 \alpha_0^{2p} + (\|\tilde{v}\|_{X_1} \delta + \Delta)^2 \right), \quad (12)$$

$$y_2 > \frac{\alpha_0}{2c_2}. \quad (13)$$

It is easily checked that each solution $y = y(t), t \geq 0$ of equation (8) with an initial state $y(0) \in (y_1, y_2)$ decreases and, moreover,

$$\lim_{t \to +\infty} y(t) = y_1. \quad (14)$$

On the other hand, if $y(0) \in [0, y_1]$ then $y(t), t \geq 0$ is a nondecreasing function, and again $y(t)$ satisfies (14). Thus (14) is valid for each solution $y(t)$ to (8) with $y(0) \in [0, y_2)$. Next, suppose

$$\|x_0 - x^*\|_{X_1} \leq \sqrt{\frac{\alpha_0}{c_2}}, \quad \alpha_0 < c_2 R^2. \tag{15}$$

From (9), (10), and (13) it follows that in the case where

$$y_1 < u(0) = \frac{1}{2}\|x_0 - x^*\|_{X_1}^2,$$

the conditions (15) imply

$$\|x(t) - x^*\|_{X_1} = \sqrt{2u(t)} \leq \sqrt{2y(t)} \leq \sqrt{2y(0)} =$$
$$= \sqrt{2u(0)} = \|x_0 - x^*\|_{X_1}, \quad 0 \leq t \leq \tilde{T}.$$

If $0 \leq u(0) \leq y_1$, then again we have

$$\|x(t) - x^*\|_{X_1} \leq \sqrt{2y(t)}, \quad 0 \leq t \leq \tilde{T}.$$

Since in this case the function $y(t)$ is nondecreasing, from (11) we get

$$y(t) \leq \lim_{t \to +\infty} y(t) = y_1 <$$
$$< 2c_3 \left(\frac{\delta^2}{\alpha_0} + \|\tilde{v}\|_{X_1}^2 \alpha_0^{2p} + (\|\tilde{v}\|_{X_1}\delta + \Delta)^2 \right) < \frac{\alpha_0}{2c_2},$$
$$\|x(t) - x^*\|_{X_1} < \sqrt{\frac{\alpha_0}{c_2}}, \quad 0 \leq t \leq \tilde{T}.$$

Therefore each trajectory $x = x(t)$ with

$$x(0) = x_0 \in \Omega_{\sqrt{\alpha_0/c_2}}(x^*)$$

remains in the ball $\Omega_{\sqrt{\alpha_0/c_2}}(x^*)$ for all $0 \leq t \leq \tilde{T}$. Let $\tilde{T}^* = \tilde{T}^*(x_0) \geq \tilde{T}$ be the full time of existence of the solution $x = x(t)$ to (4)–(5), that is, $x(t)$ is uniquely determined for $t \in [0, \tilde{T}^*)$ but $x(t)$ can't be uniquely continued forward by time through $t = \tilde{T}^*$. Assume that $\tilde{T}^* < \infty$. Then we have

$$\sup\{\|x(t) - x^*\|_{X_1} : t \in [0, \tilde{T}^*)\} < R \tag{16}$$

and hence

$$\sup\{\|\dot{x}(t)\|_{X_1} : t \in [0, T^*)\} < \infty.$$

As in Section 4.5, from (16) we deduce that there exists $\tilde{x} \in \Omega_R^0(x^*)$ such that

$$\lim_{t \to \tilde{T}^* - 0} \|x(t) - \tilde{x}\|_{X_1} = 0.$$

Therefore the solution $x(t)$ can be continued forward by time through $t = \tilde{T}^*$, and we conclude that $\tilde{T}^* = \infty$ ([55]). Thus we come to the following proposition.

Theorem 5.6. *Suppose Conditions 4.2–4.5 and inequalities (11), (15) are satisfied. Assume that the sourcewise representation (4.1.29) is fulfilled. Then a solution to the problem (4)–(5) is uniquely determined for all $t \geq 0$.*

In the conditions of Theorem 5.6, estimate (10) is valid for all $t \geq 0$. Therefore,

$$\|x(t) - x^*\|_{X_1} = \sqrt{2u(t)} \leq \sqrt{2y(t)}, \quad t \geq 0. \tag{17}$$

Since

$$\lim_{t \to +\infty} y(t) = y_1,$$

by (12) and (17) we obtain the estimate (6):

$$\limsup_{t \to +\infty} \|x(t) - x^*\|_{X_1} <$$

$$< c_1 \left(\frac{\delta}{\sqrt{\alpha_0}} + \|\tilde{v}\|_{X_1} \alpha_0^p + \|\tilde{v}\|_{X_1} \delta + \Delta \right), \quad c_1 = 2\sqrt{c_3}. \tag{18}$$

Moreover, (15) implies

$$\|x(t) - x^*\|_{X_1} \leq \sqrt{\frac{\alpha_0}{c_2}}, \quad t \geq 0. \tag{19}$$

Let us summarize the obtained results.

Theorem 5.7. *Suppose Conditions 4.2–4.4 and inequalities (11), (15) are satisfied. Assume that the sourcewise representation (4.1.29) is fulfilled. Then estimates (18) and (19) are valid.*

Further, we denote

$$r_1(\delta, \Delta, \alpha_0) = c_1 \left(\frac{\delta}{\sqrt{\alpha_0}} + \|\tilde{v}\|_{X_1} \alpha_0^p + \|\tilde{v}\|_{X_1} \delta + \Delta \right),$$

$$r_2(\alpha_0) = \sqrt{\frac{\alpha_0}{c_2}}. \tag{20}$$

According to Theorem 5.7, the ball $\Omega^0_{r_1(\delta,\Delta,\alpha_0)}(x^*)$ is an attracting subset for trajectories of the system (4) with starting points $x_0 \in \Omega_{r_2(\alpha_0)}(x^*)$. If the values $\delta, \Delta, \|\widetilde{v}\|_{X_1}$, and α_0 satisfy a slightly more stronger condition than (11), namely,

$$\frac{\delta^2}{\alpha_0^2} + \frac{(\|\widetilde{v}\|_{X_1}\delta + \Delta)^2}{\alpha_0} + \|\widetilde{v}\|^2_{X_1}\alpha_0^{2p-1} < \frac{1}{12c_2c_3}, \qquad (21)$$

then from (19) and (20) we obtain

$$\Omega_{r_1(\delta,\Delta,\alpha_0)}(x^*) \subset \Omega_{r_2(\alpha_0)}(x^*).$$

Notice that if $\alpha_0 \in (0,1)$, the value p is essentially greater than $1/2$, and the error levels δ, Δ are sufficiently small, then (21) is satisfied and $r_1(\delta, \Delta, \alpha_0)$ is much less than $r_2(\alpha_0)$. Moreover, for each point $x_0 \in \Omega_{r_2(\alpha_0)}(x^*)$ there exists $t = t(x_0)$ such that $x(t) \in \Omega_{r_1(\delta,\Delta,\alpha_0)}(x^*)$ for all $t \geq t(x_0)$. Therefore the points $x(t)$ for large t can be considered as approximations to x^* adequate to the error levels δ and Δ.

On the other hand, a behavior of the trajectory $x = x(t)$ after $x(t)$ reaches the ball $\Omega_{r_1(\delta,\Delta,\alpha_0)}(x^*)$ is also of interest. Let us discuss this subject with special emphasis on dependence of approximating points $x(t), t \to \infty$ on the initial state $x(0) = x_0$.

Suppose (21) and conditions of Theorem 5.7 are satisfied. Let us consider the metric space (X, ρ), where

$$X = \Omega_{r_2(\alpha_0)}(x^*); \quad \rho(x_1, x_2) = \|x_1 - x_2\|_{X_1}, \quad x_1, x_2 \in X.$$

Let $V_t : X \to X, t \geq 0$ be a nonlinear semigroup associated with the dynamical system (4). By definition, $V_t(x_0) = x(t); x_0 \in X, t \geq 0$, where $x = x(t)$ is the solution of (4)–(5). It can easily be shown that the semigroup V_t is continuous, i.e., the mapping $(t, y) \to V_t(y)$ is continuous in $(t, y) \in \{t : t \geq 0\} \times X$. For the sequel we need some notation.

Following [99, 100], we say that $B_0 \subset X$ attracts a subset $B \subset X$, if for each $\varepsilon > 0$ there exists $t_1(\varepsilon, B)$ such that for all $t \geq t_1(\varepsilon, B)$,

$$V_t(B) \equiv \{z \in X : z = V_t(y), y \in B\} \subset \bigcup_{y \in B_0} \Omega_\varepsilon(y).$$

The minimal (in the sense of inclusion) nonempty subset $\mathfrak{M} \subset X$ attracting each $B \subset X$ is called the minimal global (on X) B–attractor of the semigroup V_t. A semigroup V_t is said to be pointwise dissipative if there exists a subset $B_0 \subset X$ attracting each point of X. From Theorem 5.7 it follows that the semigroup V_t defined by (4) is pointwise dissipative with $B_0 = \Omega^0_{r_1(\delta,\Delta,\alpha_0)}(x^*)$. The aim of forthcoming arguing is to prove that this semigroup has a minimal global B–attractor and to establish its properties. In recent years many conditions have been derived which imply that a semigroup has a minimal global B–attractor (see, e.g., [5, 51, 99, 100] and references therein). Below we shall use a criterion given in [99].

Lemma 5.5. *([99]) Suppose a pointwise dissipative semigroup V_t on $X = \Omega_{r_2(\alpha_0)}(x^*)$ has the form $V_t = W_t + U_t$, $t \geq 0$, where*

$$\sup_{y \in X} \|W_t(y)\|_{X_1} \leq m(t), \quad t \geq 0; \quad \lim_{t \to +\infty} m(t) = 0, \quad (22)$$

and the operators $U_t : X \to X_1$, $t \geq 0$ take each subset of X to a compact subset of X_1. Then the semigroup V_t has a unique minimal global B–attractor $\mathfrak{M}$. The set $\mathfrak{M}$ is connected, compact, and invariant with respect to the semigroup V_t.

Recall that the invariance of $\mathfrak{M}$ means $V_t(\mathfrak{M}) = \mathfrak{M} \; \forall t \geq 0$.

Let us show that the semigroup V_t generated by (4) satisfies conditions of Lemma 5.5.

Recall that an operator $G : X_1 \to X_2$ is said to be strongly continuous on a closed convex subset $Y \subset X_1$ if for each sequence $\{x_n\} \subset Y$ weakly converging to x in X_1, the sequence $\{G(x_n)\}$ strongly converges to $G(x)$ in X_2 ([142]).

Suppose $\widetilde{F}(x)$ satisfies the following condition.

Condition 5.6. *The operator $\widetilde{F}(x)$ is strongly continuous on $\Omega_r(x^*)$, where $r > r_2(\alpha_0)$.*

Remark 5.8. Condition 5.6 yields that the derivative $\widetilde{F}'(x) \in L(X_1, X_2)$ is a compact operator for each $x \in \Omega_{r_2(\alpha)}^0(x^*)$. Besides, $\widetilde{F}'(x)$ is strongly continuous as a mapping from $\Omega_{r_2(\alpha)}(x^*)$ into $L(X_1, X_2)$ ([142]).

Lemma 5.6. *Suppose Condition 5.6 is satisfied. Then the operator*

$$\widetilde{\Phi} : X \to X_1,$$

$$\widetilde{\Phi}(x) = \xi - \Theta(\widetilde{F}'^*(x)\widetilde{F}'(x), \alpha_0)\widetilde{F}'^*(x)(\widetilde{F}(x) - \widetilde{F}'(x)(x - \xi)) - x$$

takes each subset of X to a compact subset of X_1.

Proof. Since $\widetilde{\Phi}(x)$ is continuous, it suffices to prove that each sequence $\{x_n\} \subset \Omega_{r_2(\alpha)}(x^*)$ has a subsequence $\{x_{n_k}\}$ for which $\{\widetilde{\Phi}(x_{n_k})\}$ converges strongly in X_1. Let $\{x_{n_k}\}$ be a weakly converging subsequence of $\{x_n\}$ with the weak limit $\widetilde{x}$. Taking into account Condition 5.6 and Remark 5.8, we conclude that $\{\widetilde{F}(x_{n_k})\}$ has a strong limit $\widetilde{f} \in X_2$. Moreover, the sequences of compact operators $\{\widetilde{F}'(x_{n_k})\}$ and $\{\widetilde{F}'^*(x_{n_k})\}$ converge in $L(X_1, X_2)$ and $L(X_2, X_1)$ to compact operators $\widetilde{A}$ and $\widetilde{A}^*$ respectively. It now follows that the sequence

$$\{\widetilde{F}'^*(x_{n_k})(\widetilde{F}(x_{n_k}) - \widetilde{F}'(x_{n_k})(x_{n_k} - \xi))\}$$

has a strong limit $\widetilde{A}^*(\widetilde{f} - \widetilde{A}(\widetilde{x} - \xi))$. To conclude the proof it remains to note that $\{\Theta(\widetilde{F}'^*(x_{n_k})\widetilde{F}'(x_{n_k}), \alpha_0)\}$ converges to $\Theta(\widetilde{A}^*\widetilde{A}, \alpha_0)$ in $L(X_1)$. This immediately follows by the representation

$$\Theta(\widetilde{F}'^*(x_{n_k})\widetilde{F}'(x_{n_k}), \alpha_0) = \frac{1}{2\pi i} \int_{\Gamma_{\alpha_0}} \Theta(\lambda, \alpha_0) R(\lambda, \widetilde{F}'^*(x_{n_k})\widetilde{F}'(x_{n_k})) d\lambda$$

and the equality

$$\lim_{k \to \infty} \|R(\lambda, \widetilde{F}'^*(x_{n_k})\widetilde{F}'(x_{n_k})) - R(\lambda, \widetilde{A}^*\widetilde{A})\|_{L(X_1)} = 0 \quad \forall \lambda \in \Gamma_{\alpha_0}.$$

The lemma is proved.   □

Let us now turn to the original semigroup V_t. Consider the semigroup of linear operators $W_t(x) = \exp(-t)x$, $t \geq 0$ associated with the equation

$$\dot{x} = -x.$$

We see that condition (22) is fulfilled with

$$m(t) = (\|x^*\|_{X_1} + r_2(\alpha)) \exp(-t).$$

Lemma 5.6 yields that the operator $U_t = V_t - W_t$, $t \geq 0$ takes each subset of X to a compact subset (see [93] for details). As a direct consequence of Lemma 5.5, we have the following result.

Theorem 5.8. *Suppose Conditions 4.2–4.4, 5.6 and inequalities (15), (21) are satisfied. Then the dynamical system (4) has on $\Omega^0_{r_2(\alpha_0)}(x^*)$ a minimal global B–attractor $\mathfrak{M} \subset \Omega_{r_1(\delta, \Delta, \alpha_0)}(x^*)$, which is a connected compact and invariant subset.*

With respect to the solution x^*, the B–attractor $\mathfrak{M}$ can be considered as an approximating set independent of an initial point x_0, unlike an individual trajectory $x = x(t)$ of the system (4)–(5).

Example 5.3. Let D be a finite–dimensional domain with a regular boundary. Let $\widetilde{K} \in C(\overline{D} \times \overline{D})$ and $\widetilde{g} \in L_2(D)$. Then the operator

$$\widetilde{F} : H^1(D) \to L_2(D),$$

$$[\widetilde{F}(x)](s) = \int_D \widetilde{K}(s,\sigma)x^2(\sigma)d\sigma - \widetilde{g}(s), \quad s \in D$$

satisfies all the conditions imposed on $\widetilde{F}(x)$ in Theorem 5.8. Indeed, since the embedding $H^1(D) \subset L_2(D)$ is compact, $\widetilde{F}$ is strongly continuous; the check of (4.1.2) is straightforward.

5.6 Iteratively Regularized Gradient Method and Its Continuous Version

In this section we study a regularized gradient method and its continuous analog. Unlike stable iterative and continuous processes proposed in Sections 5.1–5.5, the iteratively regularized gradient methods require a stopping to obtain an acceptable approximation to a solution.

Consider an operator equation

$$F(x) = 0, \quad x \in X_1, \tag{1}$$

where $F : X_1 \to X_2$ is a nonlinear operator and X_1, X_2 are Hilbert spaces. As in the previous section, we assume that the operator $F(x)$ and its approximation $\widetilde{F}(x)$ satisfy condition (4.1.2) and inequalities (4.2) and (4.3).

In Section 1.2, we have pointed out that minimization of the Tikhonov functional

$$\widetilde{\Psi}_\alpha(x) = \frac{1}{2}\|\widetilde{F}(x)\|_{X_2}^2 + \frac{1}{2}\alpha\|x - \xi\|_{X_1}^2, \quad x \in X_1 \tag{2}$$

until the present time remains the sole universal approach to constructing solution methods for general nonlinear ill–posed problems. In (2), $\alpha > 0$ is a regularization parameter and $\xi \in X_1$ an initial estimate of the solution x^* to (1). Under nonrestrictive additional conditions on the operator $\widetilde{F}(x)$, the functional $\widetilde{\Psi}_\alpha(x), x \in X_1$ has a global minimizer x_α. This is so, e.g., if $\widetilde{F}(x)$ is weakly continuous (see Section 4.2). Further, it has been proved that the minimizers x_α with $\alpha = \alpha(\delta)$ strongly converge to x^* as $\delta \to 0$, provided the regularization parameter α is properly coordinated with the error level δ ([17, 42, 137, 139]). However an implementation of similar computational schemes in application to general nonlinear operators $\widetilde{F}(x)$ is considerably hampered by the necessity of finding an exact or approximate solution to the complicated global optimization problem

$$\min_{x \in X_1} \widetilde{\Psi}_\alpha(x).$$

This fact stimulates a permanent interest to iterative realizations of Tikhonov's scheme where an iteration of a chosen basic minimization method alternates with a decrease of the regularization parameter α. In particular, taking as a basic minimization method the standard gradient process, we obtain the iteratively regularized gradient method

$$x_0 \in \Omega_R(x^*), \quad x_{n+1} = x_n - \mu_n(\widetilde{F}'^*(x_n)\widetilde{F}(x_n) + \alpha_n(x_n - \xi)). \qquad (3)$$

Here μ_n and α_n are a priori prescribed values of the stepsize and the regularization parameter at n–th iteration;

$$\mu_n > 0, \quad 0 < \alpha_{n+1} \leq \alpha_n, \quad \lim_{n \to \infty} \alpha_n = 0.$$

It is well known that in case of $\delta > 0$, the processes of type (3) usually diverge as $n \to \infty$. Therefore to get an acceptable approximation of x^*, we should terminate iterations (3) at an appropriate step $n = n(\delta)$ such that $\lim_{\delta \to 0} n(\delta) = \infty$.

If the functional

$$\Psi(x) = \frac{1}{2}\|F(x)\|_{X_2}^2$$

is convex, then the convergence $x_{n(\delta)} \to x^*(\delta \to 0)$ can be established in the context of iterative regularization theory ([17]). Unfortunately, such convexity conditions arise to be too restrictive for most applied nonlinear inverse problems. In [54, 127], instead of the convexity of $\Psi(x)$ the following condition on $F(x)$ is used:

$$\|F(\overline{x}) - F(x) - F'(x)(x - \overline{x})\|_{X_2} \leq$$
$$\leq C\|F(\overline{x}) - F(x)\|_{X_2} \quad \forall x, \overline{x} \in \Omega_R(x^*). \qquad (4)$$

Let us remark that practical verification of conditions of type (4) for most nonlinear inverse problems is highly complicated, as well.

In this section we study the iterative process (3) and its continuous variant without any structural conditions concerning the functional $\Psi(x)$ or the operator $F(x)$. The continuous version of (3) has the form

$$\dot{x} = -(\widetilde{F}'^*(x)\widetilde{F}(x) + \alpha(t)(x - \xi)), \quad x(0) = x_0. \qquad (5)$$

Here $x = x(t) \in X_1, t \geq 0; \alpha(t), t \geq 0$ is a differentiable function such that

$$\alpha(t) > 0, \quad \dot{\alpha}(t) \leq 0, \quad t \geq 0; \quad \lim_{t \to +\infty} \alpha(t) = 0.$$

The following approximate sourcewise representation condition plays a key role in subsequent considerations.

Condition 5.7.

$$x^* - \xi = F'^*(x^*)v + w, \quad v \in X_2, w \in X_1; \quad \|w\|_{X_1} \leq \Delta. \qquad (6)$$

Below we shall justify a stopping rule $n = n(\delta, \Delta)$ such that approximating elements $x_{n(\delta,\Delta)}$ obtained by the process (3) converge to x^* as $\delta, \Delta \to 0$. In the case of continuous process (5), a stopping criterion has the form $t = t(\delta, \Delta)$, so that the solution x^* is approximated by the point $x(t(\delta, \Delta))$ of the trajectory $x = x(t)$. The stopping rule $t = t(\delta, \Delta)$ that we propose below ensures the convergence $x(t(\delta, \Delta)) \to x^*$ as $\delta, \Delta \to 0$.

Let us return for a short while to the classical Tikhonov's method

$$x_\alpha^\delta \in X_1, \quad \tilde{\Psi}_\alpha(x_\alpha^\delta) = \min_{x \in X_1} \tilde{\Psi}_\alpha(x). \tag{7}$$

Suppose $\alpha = \alpha(\delta) = O(\delta)$ and Condition 5.7 is satisfied with $\Delta = 0$; then the estimate is valid ([42]):

$$\|x_\alpha^\delta - x^*\|_{X_1} = O(\delta^{1/2}).$$

We shall prove that the stopping rules $n = n(\delta, \Delta)$ and $t = t(\delta, \Delta)$ ensure analogous estimates:

$$\|x_{n(\delta,\Delta)} - x^*\|_{X_1} = O((\delta + \Delta)^{1/2}),$$
$$\|x(t(\delta, \Delta)) - x^*\|_{X_1} = O((\delta + \Delta)^{1/2}). \tag{8}$$

From (8) it follows that the schemes (3) and (5) supplied with corresponding stopping criterions guarantee the same error order in δ as the regularization procedure (7) based on a complicated global optimization technique.

First, let us establish necessary auxiliary estimates. For $x \in \Omega_R(x^*)$ we have

$$\tilde{F}(x) = \tilde{F}(x^*) + \tilde{F}'(x)(x - x^*) + \tilde{G}(x, x^*),$$
$$\|\tilde{G}(x, x^*)\|_{X_2} \le \frac{1}{2} L \|x - x^*\|_{X_1}^2. \tag{9}$$

Let $x_n \in \Omega_R(x^*)$. From (3) it follows that

$$\|x_{n+1} - x^*\|_{X_1}^2 =$$
$$= \|(x_n - x^*) - \mu_n(\tilde{F}'^*(x_n)\tilde{F}(x_n) + \alpha_n(x_n - \xi))\|_{X_1}^2 =$$
$$= \|x_n - x^*\|_{X_1}^2 - 2\mu_n(x_n - x^*, \tilde{F}'^*(x_n)\tilde{F}(x_n) + \alpha_n(x_n - \xi))_{X_1} +$$
$$+ \mu_n^2 \|\tilde{F}'^*(x_n)\tilde{F}(x_n) + \alpha_n(x_n - \xi)\|_{X_1}^2. \tag{10}$$

By (6) and (9) we get

$$A_n \equiv -2\mu_n(x_n - x^*, \tilde{F}'^*(x_n)\tilde{F}(x_n) + \alpha_n(x_n - \xi))_{X_1} =$$
$$= -2\alpha_n\mu_n\|x_n - x^*\|_{X_1}^2 - 2\mu_n(x_n - x^*, \tilde{F}'^*(x_n)\tilde{G}(x_n, x^*) +$$
$$+ \tilde{F}'^*(x_n)\tilde{F}(x^*) + \tilde{F}'^*(x_n)\tilde{F}'(x_n)(x_n - x^*) +$$
$$+ \alpha_n\tilde{F}'^*(x_n)v + \alpha_n(F'^*(x^*) - \tilde{F}'^*(x_n))v + \alpha_n w)_{X_1}. \tag{11}$$

According to (4.1.2) and (4.3),

$$\|\widetilde{F}'^*(x) - F'^*(x^*)\|_{L(X_2, X_1)} = \|\widetilde{F}'(x) - F'(x^*)\|_{L(X_1, L_2)} \leq$$
$$\leq L\|x - x^*\|_{X_1} + \delta. \tag{12}$$

Now suppose

$$\|v\|_{X_2} \leq \min\left\{\frac{1}{2L}, 1\right\}. \tag{13}$$

By (11)–(13) we obtain the estimate

$$A_n \leq -2\alpha_n\mu_n\|x_n - x^*\|_{X_1}^2 + 2\mu_n\left(-\|\widetilde{F}'(x_n)(x_n - x^*)\|_{X_2}^2 + \right.$$
$$+ \|\widetilde{F}'(x_n)(x_n - x^*)\|_{X_2}\left(\|\widetilde{F}(x^*)\|_{X_2} + \|\widetilde{G}(x_n, x^*)\|_{X_2} + \|v\|_{X_2}\alpha_n\right)\right) +$$
$$+ 2\alpha_n\mu_n\left(\|v\|_{X_2}(L\|x_n - x^*\|_{X_1} + \delta) + \Delta\right)\|x_n - x^*\|_{X_1} \leq$$
$$\leq -2\alpha_n\mu_n\|x_n - x^*\|_{X_1}^2 + 2L\|v\|_{X_2}\alpha_n\mu_n\|x_n - x^*\|_{X_1}^2 +$$
$$+ \frac{1}{2}\mu_n\left(\|\widetilde{F}(x^*)\|_{X_2} + \|\widetilde{G}(x_n, x^*)\|_{X_2} + \|v\|_{X_2}\alpha_n\right)^2 +$$
$$+ 2\alpha_n\mu_n(\|v\|_{X_2}\delta + \Delta)\|x_n - x^*\|_{X_1} \leq$$
$$\leq -\alpha_n\mu_n\|x_n - x^*\|_{X_1}^2 + 2\alpha_n\mu_n(\delta + \Delta)\|x_n - x^*\|_{X_1} +$$
$$+ \frac{1}{2}\mu_n\left(\delta + \frac{1}{2}L\|x_n - x^*\|_{X_1}^2 + \|v\|_{X_2}\alpha_n\right)^2. \tag{14}$$

Using inequality (4.5.24), we get

$$(\delta + \Delta)\|x_n - x^*\|_{X_1} \leq \frac{1}{4}\|x_n - x^*\|_{X_1}^2 + (\delta + \Delta)^2. \tag{15}$$

Taking into account (4.5.23), by (14) and (15) we deduce

$$A_n \leq -\frac{1}{2}\alpha_n\mu_n\|x_n - x^*\|_{X_1}^2 + \frac{3}{8}L^2\mu_n\|x_n - x^*\|_{X_1}^4 +$$
$$+ \frac{3}{2}\mu_n(\delta^2 + \|v\|_{X_2}^2\alpha_n^2) + 2\alpha_n\mu_n(\delta + \Delta)^2. \tag{16}$$

Further, by (4.1.2) and (4.2),

$$\|\widetilde{F}'(x)\|_{L(X_1, X_2)} \leq N_1 \quad \forall x \in \Omega_R(x^*). \tag{17}$$

Therefore,

$$\|\widetilde{F}(x)\|_{X_2} \leq \|\widetilde{F}(x) - \widetilde{F}(x^*)\|_{X_2} + \|\widetilde{F}(x^*)\|_{X_2} \leq$$
$$\leq N_1\|x - x^*\|_{X_1} + \delta \quad \forall x \in \Omega_R(x^*). \tag{18}$$

From (6), (17), (18), and (4.5.23) we obtain

$$B_n \equiv \mu_n^2 \|\tilde{F}'^*(x_n)\tilde{F}(x_n) + \alpha_n(x_n - \xi)\|_{X_1}^2 \leq$$
$$\leq 3\mu_n^2 \left(\|\tilde{F}'^*(x_n)\tilde{F}(x_n)\|_{X_1}^2 + \alpha_n^2 \|x_n - x^*\|_{X_1}^2 + \alpha_n^2 \|x^* - \xi\|_{X_1}^2 \right) \leq$$
$$\leq 3\mu_n^2 \left(N_1^2(N_1\|x_n - x^*\|_{X_1} + \delta)^2 + \alpha_n^2 \|x_n - x^*\|_{X_1}^2 + \tag{19} \right.$$
$$\left. + \alpha_n^2 \|F'^*(x^*)v + w\|_{X_1}^2 \right) \leq 3\mu_n^2 \left((2N_1^4 + \alpha_n^2)\|x_n - x^*\|_{X_1}^2 + \right.$$
$$\left. + 2N_1^2\delta^2 + 2\alpha_n^2(N_1^2\|v\|_{X_2}^2 + \Delta^2) \right).$$

Combining (10), (16), and (19), we get

$$\|x_{n+1} - x^*\|_{X_1}^2 = \|x_n - x^*\|_{X_1}^2 + A_n + B_n \leq$$
$$\leq \left(1 - \frac{1}{2}\alpha_n\mu_n + 3(2N_1^4 + \alpha_n^2)\mu_n^2 \right) \|x_n - x^*\|_{X_1}^2 +$$
$$+ \frac{3}{8}L^2\mu_n\|x_n - x^*\|_{X_1}^4 + \frac{3}{2}\mu_n(\delta^2 + \|v\|_{X_2}^2\alpha_n^2) + \tag{20}$$
$$+ 2\alpha_n\mu_n(\delta + \Delta)^2 + 6\mu_n^2 \left(N_1^2\delta^2 + \alpha_n^2(N_1^2\|v\|_{X_2}^2 + \Delta^2) \right).$$

Suppose the stepsize in (3) is proportional to the regularization parameter, i.e.,

$$\mu_n = \varepsilon\alpha_n, \quad \varepsilon > 0; \tag{21}$$

then the inequality

$$0 < \varepsilon \leq \frac{1}{12(2N_1^4 + \alpha_0^2)} \tag{22}$$

yields

$$1 - \frac{1}{2}\alpha_n\mu_n + 3(2N_1^4 + \alpha_n^2)\mu_n^2 \leq 1 - \frac{1}{4}\varepsilon\alpha_n^2. \tag{23}$$

Using (20), (21), and (23), we get

$$\|x_{n+1} - x^*\|_{X_1}^2 \leq$$
$$\leq \left(1 - \frac{1}{4}\varepsilon\alpha_n^2 \right) \|x_n - x^*\|_{X_1}^2 + \frac{3}{8}L^2\varepsilon\alpha_n\|x_n - x^*\|_{X_1}^4 +$$
$$+ \frac{3}{2}\varepsilon\alpha_n(\delta^2 + \|v\|_{X_2}^2\alpha_n^2) + 2(\delta + \Delta)^2\varepsilon\alpha_n^2 + \tag{24}$$
$$+ 6\varepsilon^2\alpha_n^2 \left(N_1^2\delta^2 + \alpha_n^2(N_1^2\|v\|_{X_2}^2 + \Delta^2) \right).$$

Thus we proved the following statement.

Theorem 5.9. *Suppose Condition 5.7 and conditions (13), (21), and (22) are satisfied. Assume that $x_n \in \Omega_R(x^*)$. Then estimate (24) is fulfilled.*

Let us now describe a stopping rule for the process (3). Pick a number $m > 0$ such that

$$m(\delta + \Delta) < \alpha_0 \tag{25}$$

and define the stopping point $n = n(\delta, \Delta) \geq 0$ as follows:

$$n(\delta, \Delta) = \max\{n = 0, 1, \cdots : \alpha_n \geq m(\delta + \Delta)\}. \tag{26}$$

Since $\lim_{n \to \infty} \alpha_n = 0$ and $\alpha_{n+1} \leq \alpha_n, n = 0, 1, \ldots$, the number $n(\delta, \Delta)$ is well defined. Besides, we have

$$\alpha_n \geq m(\delta + \Delta), \quad n = 0, 1, \ldots, n(\delta, \Delta); \quad \lim_{\delta, \Delta \to 0} n(\delta, \Delta) = \infty, \tag{27}$$

$$\alpha_{n(\delta, \Delta)+1} < m(\delta + \Delta). \tag{28}$$

From (24) and (27) it follows that under the conditions of Theorem 5.9,

$$\|x_{n+1} - x^*\|_{X_1}^2 \leq$$

$$\leq \left(1 - \frac{1}{4}\varepsilon\alpha_n^2\right)\|x_n - x^*\|_{X_1}^2 + \frac{3}{8}L^2\varepsilon\alpha_n\|x_n - x^*\|_{X_1}^4 + \tag{29}$$

$$+ \varepsilon\alpha_n^3 D(v, m), \quad 0 \leq n \leq n(\delta, \Delta) - 1,$$

where

$$D(v, m) = \frac{1}{m^2}\left(\frac{3}{2} + 2\alpha_0 + 6N_1^2\alpha_0\varepsilon + 6\alpha_0^3\varepsilon\right) + \left(\frac{3}{2} + 6N_1^2\alpha_0\varepsilon\right)\|v\|_{X_2}. \tag{30}$$

Fix a constant C such that

$$0 < C < \frac{R}{\sqrt{\alpha_0}}. \tag{31}$$

Let us prove by induction the estimate

$$\|x_n - x^*\|_{X_1} \leq C\sqrt{\alpha_n}, \quad 0 \leq n \leq n(\delta, \Delta). \tag{32}$$

To this end suppose the starting point x_0 satisfies

$$\|x_0 - x^*\|_{X_1} \leq C\sqrt{\alpha_0}. \tag{33}$$

Let $0 < k \leq n(\delta, \Delta) - 1$, $x_k \in \Omega_R(x^*)$, and (32) is valid for $n = k$. Now notice that (31) and (33) imply $x_0 \in \Omega_R(x^*)$. Next, from (29) and (32) we get

$$\|x_{k+1} - x^*\|_{X_1}^2 \leq$$

$$\leq C^2\alpha_{k+1}\left\{\left(1 - \frac{1}{4}\varepsilon\alpha_k^2\right)\frac{\alpha_k}{\alpha_{k+1}} + \left(\frac{3}{8}L^2C^2\varepsilon + \frac{D(v, m)}{C^2}\varepsilon\right)\frac{\alpha_k^3}{\alpha_{k+1}}\right\}.$$

It is readily seen that the right part of this inequality is estimated from above by $C\alpha_{k+1}$, provided $\{\alpha_n\}$ and C, ε satisfy

$$\alpha_n - \alpha_{n+1} \leq \left(\frac{1}{4} - \frac{3}{8}L^2C^2 - \frac{D(v,m)}{C^2}\right)\varepsilon\alpha_n^3, \quad n = 0, 1, \ldots . \quad (34)$$

Simple calculations prove that the following conditions are sufficient for (34):

$$0 < C \leq \left\{\frac{R}{\sqrt{\alpha_0}}, \frac{1}{L\sqrt{6}}\right\}, \quad D(v,m) \leq \frac{C^2}{16},$$

$$\sup_{n=0,1,\ldots} \frac{\alpha_n - \alpha_{n+1}}{\alpha_n^3} \leq \frac{\varepsilon}{8}. \quad (35)$$

As an example of $\{\alpha_n\}$ satisfying (35), we can take the sequence

$$\alpha_n = \frac{\alpha_0}{(n+1)^s}, \quad 0 < s < \min\left\{\frac{1}{2}, \frac{\alpha_0^2\varepsilon}{8}\right\}.$$

Therefore under the above conditions, estimate (32) is valid for the number $n = k + 1$. Finally note that (31), (33), and the monotonicity of $\{\alpha_n\}$ yield

$$\|x_{k+1} - x^*\|_{X_1} \leq C\sqrt{\alpha_{k+1}} \leq C\sqrt{\alpha_0} < R,$$

i.e., $x_{k+1} \in \Omega_R(x^*)$. This completes the proof of (32).

In addition, suppose

$$\sup_{n=0,1,\ldots} \frac{\alpha_n}{\alpha_{n+1}} \equiv r < \infty. \quad (36)$$

The sequence $\{\alpha_n\}$ from the above example obviously satisfies (36). Letting in (32) $n = n(\delta, \Delta)$ and using (28) and (36), we get

$$\|x_{n(\delta,\Delta)} - x^*\|_{X_1} \leq C\sqrt{\alpha_{n(\delta,\Delta)+1}}\sqrt{\frac{\alpha_{n(\delta,\Delta)}}{\alpha_{n(\delta,\Delta)+1}}} < C\sqrt{mr(\delta + \Delta)}.$$

The following theorem summarizes the preceding arguing.

Theorem 5.10. *Suppose Condition 5.7 and conditions (13), (21), (22), (25), (33), (35), and (36) are satisfied. Then estimate (32) is fulfilled. Moreover,*

$$\|x_{n(\delta,\Delta)} - x^*\|_{X_1} < C\sqrt{mr(\delta + \Delta)}. \quad (37)$$

Let us discuss the conditions of Theorem 5.10. From (30) we conclude that the second inequality in (35) is valid if

$$m \geq 4\sqrt{3 + 4\alpha_0 + 12N_1^2\alpha_0\varepsilon + 12\alpha_0^3\varepsilon} \Big/ C,$$

$$\|v\|_{X_2} \leq \frac{C^2}{48(1 + 4N_1^2\alpha_0\varepsilon)}. \tag{38}$$

The first inequality in (38) and the condition (25) impose a restriction on error levels δ and Δ; the second inequality in (38) and condition (13) establish an upper bound for the norm of v in (6). At last, condition (33) means that the starting point x_0 must be sufficiently close to the solution x^*.

Let us now turn to the continuous analog (5) of the iteration (3).

Using (4.1.2), as in Section 4.5 we conclude that there exists $\widetilde{T} > 0$ such that a solution $x = x(t)$ to the Cauchy problem (5) is uniquely defined for all $t \in [0, \widetilde{T}]$. Without loss of generality we can assume that $x_0 \in \Omega_R^0(x^*)$ and $x(t) \in \Omega_R^0(x^*)$ for all $t \in [0, \widetilde{T}]$.

From (5), (6), and (9) we obtain

$$\frac{d}{dt}\left(\frac{1}{2}\|x - x^*\|_{X_1}^2\right) =$$

$$= (x - x^*, -\widetilde{F}'^*(x)\widetilde{F}(x^*) - \widetilde{F}'^*(x)\widetilde{F}'(x)(x - x^*) - $$

$$- \widetilde{F}'^*(x)\widetilde{G}(x, x^*) - \alpha(t)(x - x^*) - \alpha(t)\widetilde{F}'^*(x)v + \tag{39}$$

$$+ \alpha(t)(\widetilde{F}'^*(x) - F'^*(x^*))v - \alpha(t)w)_{X_1}, \quad 0 \leq t \leq \widetilde{T}.$$

Further, by (9), (13), and (39) we deduce

$$\frac{d}{dt}\left(\frac{1}{2}\|x - x^*\|_{X_1}^2\right) \leq$$

$$\leq \|\widetilde{F}'(x)(x - x^*)\|_{X_2}\delta - \|\widetilde{F}'(x)(x - x^*)\|_{X_2}^2 + $$

$$+ \|\widetilde{F}'(x)(x - x^*)\|_{X_2}\|\widetilde{G}(x, x^*)\|_{X_2} - \alpha(t)\|x - x^*\|_{X_1}^2 + $$

$$+ \|\widetilde{F}'(x)(x - x^*)\|_{X_2}\|v\|_{X_2}\alpha(t) + \|x - x^*\|_{X_1}\alpha(t)\Delta + $$

$$+ (L\|x - x^*\|_{X_1} + \delta)\|x - x^*\|_{X_1}\|v\|_{X_2}\alpha(t) \leq \tag{40}$$

$$\leq \alpha(t)(\delta + \Delta)\|x - x^*\|_{X_1} - \alpha(t)(1 - L\|v\|_{X_2})\|x - x^*\|_{X_1}^2 + $$

$$+ \left(\frac{1}{2}L\|x - x^*\|_{X_1}^2 + \|v\|_{X_2}\alpha(t) + \delta\right)\|\widetilde{F}'(x)(x - x^*)\|_{X_2} - $$

$$- \|\widetilde{F}'(x)(x - x^*)\|_{X_2}^2.$$

Let us remember that from (4.5.24) it follows

$$(\delta + \Delta)\|x - x^*\|_{X_1} \leq \frac{1}{4}\|x - x^*\|_{X_1}^2 + (\delta + \Delta)^2. \tag{41}$$

Besides,

$$\left(\frac{1}{2}L\|x - x^*\|_{X_1}^2 + \|v\|_{X_2}\alpha(t) + \delta\right)\|\tilde{F}'(x)(x - x^*)\|_{X_2} -$$

$$- \|\tilde{F}'(x)(x - x^*)\|_{X_2}^2 \leq$$

$$\leq \frac{1}{4}\left(\frac{1}{2}L\|x - x^*\|_{X_1}^2 + (\|v\|_{X_2}\alpha(t) + \delta)\right)^2 \leq \qquad (42)$$

$$\leq \frac{1}{8}L^2\|x - x^*\|_{X_1}^4 + \|v\|_{X_2}^2\alpha^2(t) + \delta^2.$$

Combining the inequalities (40)–(42), we get the estimate

$$\frac{d}{dt}\left(\frac{1}{2}\|x - x^*\|_{X_1}^2\right) \leq -\frac{1}{4}\alpha(t)\|x - x^*\|_{X_1}^2 +$$

$$+ \frac{1}{8}L^2\|x - x^*\|_{X_1}^4 + \left(\alpha(t)(\delta + \Delta)^2 + \|v\|_{X_2}^2\alpha^2(t) + \delta^2\right),$$

$$0 \leq t \leq \tilde{T}.$$

Thus with the notation

$$u(t) = \frac{1}{2}\|x(t) - x^*\|_{X_1}^2$$

we have the following result.

Theorem 5.11. *Suppose Condition 5.7 and condition (13) are satisfied. Then*

$$\dot{u} \leq -\frac{1}{2}\alpha(t)u + \frac{1}{2}L^2u^2 +$$

$$+ \left(\alpha(t)(\delta + \Delta)^2 + \|v\|_{X_2}^2\alpha^2(t) + \delta^2\right) \quad \forall t \in [0, \tilde{T}]. \qquad (43)$$

By analogy with (25) and (26), pick a constant $m > 0$ such that

$$m(\delta + \Delta) < \alpha(0) \qquad (44)$$

and define the stopping point $t = t(\delta, \Delta) > 0$ as a root of the equation

$$\alpha(t) = m(\delta + \Delta). \qquad (45)$$

Since $\lim_{t \to +\infty} \alpha(t) = 0$ and the function $\alpha = \alpha(t), t \geq 0$ is continuous and nonincreasing, the equation (45) has a solution. Besides, from (45) it follows that

$$\alpha(t) \geq m(\delta + \Delta), \quad 0 \leq t \leq t(\delta, \Delta). \qquad (46)$$

Our task is now to establish a unique solvability of the problem (5), at least for $0 \le t \le t(\delta, \Delta)$. Let us denote

$$H(v, m) = \frac{\alpha(0) + 1}{m^2} + \|v\|_{X_2}^2. \tag{47}$$

Combining (43), (46), and (47), we obtain

$$\dot{u} \le -\frac{1}{2}\alpha(t)u + \frac{1}{2}L^2u^2 + H(v, m)\alpha^2(t), \quad 0 \le t \le \min\{t(\delta, \Delta), \tilde{T}\}. \tag{48}$$

Now, we need the following lemma.

Lemma 5.7. *([2]) Let functions $\gamma(t)$, $\sigma(t)$, and $\beta(t)$ be continuous on the positive semiaxis. Assume that continuously differentiable functions $\mu(t)$ and $w(t)$ satisfy the conditions*

$$\mu(0)w(0) < 1; \quad \mu(t) > 0, \quad w(t) \ge 0,$$
$$\dot{w} \le -\gamma(t)w + \sigma(t)w^2 + \beta(t),$$
$$0 \le \sigma(t) \le \frac{\mu(t)}{2}\left(\gamma(t) - \frac{\dot{\mu}(t)}{\mu(t)}\right),$$
$$\beta(t) \le \frac{1}{2\mu(t)}\left(\gamma(t) - \frac{\dot{\mu}(t)}{\mu(t)}\right) \quad \forall t \ge 0.$$

Then for all t $\ge$ 0,

$$w(t) < \frac{1}{\mu(t)}.$$

It can easily be checked that the functions

$$w(t) = u(t), \quad \gamma(t) = \frac{1}{2}\alpha(t), \quad \sigma(t) = \frac{1}{2}L^2,$$

$$\beta(t) = H(v, m)\alpha^2(t), \quad \mu(t) = \frac{d}{\alpha(t)}; \quad t \ge 0$$

satisfy all the conditions of Lemma 5.7 if the following inequalities are valid:

$$\frac{\dot{\alpha}(t)}{\alpha^2(t)} \ge -\frac{1}{4}, \quad t \ge 0, \tag{49}$$

$$d \ge \max\left\{4L^2, \frac{2\alpha(0)}{R^2}\right\}, \tag{50}$$

$$H(v, m) \le \frac{1}{8d}, \tag{51}$$

$$\|x_0 - x^*\|_{X_1} < \sqrt{\frac{2\alpha(0)}{d}}. \tag{52}$$

As an example of a function $\alpha(t)$ satisfying (49) we can take

$$\alpha(t) = \frac{\alpha(0)}{(t+1)^s}; \quad s \in (0,1), \quad \alpha(0) = 4s.$$

Let us now analyze conditions (50)–(52). Inequality (51) is valid if

$$m \geq 4\sqrt{d(\alpha(0) + 1)}, \tag{53}$$

$$\|v\|_{X_2} \leq \frac{1}{4\sqrt{d}}. \tag{54}$$

Conditions (44) and (53) impose restrictions on error levels δ and Δ; (54) and (13) determine requirements on the element v in representation (6). Finally, (52) establishes an upper bound for deviations of the starting point x_0 from the solution x^*.

If conditions (49)–(52) are satisfied, then by Lemma 5.7,

$$\|x(t) - x^*\|_{X_1} < \sqrt{\frac{2\alpha(t)}{d}}, \quad 0 \leq t \leq \min\{t(\delta, \Delta), \tilde{T}\}. \tag{55}$$

According to (50) and (55), the trajectory $x = x(t)$ lies in the ball

$$\Omega^0_{\sqrt{2\alpha(0)/d}}(x^*) \subset \Omega_R(x^*)$$

for all $0 \leq t \leq \min\{t(\delta, \Delta), \tilde{T}\}$. It now follows in the standard way ([55]) that $\tilde{T} \geq t(\delta, \Delta)$. Therefore inequality (55) is valid for all $0 \leq t \leq t(\delta, \Delta)$. Letting in (55) $t = t(\delta, \Delta)$ and using (44), we get the following statement.

Theorem 5.12. *Suppose Condition 5.7 and conditions (13), (44), (49), (50), and (52)–(55) are satisfied. Then a solution to the problem (5) is uniquely determined for all $0 \leq t \leq t(\delta, \Delta)$, where $t(\delta, \Delta)$ is specified by (45). Moreover,*

$$\|x(t) - x^*\|_{X_1} < \sqrt{\frac{2\alpha(t)}{d}}, \quad 0 \leq t \leq t(\delta, \Delta); \tag{56}$$

$$\|x(t(\delta, \Delta)) - \dot{x}^*\|_{X_1} < \sqrt{\frac{2m(\delta + \Delta)}{d}}. \tag{57}$$

Let us denote

$$\mathbb{M}(x^*) = \left\{ x^* + F'^*(x^*)v : \|v\|_{X_2} \leq \min\left\{1, \frac{1}{2L}, \frac{1}{4\sqrt{d}}\right\}\right\}.$$

By (37) and (57) we conclude that in the case where $\Delta = 0$, that is,

$$\xi \in \mathbb{M}(x^*) \tag{58}$$

the mapping that takes each approximate operator $\widetilde{F}(x)$ to the element $x_{n(\delta,0)}$ or the element $x(t(\delta,0))$, defines a regularization algorithm for the original equation (1). In general situation where $\Delta > 0$, estimates (37) and (57) mean that this algorithm is stable with respect to small perturbations of representation (58). Finally, if $\delta = \Delta = 0$, then in (32) and (56) we can set $n(\delta,\Delta) = t(\delta,\Delta) = \infty$. Therefore in this case we have the estimates

$$\|x_n - x^*\|_{X_1} = O(\sqrt{\alpha_n}), n = 0, 1, \ldots; \quad \|x(t) - x^*\|_{X_1} = O(\sqrt{\alpha(t)}), t \geq 0.$$

5.7 On Construction of Stable Iterative Methods for Smooth Irregular Equations and Equations with Discontinuous Operators

This section is devoted to justification of a general scheme of constructing stable iterative processes for nonlinear equations under different conditions on their operators. At the beginning, we consider irregular equations with smooth operators in a Hilbert space and propose an approach to constructing stable iterative methods on the basis of classical minimization processes for strongly convex finite–dimensional optimization problems. It is shown that the gradient projection method (2.1) can be obtained in this context as a simple special case. Then we explain how to extend the scheme to equations with discontinuous operators. We also discuss an extension of the methods of parametric approximations to discontinuous equations.

We start with equation (6.1) in the assumption that $F : X_1 \to X_2$ is a nonlinear Fréchet differentiable operator in a pare of Hilbert spaces (X_1, X_2). Suppose the Lipschitz condition (4.1.2) is satisfied so that

$$\|F'(x) - F'(y)\|_{L(X_1,X_2)} \leq L\|x - y\|_{X_1} \quad \forall x, y \in \Omega_R(x^*). \tag{1}$$

Let $\widetilde{F} : X_1 \to X_2$ be an approximation of $F(x)$. Assume that $\widetilde{F}(x)$ satisfies (1) with the same constant L. Furthermore, let

$$\|\widetilde{F}(x) - F(x)\|_{X_2} \leq \delta, \quad \|\widetilde{F}'(x) - F'(x)\|_{L(X_1,X_2)} \leq \delta \quad \forall x \in \Omega_R(x^*). \tag{2}$$

According to the above conditions,

$$\|F'(x)\|_{L(X_1,X_2)}, \quad \|\widetilde{F}'(x)\|_{L(X_1,X_2)} \leq N_1 \quad \forall x \in \Omega_R(x^*).$$

Now pick a finite–dimensional subspace $\mathcal{M} \subset X_1$ and assume that the following condition is satisfied (compare with Condition 5.3):

Condition 5.8.

$$N(F'(x^*)) \cap \mathcal{M} = \{0\}.$$

We shall consider $\mathcal{M}$ as a Euclidean space supplied with the norm and scalar product of X_1.

Also, choose an element $\xi \in X_1$ satisfying Condition 5.4 with some $\Delta \geq 0$. Let us construct for equation (6.1) the approximating optimization problem

$$\min_{x \in \mathcal{M}} \widetilde{\Psi}_{\mathcal{M}, \xi}(x), \tag{3}$$

where

$$\widetilde{\Psi}_{\mathcal{M}, \xi}(x) = \frac{1}{2}\|\widetilde{F}_{\mathcal{M}, \xi}(x)\|_{X_2}^2; \quad \widetilde{F}_{\mathcal{M}, \xi}(x) = \widetilde{F}(x + \xi - P_{\mathcal{M}}\xi), \quad x \in \mathcal{M}.$$

With the notation

$$\mathcal{M}_\xi = \{x \in X_1 : x = u + \xi, u \in \mathcal{M}\},$$

we can rewrite the problem (3) as

$$\min_{x \in \mathcal{M}_\xi} \frac{1}{2}\|\widetilde{F}(x)\|_{X_2}^2.$$

It can easily be checked that the functional $\widetilde{\Psi}_{\mathcal{M}, \xi} : X_1 \to \mathbf{R}$ has the Fréchet derivative $\widetilde{F}'^*(x + \xi - P_{\mathcal{M}}\xi)\widetilde{F}(x + \xi - P_{\mathcal{M}}\xi), x \in \Omega_R(x^*)$ but if we consider $\widetilde{\Psi}_{\mathcal{M}, \xi}(x)$ as a functional on $\mathcal{M}$, then the derivative takes the form

$$\widetilde{\Psi}'_{\mathcal{M}, \xi}(x) = P_{\mathcal{M}}\widetilde{F}'^*(x + \xi - P_{\mathcal{M}}\xi)\widetilde{F}(x + \xi - P_{\mathcal{M}}\xi), \quad x \in \mathcal{M}. \tag{4}$$

Further, let

$$\Delta \leq \frac{R}{2}. \tag{5}$$

Let us remark that by (5) and Condition 5.4,

$$\|(x + \xi - P_{\mathcal{M}}\xi) - x^*\|_{X_1} \leq$$
$$\leq \|(P_{\mathcal{M}} - E)(x^* - \xi)\|_{X_1} + \|x - P_{\mathcal{M}}x^*\|_{X_1} \leq R$$

for all $x \in \Omega_{R/2}(P_{\mathcal{M}}x^*) \cap \mathcal{M}$. This allows to use (1) and (2) for points $x + \xi - P_{\mathcal{M}}\xi$ with $x \in \Omega_{R/2}(P_{\mathcal{M}}x^*) \cap \mathcal{M}$.

The aim of forthcoming arguing is to estimate from below the norm $\|\widetilde{\Psi}'_{\mathcal{M},\xi}(x)\|_x$
From (1.1.14), (1), (2), (4), and Condition 5.4 we obtain

$$\|\widetilde{\Psi}'_{\mathcal{M},\xi}(x)\|_{X_1} = \|P_{\mathcal{M}}\widetilde{F}'^*(x+\xi-P_{\mathcal{M}}\xi)\widetilde{F}(x+\xi-P_{\mathcal{M}}\xi)\|_{X_1} =$$
$$= \|P_{\mathcal{M}}\widetilde{F}'^*(x+\xi-P_{\mathcal{M}}\xi)[\widetilde{F}(x+\xi-P_{\mathcal{M}}\xi)-$$
$$- \widetilde{F}(P_{\mathcal{M}}x^*+\xi-P_{\mathcal{M}}\xi) - \widetilde{F}'(x+\xi-P_{\mathcal{M}}\xi)(x-P_{\mathcal{M}}x^*)]+$$
$$+ P_{\mathcal{M}}\widetilde{F}'^*(x+\xi-P_{\mathcal{M}}\xi)[\widetilde{F}(P_{\mathcal{M}}x^*+\xi-P_{\mathcal{M}}\xi)-$$
$$- \widetilde{F}(x^*) - \widetilde{F}'(x^*)(P_{\mathcal{M}}x^*+\xi-P_{\mathcal{M}}\xi-x^*)]+$$
$$+ P_{\mathcal{M}}\widetilde{F}'^*(x+\xi-P_{\mathcal{M}}\xi)\widetilde{F}'(x+\xi-P_{\mathcal{M}}\xi)(x-P_{\mathcal{M}}x^*)+$$
$$+ P_{\mathcal{M}}\widetilde{F}'^*(x+\xi-P_{\mathcal{M}}\xi)(\widetilde{F}(x^*)-F(x^*))+$$
$$+ P_{\mathcal{M}}\widetilde{F}'^*(x+\xi-P_{\mathcal{M}}\xi)\widetilde{F}'(x^*)(P_{\mathcal{M}}x^*+\xi-P_{\mathcal{M}}\xi-x^*)\|_{X_1} \geq$$
$$\geq \|P_{\mathcal{M}}\widetilde{F}'^*(x+\xi-P_{\mathcal{M}}\xi)\widetilde{F}'(x+\xi-P_{\mathcal{M}}\xi)P_{\mathcal{M}}(x-P_{\mathcal{M}}x^*)\|_{X_1}-$$
$$- \frac{1}{2}N_1 L\|x-P_{\mathcal{M}}x^*\|^2_{X_1} - N_1^2\Delta - \frac{1}{2}N_1 L\Delta^2 - N_1\delta$$
$$\forall x \in \Omega_{R/2}(P_{\mathcal{M}}x^*)\cap\mathcal{M}.$$

$$(6)$$

As in Section 5.2, from Condition 5.8 we deduce

$$\rho \equiv \inf\{\lambda : \lambda \in \sigma(P_{\mathcal{M}}F'^*(x^*)F'(x^*)P_{\mathcal{M}})\} > 0.$$

We shall need the following analog of Lemma 5.3.

Lemma 5.8. *Let Conditions 5.4 and 5.8 be satisfied. Assume that*

$$\delta < \frac{\rho}{4N_1}; \quad \|(x+\xi-P_{\mathcal{M}}\xi)-x^*\|_{X_1} \leq \frac{\rho-4N_1\delta}{4N_1 L},$$
$$x \in \Omega_{R/2}(P_{\mathcal{M}}x^*)\cap\mathcal{M}.$$

$$(7)$$

Then

$$\left(-\infty,\frac{\rho}{2}\right)\cap\sigma(P_{\mathcal{M}}\widetilde{F}'^*(x+\xi-P_{\mathcal{M}}\xi)\widetilde{F}'(x+\xi-P_{\mathcal{M}}\xi)P_{\mathcal{M}}) = \emptyset.$$

Now suppose the error levels δ and Δ satisfy the following additional condition:

$$\delta + L\Delta \leq \frac{\rho}{8N_1}.$$

$$(8)$$

Using (8), for all points $x \in \Omega_{R/2}(P_{\mathcal{M}}x^*)\cap\mathcal{M}$ such that

$$\|x-P_{\mathcal{M}}x^*\|_{X_1} \leq \frac{\rho}{8N_1 L}$$

$$(9)$$

we get

$$\|(x + \xi - P_{\mathcal{M}}\xi) - x^*\|_{X_1} \leq$$
$$\leq \|(P_{\mathcal{M}} - E)(x^* - \xi)\|_{X_1} + \|x - P_{\mathcal{M}}x^*\|_{X_1} \leq$$
$$\leq \Delta + \frac{\rho}{8N_1 L} \leq \frac{\rho - 8N_1\delta}{8N_1 L} + \frac{\rho}{8N_1 L} = \frac{\rho - 4N_1\delta}{4N_1 L}.$$

Therefore conditions (8) and (9) yield inequalities (7). For this reason, referring to Lemma 5.8 below, we shall use (8) and (9) instead of (7).

Let us now analyze the first summand in the right part of inequality (6). By Lemma 5.8, for a bounded selfadjoint operator

$$A = P_{\mathcal{M}}\widetilde{F}'^*(x + \xi - P_{\mathcal{M}}\xi)\widetilde{F}'(x + \xi - P_{\mathcal{M}}\xi)P_{\mathcal{M}}$$

we have

$$\frac{\rho}{2} \leq \lambda \leq N_1^2 \quad \forall \lambda \in \sigma(A)$$

and hence,

$$\|Ay\|_{X_1}^2 = \int_0^{N_1^2} \lambda^2 d(E_\lambda y, y)_{X_1} \geq$$

$$\geq \frac{1}{4}\rho^2 \int_0^{N_1^2} d(E_\lambda y, y)_{X_1} = \frac{1}{4}\rho^2 \|y\|_{X_1}^2 \quad \forall y \in \mathcal{M}.$$

Here $\{E_\lambda\}$ is the spectral family of A. Consequently for all points

$$x \in \Omega_{R/2}(P_{\mathcal{M}}x^*) \cap \mathcal{M}$$

satisfying condition (9),

$$\|P_{\mathcal{M}}\widetilde{F}'^*(x + \xi - P_{\mathcal{M}}\xi)\widetilde{F}'(x + \xi - P_{\mathcal{M}}\xi)P_{\mathcal{M}}(x - P_{\mathcal{M}}x^*)\|_{X_1} \geq$$

$$\geq \frac{1}{2}\rho\|x - P_{\mathcal{M}}x^*\|_{X_1}.$$

Using (6) we then obtain

$$\|\widetilde{\Psi}'_{\mathcal{M},\xi}(x)\|_{X_1} \geq \frac{1}{2}\rho\|x - P_{\mathcal{M}}x^*\|_{X_1} - \frac{1}{2}N_1 L\|x - P_{\mathcal{M}}x^*\|_{X_1}^2 - $$
$$- N_1^2\Delta - \frac{1}{2}N_1 L\Delta^2 - N_1\delta. \tag{10}$$

From (10) it follows that for all $x \in \mathcal{M}$ such that

$$\|x - P_{\mathcal{M}}x^*\|_{X_1} \leq r_0 \equiv \min\left\{\frac{R}{2}, \frac{\rho}{2N_1 L}\right\}, \tag{11}$$

the estimate is valid:

$$\|x - P_{\mathcal{M}}x^*\|_{X_1} \leq \rho^{-1}(4\|\tilde{\Psi}'_{\mathcal{M},\xi}(x)\|_{X_1} + 4N_1^2\Delta + 2N_1 L\Delta^2 + 4N_1\delta). \quad (12)$$

Thus we proved the following auxiliary result.

Theorem 5.13. *Suppose Conditions 5.4 and 5.8 and inequalities (5), (8), and (11) are satisfied. Then (12) is fulfilled.*

Let us now study local properties of the functional $\tilde{\Psi}_{\mathcal{M},\xi}(x)$ on $\mathcal{M}$ in a neighborhood of the point $P_{\mathcal{M}}x^*$. Our immediate task is to prove that $\tilde{\Psi}_{\mathcal{M},\xi}(x)$ is strongly convex near $P_{\mathcal{M}}x^*$ and to estimate the diameter of a neighborhood where $\tilde{\Psi}_{\mathcal{M},\xi}(x)$ is strongly convex. By Lemma 5.8,

$$\left(P_{\mathcal{M}}\tilde{F}'^*(x + \xi - P_{\mathcal{M}}\xi)\tilde{F}'(x + \xi - P_{\mathcal{M}}\xi)P_{\mathcal{M}}(x - y), x - y\right)_{X_1} \geq$$
$$\geq \frac{1}{2}\rho\|x - y\|_{X_1}^2 \quad \forall x, y \in \Omega_{r_0}(P_{\mathcal{M}}x^*) \cap \mathcal{M}. \quad (13)$$

In addition to (5) and (8), let

$$\delta + N_1\Delta \leq \frac{\rho}{8L}. \quad (14)$$

For all

$$x, y \in \Omega_{r_1}(P_{\mathcal{M}}x^*) \cap \mathcal{M}; \quad r_1 = \min\left\{\frac{R}{2}, \frac{\rho}{16N_1 L}\right\}$$

we have

$$\left(\tilde{\Psi}'_{\mathcal{M},\xi}(x) - \tilde{\Psi}'_{\mathcal{M},\xi}(y), x - y\right)_{X_1} =$$
$$= \left(P_{\mathcal{M}}\tilde{F}'^*(x + \xi - P_{\mathcal{M}}\xi)[\tilde{F}(x + \xi - P_{\mathcal{M}}\xi) - \tilde{F}(y + \xi - P_{\mathcal{M}}\xi) - \right.$$
$$\left. - \tilde{F}'(x + \xi - P_{\mathcal{M}}\xi)(x - y)], x - y\right)_{X_1} +$$
$$+ \left(P_{\mathcal{M}}\tilde{F}'^*(x + \xi - P_{\mathcal{M}}\xi)\tilde{F}'(x + \xi - P_{\mathcal{M}}\xi)P_{\mathcal{M}}(x - y), x - y\right)_{X_1} +$$
$$+ \left([P_{\mathcal{M}}\tilde{F}'^*(x + \xi - P_{\mathcal{M}}\xi) - \right.$$
$$\left. - P_{\mathcal{M}}\tilde{F}'^*(y + \xi - P_{\mathcal{M}}\xi)]\tilde{F}(y + \xi - P_{\mathcal{M}}\xi), x - y\right)_{X_1}.$$

Hence from (13) and (14) it follows

$$(\widetilde{\Psi}'_{\mathcal{M},\xi}(x) - \widetilde{\Psi}'_{\mathcal{M},\xi}(y), x - y)_{X_1} \geq$$

$$\geq \frac{1}{2}\rho\|x - y\|_{X_1}^2 - \frac{1}{2}N_1 L\|x - y\|_{X_1}^3 - L\|x - y\|_{X_1}^2 \cdot$$

$$\cdot \Big(\|\widetilde{F}(y + \xi - P_{\mathcal{M}}\xi) - \widetilde{F}(P_{\mathcal{M}}x^* + \xi - P_{\mathcal{M}}\xi)\|_{X_2} +$$

$$+ \|\widetilde{F}(P_{\mathcal{M}}x^* + \xi - P_{\mathcal{M}}\xi) - \widetilde{F}(x^*)\|_{X_2} + \|\widetilde{F}(x^*) - F(x^*)\|_{X_2}\Big) \geq$$

$$\geq \|x - y\|_{X_1}^2 \Big(\frac{1}{2}\rho - \frac{1}{2}N_1 L\|x - P_{\mathcal{M}}x^*\|_{X_1} - \frac{1}{2}N_1 L\|y - P_{\mathcal{M}}x^*\|_{X_1} -$$

$$- N_1 L\|y - P_{\mathcal{M}}x^*\|_{X_1} - L(\delta + N_1\Delta)\Big) \geq \frac{\rho}{4}\|x - y\|_{X_1}^2.$$

$$(15)$$

Taking into account the criterion (1.1.17), by (15) we obtain the following proposition.

Theorem 5.14. *Suppose Conditions 5.4 and 5.8 and inequalities (5), (8), and (14) are satisfied. Then the functional $\widetilde{\Psi}_{\mathcal{M},\xi}(x)$ is strongly convex on $\Omega_{r_1}(P_{\mathcal{M}}x^*) \cap \mathcal{M}$ with the strong convexity constant $\kappa = \rho/8$.*

We now intend to prove that there exists a point $\overline{x} \in \Omega_{r_1}^0(P_{\mathcal{M}}x^*) \cap \mathcal{M}$ with $\widetilde{\Psi}'_{\mathcal{M},\xi}(\overline{x}) = 0$. The point $\overline{x}$ obviously is a local unconstrained minimizer of $\widetilde{\Psi}_{\mathcal{M},\xi}(x)$ over the finite–dimensional subspace $\mathcal{M}$.

The following proposition is well known (see, e.g., [95]).

Lemma 5.9. *Let $f : D \to H$ be a continuous mapping from an open subset D of a finite–dimensional Euclidean space H into H. Let $a \in D$ and*

$$(f(x), x - a)_H > 0 \quad \forall x \in \partial D \equiv \overline{D}\backslash D.$$

Then there exists a point $\overline{x} \in D$ such that $f(\overline{x}) = 0$.

Let us apply Lemma 5.9 setting

$$f(x) = \widetilde{\Psi}'_{\mathcal{M},\xi}(x), \quad a = P_{\mathcal{M}}x^*, \quad D = \Omega_{r_1}^0(P_{\mathcal{M}}x^*) \cap \mathcal{M}, \quad H = \mathcal{M}.$$

Suppose δ and Δ are sufficiently small so that

$$2N_1^2\Delta + N_1 L\Delta^2 + 2N_1\delta \leq \frac{1}{4}\rho r_1. \tag{16}$$

Since

$$(\widetilde{\Psi}'_{\mathcal{M},\xi}(x), x - P_{\mathcal{M}}x^*)_{X_1} = (P_{\mathcal{M}}\widetilde{F}'^*(x + \xi - P_{\mathcal{M}}\xi)[\widetilde{F}(x + \xi - P_{\mathcal{M}}\xi) -$$
$$- \widetilde{F}(P_{\mathcal{M}}x^* + \xi - P_{\mathcal{M}}\xi) - \widetilde{F}'(x + \xi - P_{\mathcal{M}}\xi)(x - P_{\mathcal{M}}x^*)], x - P_{\mathcal{M}}x^*)_{X_1} +$$
$$+ (P_{\mathcal{M}}\widetilde{F}'^*(x + \xi - P_{\mathcal{M}}\xi)[\widetilde{F}(P_{\mathcal{M}}x^* + \xi - P_{\mathcal{M}}\xi) - \widetilde{F}(x^*) -$$
$$- \widetilde{F}'(x^*)(P_{\mathcal{M}}x^* + \xi - P_{\mathcal{M}}\xi - x^*)], x - P_{\mathcal{M}}x^*)_{X_1} +$$
$$+ (P_{\mathcal{M}}\widetilde{F}'^*(x + \xi - P_{\mathcal{M}}\xi)(\widetilde{F}(x^*) - F(x^*)), x - P_{\mathcal{M}}x^*)_{X_1} +$$
$$+ (P_{\mathcal{M}}\widetilde{F}'^*(x + \xi - P_{\mathcal{M}}\xi)\widetilde{F}'(x^*)(P_{\mathcal{M}}x^* + \xi - P_{\mathcal{M}}\xi - x^*), x - P_{\mathcal{M}}x^*)_{X_1} +$$
$$+ (P_{\mathcal{M}}\widetilde{F}'^*(x + \xi - P_{\mathcal{M}}\xi)\widetilde{F}'(x + \xi - P_{\mathcal{M}}\xi)P_{\mathcal{M}}(x - P_{\mathcal{M}}x^*), x - P_{\mathcal{M}}x^*)_{X_1},$$

using (13) and (16), for all $x \in \mathcal{M}$ with $\|x - P_{\mathcal{M}}x^*\|_{X_1} = r_1$ we get

$$(\widetilde{\Psi}'_{\mathcal{M},\xi}(x), x - P_{\mathcal{M}}x^*)_{X_1} \geq$$
$$\geq \frac{1}{2}\rho\|x - P_{\mathcal{M}}x^*\|^2_{X_1} - \frac{1}{2}N_1 L\|x - P_{\mathcal{M}}x^*\|^3_{X_1} -$$
$$- \left(N_1^2\Delta + \frac{1}{2}N_1 L\Delta^2 + N_1\delta\right)\|x - P_{\mathcal{M}}x^*\|_{X_1} \geq$$
$$\geq r_1\left(\frac{3}{8}\rho r_1 - \left(N_1^2\Delta + \frac{1}{2}N_1 L\Delta^2 + N_1\delta\right)\right) \geq \frac{1}{4}\rho r_1^2 > 0.$$

From Theorems 5.13, 5.14 and Lemma 5.9 it now follows

Theorem 5.15. *Suppose Conditions 5.4 and 5.8 and inequalities (5), (8), and (16) are satisfied. Then there exists a unique point $\overline{x} \in \Omega^0_{r_1}(P_{\mathcal{M}}x^*) \cap \mathcal{M}$ such that*

$$\widetilde{\Psi}'_{\mathcal{M},\xi}(\overline{x}) = 0.$$

Moreover, the estimate is valid:

$$\|\overline{x} - P_{\mathcal{M}}x^*\|_{X_1} \leq \rho^{-1}(4N_1^2\Delta + 2N_1 L\Delta^2 + 4N_1\delta).$$

Now let us choose a finite–dimensional optimization method and apply it to the problem (3). Such methods usually generate relaxational sequences of iterative points, that is, sequences $\{x_n\} \subset \mathcal{M}$ with the property

$$\widetilde{\Psi}_{\mathcal{M},\xi}(x_{n+1}) \leq \widetilde{\Psi}_{\mathcal{M},\xi}(x_n), \quad n = 0, 1, \dots.$$

Let us determine conditions under which all points of a relaxational sequence $x_0, x_1, \dots$ belong to the ball $\Omega_{r_1}(P_{\mathcal{M}}x^*) \cap \mathcal{M}$. Suppose

$$x_0 \in \Omega_{r_1}(P_{\mathcal{M}}x^*) \cap \mathcal{M}.$$

It is sufficient to establish the inclusion

$$E_{\mathcal{M},\xi}(x_0) \equiv \Omega_{r_1}(P_{\mathcal{M}}x^*) \cap \{x \in \mathcal{M} : \widetilde{\Psi}_{\mathcal{M},\xi}(x) \leq \widetilde{\Psi}_{\mathcal{M},\xi}(x_0)\} \subset$$
$$\subset \Omega^0_{r_1}(P_{\mathcal{M}}x^*) \cap \mathcal{M}.$$

By Theorem 5.14 and (1.1.18),

$$\widetilde{\Psi}_{\mathcal{M},\xi}(x) \geq \widetilde{\Psi}_{\mathcal{M},\xi}(x_0) + (\widetilde{\Psi}'_{\mathcal{M},\xi}(x_0), x - x_0)_{X_1} + \frac{\rho}{8}\|x - x_0\|^2_{X_1} \qquad (17)$$
$$\forall x \in \Omega_{r_1}(P_{\mathcal{M}}x^*) \cap \mathcal{M}.$$

From (17) it follows that for each $x \in E_{\mathcal{M},\xi}(x_0)$,

$$\frac{\rho}{8}\|x - x_0\|^2_{X_1} \leq (\widetilde{\Psi}'_{\mathcal{M},\xi}(x_0), x_0 - x)_{X_1} =$$
$$= (P_{\mathcal{M}}\widetilde{F}'^*(x_0 + \xi - P_{\mathcal{M}}\xi)\widetilde{F}(x_0 + \xi - P_{\mathcal{M}}\xi), x_0 - x)_{X_1} \leq$$
$$\leq N_1\Big(\|\widetilde{F}(x_0 + \xi - P_{\mathcal{M}}\xi) - \widetilde{F}(P_{\mathcal{M}}x^* + \xi - P_{\mathcal{M}}\xi)\|_{X_2} +$$
$$+ \|\widetilde{F}(P_{\mathcal{M}}x^* + \xi - P_{\mathcal{M}}\xi) - \widetilde{F}(x^*)\|_{X_2} +$$
$$+ \|\widetilde{F}(x^*) - F(x^*)\|_{X_2}\Big)\|x_0 - x\|_{X_1} \leq$$
$$\leq (N_1^2\|x_0 - P_{\mathcal{M}}x^*\|_{X_1} + N_1^2\Delta + N_1\delta)\|x_0 - x\|_{X_1}.$$

Consequently,

$$\|x_0 - x\|_{X_1} \leq$$
$$\leq 8\rho^{-1}(N_1^2\|x_0 - P_{\mathcal{M}}x^*\|_{X_1} + N_1^2\Delta + N_1\delta) \quad \forall x \in E_{\mathcal{M},\xi}(x_0).$$

Let δ and Δ satisfy the condition

$$16N_1^2\Delta + 16N_1\delta \leq \rho r_1, \qquad (18)$$

and the starting point $x_0 \in \mathcal{M}$ be chosen such that

$$\|x_0 - P_{\mathcal{M}}x^*\|_{X_1} < r_2 \equiv \frac{\rho r_1}{16N_1^2 + 2\rho}. \qquad (19)$$

Then by (19) we get

$$\|x - P_{\mathcal{M}}x^*\|_{X_1} \leq \|x_0 - x\|_{X_1} + \|x_0 - P_{\mathcal{M}}x^*\|_{X_1} \leq$$
$$\leq \rho^{-1}((8N_1^2 + \rho)\|x_0 - P_{\mathcal{M}}x^*\|_{X_1} + 8N_1^2\Delta + 8N_1\delta) < r_1.$$

Therefore,

$$E_{\mathcal{M},\xi}(x_0) \subset \Omega^0_{r_1}(P_{\mathcal{M}}x^*) \cap \mathcal{M}.$$

Thus we proved that if $x_0 \in \mathcal{M}$ is taken subject to (19), then all points of a relaxational sequence $\{x_n\}_{n=0}^{\infty}$ lie in a region of strong convexity of the functional $\widetilde{\Psi}_{\mathcal{M},\xi}(x)$.

Consider an iterative method for solving an optimization problem

$$\min_{x \in \Lambda(x_0)} \psi(x),$$

where

$$x_0 \in \mathcal{M}, \quad \Lambda(x_0) = \{x \in \mathcal{M} : \psi(x) \leq \psi(x_0)\},$$

and a functional $\psi : \mathcal{M} \to \mathbf{R}$ is smooth and strongly convex on $\Lambda(x_0)$. Assume that for each ψ satisfying the above conditions, the chosen method generates a relaxational sequence $\{x_n\} \subset \mathcal{M}$ ($\psi(x_{n+1}) \leq \psi(x_n), n = 0, 1, \dots$) such that

$$\lim_{n \to \infty} \|x_n - \widetilde{x}\|_{X_1} = 0;$$

$$\widetilde{x} \in \Lambda(x_0), \quad \psi(\widetilde{x}) = \min_{x \in \Lambda(x_0)} \psi(x). \tag{20}$$

Then we shall say that the method is a converging relaxational method.

Combining (20) with (12), we get the following result.

Theorem 5.16. *Suppose Conditions 5.4 and 5.8 are satisfied. Let δ and Δ satisfy conditions (5), (8), (14), (16), and (18), and the initial point $x_0 \in \mathcal{M}$ is chosen subject to (19). Let a sequence $\{x_n\}$, $n = 0, 1, \dots$ be constructed by a converging relaxational method in application to the problem (3). Then*

$$\lim_{n \to \infty} \left\|P_{\mathcal{M}_\xi}(x_n) - x^*\right\|_{X_1} = \left\|P_{\mathcal{M}_\xi}(\widetilde{x}) - x^*\right\|_{X_1} \leq$$
$$\leq \rho^{-1}\left((4N_1^2 + \rho)\Delta + 2N_1 \dot{L}\Delta^2 + 4N_1\delta\right); \tag{21}$$

$$\widetilde{x} \in \Omega_{r_1}^0(P_{\mathcal{M}}x^*) \cap \mathcal{M}, \quad \widetilde{\Psi}_{\mathcal{M},\xi}(\widetilde{x}) = \min_{x \in \Omega_{r_1}^0(P_{\mathcal{M}}x^*) \cap \mathcal{M}} \widetilde{\Psi}_{\mathcal{M},\xi}(x).$$

According to (21), the points $P_{\mathcal{M}_\xi}(x_n) = x_n + \xi - P_{\mathcal{M}}\xi$ are attracted to a neighborhood of x^* with diameter of order $O(\delta + \Delta)$. If $\xi \in \mathbb{M}(x^*) \equiv x^* + \mathcal{M}$ and $\delta = 0$, then by (21),

$$\lim_{n \to \infty} \left\|P_{\mathcal{M}_\xi}(x_n) - x^*\right\|_{X_1} = 0.$$

Clearly, the problem of finding such ξ is of the same order of complexity as the original equation. Estimate (21) now implies that the iterative process $\{x_n\}$ is stable with respect to errors in $F(x)$ and to deviations of the controlling element ξ from its admissible subset $\mathbb{M}(x^*)$. Theorems 5.15 and 5.16 justify the following general algorithm of constructing stable iterative processes for equations (6.1) with smooth operators.

Algorithm 5.1

1) Pick a finite–dimensional subspace $\mathcal{M} \subset X_1$ and an element $\xi \in X_1$ satisfying Conditions 5.4 and 5.8.
2) Construct the optimization problem (3).
3) Choose a converging relaxational method.
4) Choose an initial point $x_0 \in \mathcal{M}$ subject to condition (19).
5) Apply the chosen method to the problem (3) with the starting point x_0 and obtain a sequence $\{x_n\}$.
6) Obtain approximations $P_{\mathcal{M}_\xi}(x_n) = x_n + \xi - P_{\mathcal{M}}\xi, n \geq 0$ to the solution x^*.

Example 5.4. Let us apply to (3) the standard method of gradient descent

$$x_0 \in \mathcal{M}, \quad x_{n+1} = x_n - \gamma \widetilde{\Psi}'_{\mathcal{M},\xi}(x_n) \tag{22}$$

with sufficiently small stepsize $\gamma > 0$. Since $x_n \in \mathcal{M}, n \geq 0$, from (4) we get

$$x_{n+1} = x_n - \gamma P_{\mathcal{M}} \widetilde{F}'^*(x_n + \xi - P_{\mathcal{M}}\xi)\widetilde{F}(x_n + \xi - P_{\mathcal{M}}\xi) =$$
$$= P_{\mathcal{M}}(x_n - \gamma \widetilde{F}'^*(x_n + \xi - P_{\mathcal{M}}\xi)\widetilde{F}(x_n + \xi - P_{\mathcal{M}}\xi)).$$

Setting $\widetilde{x}_n = x_n + \xi - P_{\mathcal{M}}\xi$ we can rewrite this process as follows:

$$\widetilde{x}_0 \in \mathcal{M}_\xi,$$
$$\widetilde{x}_{n+1} = P_{\mathcal{M}}\big(x_n - \gamma \widetilde{F}'^*(x_n + \xi - P_{\mathcal{M}}\xi)\widetilde{F}(x_n + \xi - P_{\mathcal{M}}\xi)\big) +$$
$$+ \xi - P_{\mathcal{M}}\xi = P_{\mathcal{M}}\big(\widetilde{x}_n - \xi - \gamma \widetilde{F}'^*(\widetilde{x}_n)\widetilde{F}(\widetilde{x}_n)\big) + \xi. \tag{23}$$

The iterative process (23) coincides with the method (2.1) studied in Section 5.2 under slightly different conditions. Under the conditions of Theorem 5.16, the iteration (22) converges at a geometrical rate to a local minimizer $\overline{x}$ of the functional $\widetilde{\Psi}_{\mathcal{M},\xi}(x)$ [113, 115]. Therefore we get the following refined version of Theorem 5.2.

Theorem 5.17. *Suppose Conditions 5.4 and 5.8 and inequalities (5), (8), (14), (16), and (18) are satisfied. Let $x_0 \in \mathcal{M}$ be chosen subject to (19). Then there exists $\gamma_0 > 0$ such that for the process (23) with $\gamma \in (0, \gamma_0)$ we have*

$$\|\widetilde{x}_n - \widetilde{x}\|_{X_1} \leq C q_0^n, \quad n = 0, 1, \ldots; \quad q_0 \in (0, 1),$$

where $\widetilde{x} = P_{\mathcal{M}_\xi}(\overline{x})$. Moreover,

$$\|\widetilde{x} - x^*\|_{X_1} \leq \rho^{-1}\big((4N_1^2 + \rho)\Delta + 2N_1 L\Delta^2 + 4N_1\delta\big).$$

Further examples of converging relaxational methods are conjugate gradient type processes, Newton's method and various quasi–Newton schemes [113, 115].

Example 5.5. Fletcher and Reeves' conjugate gradient method is a converging relaxational method [34]. The method is defined by the formulae:

$$x_{n+1} = x_n + \gamma_n p_n; \tag{24}$$

$$\gamma_n > 0, \quad \widetilde{\Psi}_{\mathcal{M},\xi}(x_n + \gamma_n p_n) = \min_{\gamma \geq 0} \widetilde{\Psi}_{\mathcal{M},\xi}(x_n + \gamma p_n);$$

$$p_n = \begin{cases} -\widetilde{\Psi}'_{\mathcal{M},\xi}(x_n), & n = 0, K, 2K \ldots, \\ -\widetilde{\Psi}'_{\mathcal{M},\xi}(x_n)+ \\ +\|\widetilde{\Psi}'_{\mathcal{M},\xi}(x_n)\|^2_{X_1}\|\widetilde{\Psi}'_{\mathcal{M},\xi}(x_{n-1})\|^{-2}_{X_1}p_{n-1}, & n \neq 0, K, 2K \ldots, \end{cases}$$

$$K = \dim\mathcal{M}.$$

Let us now present an approach to relaxation of smoothness conditions imposed above on the operator $F(x)$. Consider an operator equation

$$F(x) = 0, \quad x \in D(F), \tag{25}$$

where $F : D(F) \subset X_1 \to X_2$ is a nonlinear mapping and X_1, X_2 are Hilbert spaces. We don't assume that $F(x)$ is everywhere defined. Moreover, the domain of definition $D(F)$ doesn't need to contain any open subset of X_1. Also, no conditions on the structure of $D(F)$ are a priori imposed. We only assume that $x^* \in D(F)$ and the pare $(\mathcal{M}, \xi)$ is chosen such that

$$\mathcal{M}_\xi \subset D(F).$$

Now, define for $F(x)$ the approximating operator

$$F_{\mathcal{M},\xi}(x) = F(x + \xi - P_\mathcal{M}\xi), \quad x \in \mathcal{M}.$$

Suppose the operator $F_{\mathcal{M},\xi} : \mathcal{M} \to X_2$ is Fréchet differentiable, and $F'_{\mathcal{M},\xi}(x)$ satisfies the Lipschitz condition

$$\|F'_{\mathcal{M},\xi}(x) - F'_{\mathcal{M},\xi}(y)\|_{L(\mathcal{M},X_2)} \leq L\|x - y\|_{X_1} \tag{26}$$
$$\forall x, y \in \Omega_R(P_\mathcal{M}x^*) \cap \mathcal{M}.$$

We stress that $F'_{\mathcal{M},\xi}(x), x \in \Omega_R(P_\mathcal{M}x^*) \cap \mathcal{M}$ is considered as a mapping from $\mathcal{M}$ into X_2 but not from X_1 into X_2. Moreover, nigher a differentiability of

$F_{\mathcal{M},\xi}(x)$ nor even definiteness of this operator on an open neighborhood of $P_{\mathcal{M}}x^*$ is assumed. In the case where $F(x)$ is well defined and differentiable on X_1 so that

$$F'_{\mathcal{M},\xi}(x) = F'(x + \xi - P_{\mathcal{M}}\xi), \quad x \in X_1,$$

we make no assumptions concerning regularity of $F(x)$.

Suppose instead of the true operator $F(x)$ we have an approximation

$$\widetilde{F} : D(\widetilde{F}) \subset X_1 \to X_2.$$

Then the true operator $F_{\mathcal{M},\xi}(x)$ is not accessible but we can use

$$\widetilde{F}_{\mathcal{M},\xi}(x) = \widetilde{F}(x + \xi - P_{\mathcal{M}}\xi), \quad x \in \mathcal{M}$$

as its approximation. Suppose that

$$\mathcal{M}_\xi \subset D(\widetilde{F}).$$

Let the operator $\widetilde{F}_{\mathcal{M},\xi} : \mathcal{M} \to X_2$ be Fréchet differentiable and satisfy condition (26) with the same constant L. Also, let

$$\|\widetilde{F}_{\mathcal{M},\xi}(x) - F_{\mathcal{M},\xi}(x)\|_{X_2} \le \delta, \quad \|\widetilde{F}'_{\mathcal{M},\xi}(x) - F'_{\mathcal{M},\xi}(x)\|_{L(\mathcal{M},X_2)} \le \delta$$

$$\forall x \in \Omega_R(P_{\mathcal{M}}x^*) \cap \mathcal{M}. \tag{27}$$

From (26) it follows that

$$\|F'_{\mathcal{M},\xi}(x)\|_{L(\mathcal{M},X_2)}, \quad \|\widetilde{F}'_{\mathcal{M},\xi}(x)\|_{L(\mathcal{M},X_2)} \le N_1$$

$$\forall x \in \Omega_R(P_{\mathcal{M}}x^*) \cap \mathcal{M}.$$

We can now approximate equation (25) by a variational problem of the form (3):

$$\min_{x \in \mathcal{M}} \widetilde{\Psi}_{\mathcal{M},\xi}(x); \quad \widetilde{\Psi}_{\mathcal{M},\xi}(x) = \frac{1}{2}\|\widetilde{F}_{\mathcal{M},\xi}(x)\|^2_{X_2}, \quad x \in \mathcal{M}. \tag{28}$$

Let us introduce the operator

$$F_{\mathcal{M},x^*} : \mathcal{M} \to X_2; \quad F_{\mathcal{M},x^*}(x) = F(x + x^* - P_{\mathcal{M}}x^*), \quad x \in \mathcal{M}. \tag{29}$$

In essence, $F_{\mathcal{M},x^*}(x)$ is the restriction of $F(x)$ to the affine subspace $\mathcal{M}_{x^*}$ obtained from $\mathcal{M}$ by a parallel shift at x^*. Suppose the operator $F_{\mathcal{M},x^*}(x)$ is Fréchet differentiable. We denote

$$\Delta_2 = \|F'_{\mathcal{M},\xi}(P_{\mathcal{M}}x^*) - F'_{\mathcal{M},x^*}(P_{\mathcal{M}}x^*)\|_{L(\mathcal{M},X_2)}.$$

In the sequel we assume that the following analog of Condition 5.8 is satisfied.

Condition 5.9. *The operator* $F'_{\mathcal{M}, x^*}(P_{\mathcal{M}}x^*) : \mathcal{M} \to X_2$ *is injective, i.e.,*

$$N\left(F'_{\mathcal{M}, x^*}(P_{\mathcal{M}}x^*)\right) = \{0\}.$$

As above, we denote

$$\tilde{\rho} = \inf\{\lambda : \lambda \in \sigma(F'^*_{\mathcal{M}, x^*}(P_{\mathcal{M}}x^*)F'_{\mathcal{M}, x^*}(P_{\mathcal{M}}x^*))\}.$$

Condition 5.9 then yields $\tilde{\rho} > 0$. The value $\tilde{\rho} = \tilde{\rho}(\mathcal{M})$ obviously depends on the subspace $\mathcal{M}$ but $\tilde{\rho}$ is independent of the element ξ.

Arguing as in Section 5.1 and taking into account (26), we get the following analog of Lemma 5.8.

Lemma 5.10. *Suppose Condition 5.9 is satisfied. Assume that*

$$\delta + \Delta_2 \le \frac{\tilde{\rho}}{8N_1}; \quad \|x - P_{\mathcal{M}}x^*\|_{X_1} \le \frac{\tilde{\rho}}{8N_1 L}, \quad x \in \Omega_R(P_{\mathcal{M}}x^*) \cap \mathcal{M}. \quad (30)$$

Then

$$\left(-\infty, \frac{\tilde{\rho}}{2}\right) \cap \sigma(\tilde{F}'^*_{\mathcal{M}, \xi}(x)\tilde{F}'_{\mathcal{M}, \xi}(x)) = \emptyset.$$

Notice that conditions (30) serve as analogs of inequalities (8) and (9).

Using Lemma 5.10 instead of Lemma 5.8, we obtain the following nonsmooth versions of Theorems 5.14–5.16 (with $\tilde{\rho} = \tilde{\rho}(\mathcal{M})$ in place of $\rho = \rho(\mathcal{M})$).

Theorem 5.18. *Suppose Condition 5.9 and inequalities (26) and (27) are satisfied. Let δ, Δ_2, and $\|F_{\mathcal{M}, \xi}(P_{\mathcal{M}}x^*)\|_{X_2}$ satisfy the conditions*

$$\delta + \Delta_2 \le \frac{\tilde{\rho}}{8N_1}, \quad \delta + \|F_{\mathcal{M}, \xi}(P_{\mathcal{M}}x^*)\|_{X_2} \le \frac{\tilde{\rho}}{8L}. \quad (31)$$

Then the functional $\tilde{\Psi}_{\mathcal{M}, \xi}(x)$ *is strongly convex on* $\Omega_{r_3}(P_{\mathcal{M}}x^*) \cap \mathcal{M}$, *where*

$$r_3 = \min\left\{R, \frac{\tilde{\rho}}{16N_1 L}\right\}.$$

If, in addition,

$$\delta + \|F_{\mathcal{M}, \xi}(P_{\mathcal{M}}x^*)\|_{X_2} \le \frac{\tilde{\rho} r_3}{8N_1},$$

then there exists a unique point $\overline{x} \in \Omega^0_{r_3}(P_{\mathcal{M}}x^*) \cap \mathcal{M}$ *such that* $\tilde{\Psi}'_{\mathcal{M}, \xi}(\overline{x}) = 0$.

Theorem 5.19. *Suppose Condition 5.9 and inequalities (26), (27), and (31) are satisfied. Assume that*

$$\delta + \|F_{\mathcal{M},\xi}(P_{\mathcal{M}}x^*)\|_{X_2} \leq \frac{\tilde{\rho}r_3}{16N_1}.$$

Let an initial point $x_0 \in \mathcal{M}$ be chosen subject to the condition

$$\|x_0 - P_{\mathcal{M}}x^*\|_{X_1} < \frac{\tilde{\rho}r_3}{16N_1^2 + 2\tilde{\rho}}. \tag{32}$$

Then for each sequence $\{x_n\}_{n=0}^{\infty} \subset \mathcal{M}$ such that

$$\tilde{\Psi}_{\mathcal{M},\xi}(x_{n+1}) \leq \tilde{\Psi}_{\mathcal{M},\xi}(x_n), n = 0, 1, \ldots; \quad \lim_{n\to\infty} \|\tilde{\Psi}'_{\mathcal{M},\xi}(x_n)\|_{X_1} = 0,$$

we have

$$\lim_{n\to\infty} \|P_{\mathcal{M}_\xi}(x_n) - x^*\|_{X_1} = \|P_{\mathcal{M}_\xi}(\overline{x}) - x^*\|_{X_1} \leq$$
$$\leq 4\tilde{\rho}^{-1}N_1(\|F_{\mathcal{M},\xi}(P_{\mathcal{M}}x^*)\|_{X_2} + \delta) + \Delta.$$

Under the conditions of Theorems 5.18 and 5.19, we can construct stable iterative methods for equation (25) using the following algorithm.

Algorithm 5.2

1) Pick a finite–dimensional subspace $\mathcal{M} \subset X_1$ and an element $\xi \in X_1$ such that $\mathcal{M}_\xi \subset D(F) \cap D(\tilde{F})$ and Conditions 5.4 and 5.9 are satisfied.
2) Construct the optimization problem (28).
3) Choose a converging relaxational method.
4) Choose an initial point $x_0 \in \mathcal{M}$ subject to (32).
5) Apply the chosen method to (28) with the starting point x_0 and obtain a sequence $\{x_n\}$.
6) Obtain approximations $P_{\mathcal{M}_\xi}(x_n) = x_n + \xi - P_{\mathcal{M}}\xi, n \geq 0$ to the solution x^*.

In particular, applying to (28) the method of gradient descent from Example 5.4, we obtain the process

$$x_0 \in \mathcal{M}, \quad x_{n+1} = x_n - \gamma \tilde{F}'^*_{\mathcal{M},\xi}(x_n)\tilde{F}_{\mathcal{M},\xi}(x_n); \quad \gamma > 0. \tag{33}$$

Let us make several remarks concerning a choice of $\mathcal{M}$ subject to Conditions 5.8 and 5.9.

First note that if $F(x)$ is differentiable and the derivative $F'(x^*)$ is an injective operator, then both conditions are satisfied with an arbitrary finite–dimensional subspace $\mathcal{M}$. Furthermore, let $\widehat{X}_1 \subset X_1$ be a Banach space embedded into X_1 and the operator $F : \widehat{X}_1 \to X_2$ be differentiable with the injective derivative $F'(x^*) : \widehat{X}_1 \to X_2$, where $x^* \in \widehat{X}_1$. Then Conditions 5.8 and 5.9 are satisfied with an arbitrary finite–dimensional $\mathcal{M} \subset \widehat{X}_1$. Operators $F(x)$ of this type arise, e.g., when studying acoustic inverse scattering problems ([30] and Sections 6.3–6.5).

Generally speaking, the subspace $\mathcal{M}$ can be chosen subject to an analog of Condition 5.8 involving a point $\hat{x} \neq x^*$ in place of x^*. In fact, let a differentiable operator $F(x)$ satisfy (1), and

$$N(F'(\hat{x})) \cap \mathcal{M} = \{0\}. \tag{34}$$

According to (34),

$$\omega_1 \equiv \inf_{\substack{h \in \mathcal{M} \\ \|h\|_{X_1}=1}} \|F'(\hat{x})h\|_{X_2} > 0. \tag{35}$$

Combining (1) and (35), we get the following proposition.

Theorem 5.20. *Let a differentiable operator $F(x)$ satisfy (1). Assume that (34) is fulfilled, and*

$$\|\hat{x} - x^*\|_{X_1} \leq \min\left\{\frac{R}{2}, \frac{\omega_1}{2N_1}\right\}.$$

Then Condition 5.8 is satisfied.

Due to Theorem 5.20, variations of a point $\hat{x}$ in a neighborhood of x^* provide additional means for a choice of $\mathcal{M}$.

Similar arguing is applicable to Condition 5.9, as well. By analogy with (29), consider the operator

$$F_{\mathcal{M},\hat{x}} : \mathcal{M} \to X_2; \quad F_{\mathcal{M},\hat{x}}(x) = F(x + \hat{x} - P_{\mathcal{M}}\hat{x}), \quad x \in \mathcal{M}$$

and denote

$$\Delta_3 = \|F'_{\mathcal{M},x^*}(P_{\mathcal{M}}x^*) - F'_{\mathcal{M},\hat{x}}(P_{\mathcal{M}}\hat{x})\|_{L(\mathcal{M},X_2)}.$$

Instead of Condition 5.9, assume that

$$N\left(F'_{\mathcal{M},\hat{x}}(P_{\mathcal{M}}\hat{x})\right) = \{0\}. \tag{36}$$

By (36) we deduce

$$\omega_2 \equiv \inf_{\substack{h \in \mathcal{M} \\ \|h\|_{X_1}=1}} \|F'_{\mathcal{M},\hat{x}}(P_{\mathcal{M}}\hat{x})h\|_{X_2} > 0.$$

The following assertion is analogous to Theorem 5.20.

Theorem 5.21. *Suppose a point $\hat{x} \in X_1$ and a finite–dimensional subspace $\mathcal{M} \subset X_1$ satisfy (36) and the inequality*

$$\Delta_3 \leq \frac{\omega_2}{2}.$$

Then Condition 5.9 is satisfied.

In a conclusion, let us present a scheme of extension of parametric approximations methods from Chapter 4 and Sections 5.4–5.5 and iteratively regularized methods of Section 5.6 to equation (25) with a discontinuous operator $F : D(F) \subset X_1 \to X_2$. Pick an orthoprojector $Q \in L(X_2)$ and denote $\widetilde{\mathcal{M}} = R(Q)$. Along with (25), consider the approximating equation

$$Q\widetilde{F}_{\mathcal{M},\xi}(x) - Q\widetilde{F}_{\mathcal{M},\xi}(P_{\mathcal{M}}x^*) = 0, \quad x \in \mathcal{M}. \tag{37}$$

The point $P_{\mathcal{M}}x^*$ is clearly a solution to (37). Observe that the operator

$$G : \mathcal{M} \to \widetilde{\mathcal{M}}; \quad G(x) = Q\widetilde{F}_{\mathcal{M},\xi}(x) - Q\widetilde{F}_{\mathcal{M},\xi}(P_{\mathcal{M}}x^*), \quad x \in \mathcal{M} \tag{38}$$

is Fréchet differentiable with a Lipschitz continuous derivative in a neighborhood of $P_{\mathcal{M}}x^*$. Therefore the mapping G satisfies all the conditions imposed on F in Chapter 4 and Sections 5.4–5.6. Since the last summand in (38) is not accessible, instead of G we really have the approximation

$$\widetilde{G} : \mathcal{M} \to \widetilde{\mathcal{M}}; \quad \widetilde{G}(x) = Q\widetilde{F}_{\mathcal{M},\xi}(x), \quad x \in \mathcal{M}.$$

The approximation error is estimated as follows:

$$\|\widetilde{G}(x) - G(x)\|_{X_2} \leq \delta + \|F_{\mathcal{M},\xi}(P_{\mathcal{M}}x^*)\|_{X_2}, \quad x \in \mathcal{M}.$$

Let us remark that $\widetilde{G}'(x) = G'(x), x \in \mathcal{M}$. Now, applying the iterative and continuous methods of Chapter 4 and Sections 5.4–5.6 to the smooth equation (37), we can obtain approximations to the solution $P_{\mathcal{M}}x^*$ of (37) and hence to x^*. Here the case of infinite–dimensional $\widetilde{\mathcal{M}}$ is not excluded; we can even set $Q = E$ and $\widetilde{\mathcal{M}} = X_2$.

Chapter 6

APPLICATIONS OF ITERATIVE METHODS
TO INVERSE PROBLEMS

This chapter is devoted to applications of iterative processes proposed in Chapters 4 and 5. The materials presented here primarily reflect the authors' experience in numerical solution of inverse problems of mathematical physics. Nowadays, there exists a wide range of papers with various numerical algorithms for special classes of inverse problems. These algorithms every so often are formulated at a physical level of severity, without rigorous theoretical analysis; nevertheless the use of such algorithms when solving a separate problem can give quite satisfactory results despite the lack of a comprehensive convergence theory. On the contrary, investigations presented in previous chapters are aimed at a unified formal analysis of wide classes of abstract iterative methods. The methods developed in Chapters 4 and 5 are applicable to irregular operator equations and inverse problems of various forms. Succeeding numerical examples are intended to demonstrate only a capacity for work of these formally justified methods; we don't try to obtain any kind of the best numerical solutions to the considered inverse problems. In order to get such best approximations, one should make the most use of all special features of a problem under consideration. Sections 6.1 and 6.2 are devoted to applications of iterative methods to inverse gravimetrical problems. In Sections 6.3–6.5 we describe numerical experiments with these methods when applied to inverse acoustic scattering problems in different settings.

6.1   Reconstruction of Bounded Homogeneous Inclusion

In this section we consider the problem of detecting a 3D body localized below the earth surface, by measurements of the gravitational attraction force on the surface [17, 102]. Suppose subterranean masses of density $\rho^{(0)}$ fill the half–space $\mathbf{R}^3_- = \{r = (r_1, r_2, r_3) : r_3 \leq 0\}$ and contain a bounded homogeneous inclusion D of density $\rho^{(1)} > \rho^{(0)}$; the values $\rho^{(0)}$ and $\rho^{(1)}$ are given. Suppose D is a domain in $\mathbf{R}^3_-$; then the body D creates an additional gravitational field with the potential

$$U(r) = \gamma_0(\rho^{(1)} - \rho^{(0)}) \int_D \frac{dr'_1 dr'_2 dr'_3}{\sqrt{(r'_1 - r_1)^2 + (r'_2 - r_2)^2 + (r'_3 - r_3)^2}}, \quad r \in \mathbf{R}^3,$$

where γ_0 is the gravitational constant. Due to the presence of D, on the earth surface $P = \{(r_1, r_2, r_3) : r_3 = 0\}$ arises an additional vertical component of the gravitational attraction force

$$g(r_1, r_2) = \frac{\partial}{\partial r_3} U(r)\Big|_{r_3=0} =$$

$$= \gamma_0(\rho^{(1)} - \rho^{(0)}) \int_D \frac{r'_3 dr'_1 dr'_2 dr'_3}{[(r'_1 - r_1)^2 + (r'_2 - r_2)^2 + r'^2_3]^{3/2}}. \tag{1}$$

Suppose the units of measurements are chosen such that $\gamma_0(\rho^{(1)} - \rho^{(0)}) = 1$. Let Y be a domain in the plane P. The inverse problem of reconstructing a bounded homogeneous inclusion is to determine the body D given the values $\{g(r_1, r_2) : (r_1, r_2) \in Y\}$. Unique solvability of this problem follows by Theorem 3.1.1 of [66].

Suppose the body D is star–shaped with respect to a known point $(0, 0, -H)$, where $H > 0$. Using the spherical coordinates with the origin at $(0, 0, -H)$, we can present the boundary ∂D in the parametric form as follows:

$$r_1 = \rho(\theta, \varphi) \sin \theta \cos \varphi, \quad r_2 = \rho(\theta, \varphi) \sin \theta \sin \varphi,$$

$$r_3 = -H + \rho(\theta, \varphi) \cos \theta;$$

$$\rho(\theta, \varphi) \geq 0, \quad \theta \in [0, \pi], \quad \varphi \in [0, 2\pi] \quad (r = (r_1, r_2, r_3) \in \partial D).$$

Therefore the function $\rho = \rho(\theta, \varphi)$ completely determines the boundary ∂D and hence the body D. By (1) we deduce that $\rho = \rho(\theta, \varphi)$ satisfies the nonlinear integral equation

$$\int_\Omega \Phi(\rho(\theta, \varphi), \theta, \varphi; r_1, r_2) d\theta d\varphi - g(r_1, r_2) = 0, \quad (r_1, r_2) \in Y, \tag{2}$$

where $\Omega = [0, \pi] \times [0, 2\pi]$ and

$$\Phi(\rho, \theta, \varphi; r_1, r_2) =$$

$$= \int_0^\rho \frac{t^2(H - t\cos\theta)\sin\theta \, dt}{[(r_1 - t\sin\theta\cos\varphi)^2 + (r_2 - t\sin\theta\sin\varphi)^2 + (H - t\cos\theta)^2]^{3/2}}. \quad (3)$$

Let $\rho = \rho^*(\theta, \varphi)$ be a solution to equation (2). Suppose the function $\rho^*(\theta, \varphi)$ is continuous on Ω, and

$$0 < \rho^*(\theta, \varphi) < H \quad \forall(\theta, \varphi) \in \Omega.$$

Let us denote

$$F(\rho)(r_1, r_2) = \int_\Omega \Phi(\rho(\theta, \varphi), \theta, \varphi; r_1, r_2) d\theta d\varphi - g(r_1, r_2), \quad (r_1, r_2) \in Y.$$

We shall consider F as an operator acting from $L_2(\Omega)$ into $L_2(Y)$, with the domain of definition

$$D(F) = \{\rho \in C(\Omega) : 0 < \rho(\theta, \varphi) < H \quad \forall(\theta, \varphi) \in \Omega\}.$$

As an approximate operator $\widetilde{F}(\rho)$, we take $F(\rho)$ with the true function $g(r_1, r_2)$ replaced by

$$\widetilde{g}(r_1, r_2) = g(r_1, r_2) + \delta\omega(r_1, r_2), \quad (r_1, r_2) \in Y, \quad \delta \geq 0, \quad (4)$$

where $|\omega(r_1, r_2)| \leq 1$, $(r_1, r_2) \in Y$. In computational experiments, as $\omega(r_1, r_2)$ we take realizations of a random variable uniformly distributed on the segment $[-1, 1]$.

Let us discuss methods of Chapters 4 and 5 from the point of view of their application to equation (2). Observe that the subset $D(F)$ of physically realizable functions $\rho(\theta, \varphi)$ is not open in $L_2(\Omega)$. For this reason, the operator $F : L_2(\Omega) \to L_2(Y)$ doesn't satisfy smoothness conditions required in Chapter 4 and Sections 5.1–5.6. Therefore the methods of Chapter 4 and Sections 5.1–5.6 formally are not applicable to equation (2). On the other hand, the embedding $H^s(\Omega) \subset C(\Omega)$, $s > 1$ yields that F possesses all necessary properties, at least in a small neighborhood of the solution $\rho^* \in D(F) \cap H^2(\Omega)$, if we consider F, e.g., in the pare of spaces $(H^2(\Omega), L_2(Y))$ instead of $(L_2(\Omega), L_2(Y))$. At the same time, implementation of the methods in the case of $X_1 = H^2(\Omega)$ faces a nontrivial problem of numerical approximation of the conjugate operator $F'^*(\rho) : L_2(Y) \to H^2(\Omega)$. As opposite to the case $X_1 = L_2(\Omega)$ where $F'^*(\rho)$ is an integral operator, this $F'^*(\rho)$ can't be written explicitly since evaluation of an element $F'^*(\rho)u \in H^2(\Omega)$ by a given $u \in L_2(Y)$ requires a solution of an auxiliary integro–differential equation. Let us now analyze iterative methods generated by Algorithm 5.2 of Section 5.7. Observe that the scheme of their numerical implementation doesn't depend on the space X_1, provided $\mathcal{M} \subset X_1$; a choice of X_1 affects only the constants from Theorems 5.18, 5.19, and 5.21. Having this in mind, for solving (2) we shall use the iterative process (5.7.33) derived within the framework of Algorithm 5.2.

Let $\mathcal{M}$ be the subspace of spherical functions

$$\mathcal{M}_N = \left\{ r(\theta, \varphi) = \sum_{m=0}^{N} \sum_{n=-m}^{m} a_{mn} Y_m^n(\theta, \varphi) : a_{mn} \in \mathbf{R} \right\}, \qquad (5)$$

where

$$Y_m^n(\theta, \varphi) = \begin{cases} P_m^n(\cos\theta)\cos n\varphi, & n = 0, 1, \ldots, m; \\ P_m^{|n|}(\cos\theta)\sin|n|\varphi, & n = -m, \ldots, -1 \end{cases}$$

are spherical harmonics and $\{P_m^n(\lambda)\}$ $(m = 0, 1, \ldots; n = 0, 1, \ldots, m)$ the classical and associated Legendre functions. Namely,

$$P_m^n(\lambda) = (1 - \lambda^2)^{n/2} \frac{d^n}{d\lambda^n} P_m(\lambda), \quad \lambda \in [-1, 1], \quad n = 0, 1, \ldots, m,$$

where $P_m(\lambda) = P_m^0(\lambda)$ are the Legendre polynomials (5.1.35) ([103]).

Assume that the solution $\rho^*(\theta, \varphi)$ of (2) allows for a good approximation by the subspace $\mathcal{M}_N$. Then in point 1) of Algorithm 5.2 we can simply set

$$\xi(\theta, \varphi) = 0, \quad (\theta, \varphi) \in \Omega.$$

Below we follow just this choice of ξ. Furthermore, if the approximation error $\|(E - P_{\mathcal{M}})\rho^*\|_{C(\Omega)}$ is sufficiently small then there exists $\varepsilon > 0$ such that $\Omega_\varepsilon(P_{\mathcal{M}}\rho^*) \cap \mathcal{M} \subset D(F)$ and the operator

$$F_{\mathcal{M},0} : \mathcal{M} \to L_2(Y); \quad F_{\mathcal{M},0}(\rho) = F(\rho), \quad \rho \in \mathcal{M}$$

is Fréchet differentiable on $\Omega_\varepsilon(P_{\mathcal{M}}\rho^*) \cap \mathcal{M}$. Direct calculations then prove that for all $\rho \in \Omega_\varepsilon(P_{\mathcal{M}}\rho^*) \cap \mathcal{M}$ we have

$$[F'_{\mathcal{M},0}(\rho)h](r_1, r_2) = \int_\Omega \Psi(\rho(\theta, \varphi), \theta, \varphi; r_1, r_2) h(\theta, \varphi) d\theta d\varphi;$$

$$h = h(\theta, \varphi) \in \mathcal{M}, \quad (r_1, r_2) \in Y,$$

where

$$\Psi(\rho, \theta, \varphi; r_1, r_2) = \frac{\partial}{\partial\rho} \Phi(\rho, \theta, \varphi; r_1, r_2) =$$

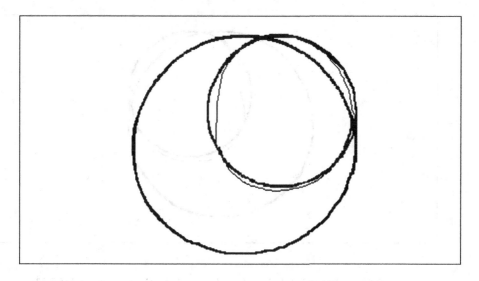

Figure 6.1. Example 6.1: $Y = (-4, 4) \times (-4, 4)$, $\delta = 0$, method (5.7.33)

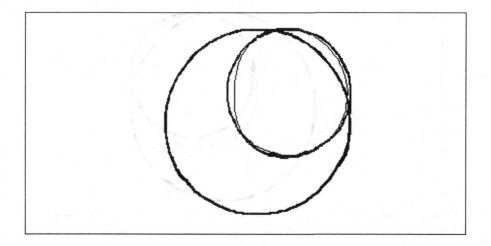

Figure 6.2. Example 6.1: $Y = (-4, 4) \times (-4, 4)$, $\delta = 0.04$, method (5.7.33)

$$= \frac{\rho^2(H - \rho \cos \theta) \sin \theta}{[(r_1 - \rho \sin \theta \cos \varphi)^2 + (r_2 - \rho \sin \theta \sin \varphi)^2 + (H - \rho \cos \theta)^2]^{3/2}}.$$

Also, it is readily seen that condition (5.7.26) is satisfied if $R = \varepsilon$ is small.

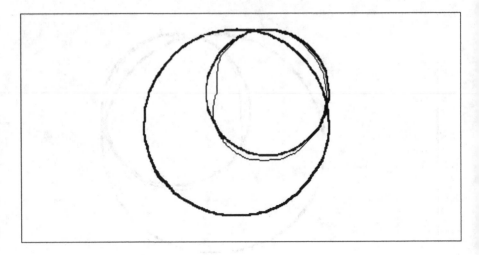

Figure 6.3.   Example 6.1: $Y = (-4, 4) \times (-4, 4)$, $\delta = 0.08$, method (5.7.33)

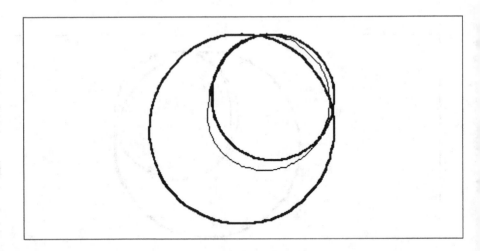

Figure 6.4.   Example 6.2: $Y = (0, 4) \times (0, 4)$, $\delta = 0$, method (5.7.33)

Let us turn to results of computational experiments. We set $H = 1.5$ and consider equation (2) with $g(r_1, r_2)$ obtained by the exact solution

$$\rho^*(\theta, \varphi) = \frac{1}{2} \sin \theta \cos \varphi + \frac{1}{2} \sqrt{\sin^2 \theta \cos^2 \varphi + 3}, \quad (\theta, \varphi) \in \Omega.$$

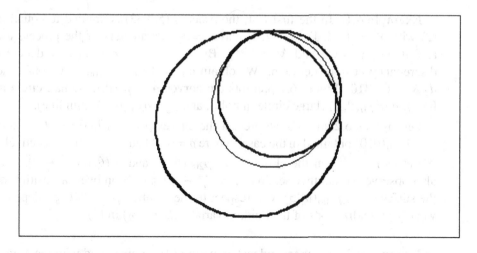

Figure 6.5. Example 6.2: $Y = (0,4) \times (0,4)$, $\delta = 0.04$, method (5.7.33)

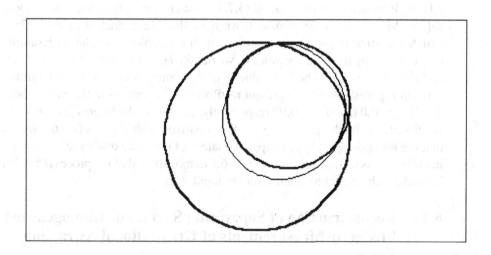

Figure 6.6. Example 6.2: $Y = (0,4) \times (0,4)$, $\delta = 0.08$, method (5.7.33)

The function $\rho = \rho^*(\theta, \varphi)$ defines a sphere of radius 1 centered at $(0.5, 0, -1.5)$. As an initial approximation, we choose a sphere of radius 1.5 centered at the point $(0, 0, -2)$. All computations are performed with the uniform grids 16×16 on Ω and 20×20 on the region of observation $\overline{Y}$. For evaluation of the integral (3) we use the Simpson rule with 5 knots.

Example 6.1. In the first test, the process (5.7.33) is applied to equation (2) with $Y = (-4, 4) \times (-4, 4)$ and $\delta = 0$. Parameters of the process are as follows: $\gamma = 0.1$ and $\mathcal{M} = \mathcal{M}_2$. By $d_n = \|\rho_n - \rho^*\|_{L_2(\Omega)}$ we denote a discrepancy at n–th iteration. We obtain $d_0 = 3.1459$, $d_{100} = 0.3095$, and $d_{200} = 0.2316$. Figure 6.1 presents the curves $\rho = \rho^*(\theta, 0)$ (small circle in bold), $\rho = \rho_0(\theta, 0)$ (large circle in bold), and $\rho = \rho_{200}(\theta, 0)$ (thin line).

On Figures 6.2 and 6.3 we present the curves $\rho = \rho^*(\theta, 0)$, $\rho_0(\theta, \varphi)$, and $\rho = \rho_{200}(\theta, 0)$ obtained in the cases where $\delta = 0.04$ and $\delta = 0.08$ respectively.

Let us remark that the closeness of $\rho_{200}(\theta, \varphi)$ and $\rho^*(\theta, \varphi)$, $\theta \in [0, \pi]$ is also observed at another sections $\varphi = \overline{\varphi} \in (0, \pi)$. Moreover, a position of the surface $\rho = \rho_{200}(\theta, \varphi)$ with respect to the solution $\rho = \rho^*(\theta, \varphi)$ depends weakly on realizations of the random variable $\omega(r_1, r_2)$ in (4).

Example 6.2. In our second test, equation (2) is considered with the region of observation $Y = (0, 4) \times (0, 4)$ so that in this experiment, the volume of observational data is essentially narrowed. The question naturally arises how will this affect the quality of approximations. Intuition suggests that reaching the same approximation error as in Example 6.1 will require more computational efforts. Parameters of the process (5.7.33) are chosen as follows: $\gamma = 0.2$ and $\mathcal{M} = \mathcal{M}_2$. In the case of $\delta = 0$ we have the discrepancy $d_{200} = 1.2785$, which is distinctly greater as compared with the analogous value of Example 6.1. Continuing the iterative process, we obtain $d_{300} = 0.9890$, $d_{400} = 0.7723$, and $d_{500} = 0.6111$. The last value is of the same order as d_{200} in Example 6.1. On Figures 6.4–6.6 we present results of 500 iterations in the cases where $\delta = 0$, $\delta = 0.04$, and $\delta = 0.08$ respectively. As above, the figures give sections of all surfaces by the plane $\varphi = 0$. A comparison with Figures 6.1–6.3 shows that the iteration $\rho_{500}(\theta, \varphi)$ is approximately of the same quality as $\rho_{200}(\theta, \varphi)$ in the previous test. Also, Figures 6.1–6.6 demonstrate that the process (5.7.33) is stable with respect to small errors in input data.

6.2 Reconstruction of Separating Surface of Homogeneous Media by Measurements of Gravitational Attraction Force

Suppose the half–space $\mathbf{R}^3_-$ is filled by two different homogeneous media of known densities $\rho^{(0)}$ and $\rho^{(1)}$; the media are separated by an unknown surface. The observable value is a vertical component of the gravitational force measured on a domain $Y \subset P = \{(r_1, r_2, r_3) : r_3 = 0\}$. Let $r_3 = z(r_1, r_2)$ be an unknown actual separating surface and $r_3 = \overline{z}(r_1, r_2)$ a reference surface, for which corresponding gravitational force on Y is a priori given; $(r_1, r_2) \in \mathbf{R}^2$. For simplicity assume that $z(r_1, r_2) = \overline{z}(r_1, r_2)$, $(r_1, r_2) \in \mathbf{R}^2 \backslash Y$. Also, suppose the units of measurements are chosen such that $\gamma_0(\rho^{(0)} - \rho^{(1)}) = 1$.

Let $U(r)$ and $\overline{U}(r)$ be potentials of the actual and reference configuration of gravitating masses respectively. Then measurements on Y give the difference $g(r_1, r_2)$ between actual and reference vertical components of the gravitational attraction force:

$$g(r_1, r_2) = \frac{\partial}{\partial r_3} U(r)\Big|_{r_3=0} - \frac{\partial}{\partial r_3} \overline{U}(r)\Big|_{r_3=0}, \quad (r_1, r_2) \in Y.$$

The inverse problem under consideration is to reconstruct the surface $r_3 = z(r_1, r_2)$ by the data $\{g(r_1, r_2) : (r_1, r_2) \in Y\}$ ([102]). As in the previous section, this problem is uniquely solvable. It is readily seen that the unknown $r_3 = z^*(r_1, r_2)$ satisfies the following nonlinear integral equation:

$$\int_Y \left(\frac{1}{\sqrt{(r_1' - r_1)^2 + (r_2' - r_2)^2 + z^2(r_1', r_2')}} - \right.$$

$$\left. - \frac{1}{\sqrt{(r_1' - r_1)^2 + (r_2' - r_2)^2 + \overline{z}^2(r_1', r_2')}} \right) dr_1' dr_2' + \tag{1}$$

$$+ g(r_1, r_2) = 0, \quad (r_1, r_2) \in Y.$$

Suppose the solution $r_3 = z^*(r_1, r_2)$ to (1) is continuous on $\overline{Y}$. Below we restrict ourselves with the simplest case of rectangular observation region $Y = (a_1, b_1) \times (a_2, b_2)$. Let us denote

$$F(z)(r_1, r_2) = \int_Y \left(\frac{1}{\sqrt{(r_1' - r_1)^2 + (r_2' - r_2)^2 + z^2(r_1', r_2')}} - \right.$$

$$\left. - \frac{1}{\sqrt{(r_1' - r_1)^2 + (r_2' - r_2)^2 + \overline{z}^2(r_1', r_2')}} \right) dr_1' dr_2' + \tag{2}$$

$$+ g(r_1, r_2), \quad (r_1, r_2) \in Y.$$

We consider F as an operator acting from $L_2(Y)$ into $L_2(Y)$, with the domain of definition

$$D(F) = \{z \in C(\overline{Y}) : z(r_1, r_2) < 0 \quad \forall (r_1, r_2) \in \overline{Y}\}.$$

Having in mind the remarks on methods of Chapters 4 and 5 made in Section 6.1, we apply to (1) iterative processes of Section 5.7, namely, the gradient process (5.7.33) and methods of parametric approximations (4.1.38), (4.1.39), and (4.1.41) (in the way outlined at the end of Section 5.7).

Let us define the subspaces

$$M_N^{(t)} = \left\{ z(r_1, r_2) = \sum_{m,n=0}^N a_{mn} \eta_m(r_1) \eta_n(r_2) : a_{mn} \in \mathbf{R} \right\}$$

and

$$\mathcal{M}_N^{(l)} = \left\{ z(r_1, r_2) = \right.$$

$$= \sum_{m,n=0}^{N} b_{mn} P_m \left(\frac{2r_1 - a_1 - b_1}{b_1 - a_1} \right) P_n \left(\frac{2r_2 - a_2 - b_2}{b_2 - a_2} \right) : b_{mn} \in \mathbf{R} \left. \right\},$$

where

$$\eta_0(r_i) = 1, \quad \eta_1(r_i) = \sin \frac{2\pi(r_i - a_i)}{b_i - a_i}, \quad \eta_2(r_i) = \cos \frac{2\pi(r_i - a_i)}{b_i - a_i},$$

$$\eta_3(r_i) = \sin \frac{4\pi(r_i - a_i)}{b_i - a_i}, \quad \eta_4(r_i) = \cos \frac{4\pi(r_i - a_i)}{b_i - a_i}, \dots; \quad i = 1, 2$$

and $\{P_n(\lambda)\}$ $(n = 0, 1, \dots)$ are the Legendre polynomials (5.1.35).

Suppose the projection $P_{\mathcal{M}} z^*$ provides a good approximation to z^* in the sense that the norm $\|(E - P_{\mathcal{M}})z^*\|_{C(\overline{Y})}$ is small; then as above we can set

$$\xi(r_1, r_2) = 0, \quad (r_1, r_2) \in Y.$$

Furthermore, there exists $\varepsilon > 0$ such that $\Omega_\varepsilon(P_{\mathcal{M}} z^*) \cap \mathcal{M} \subset D(F)$. It can easily be checked that the operator

$$F_{\mathcal{M},0} : \mathcal{M} \to L_2(Y); \quad F_{\mathcal{M},0}(z) = F(z), \quad z \in \mathcal{M}$$

is Fréchet differentiable on $\Omega_\varepsilon(P_{\mathcal{M}} z^*) \cap \mathcal{M}$, and the derivative $F'_{\mathcal{M},0}$ has the form

$$[F'_{\mathcal{M},0}(z)h](r_1, r_2) = -\int_Y \frac{z(r_1', r_2')h(r_1', r_2')dr_1' dr_2'}{[(r_1' - r_1)^2 + (r_2' - r_2)^2 + z^2(r_1', r_2')]^{3/2}};$$

$$h = h(r_1, r_2) \in \mathcal{M}, \quad (r_1, r_2) \in Y.$$

Besides, if $R = \varepsilon$ is sufficiently small, then the Lipschitz condition (5.7.26) is satisfied. Now define a noisy operator $\widetilde{F}(z)$ by (2) with the function $g(r_1, r_2)$ replaced by expression (1.4).

Let us turn to results of computational experiments. All tests are performed with the uniform grid 20×20 on $\overline{Y}$. We set $\xi(r_1, r_2) = 0, (r_1, r_2) \in Y$. In the first series of tests (Examples 6.3 and 6.4), equation (1) is solved by the iterative method (5.7.33). The separating surface to be reconstructed is given by the formula

$$r_3 = z^*(r_1, r_2) \equiv -4 + 2.5| \sin 0.7r_1 \cos 0.7r_2|, \quad (r_1, r_2) \in Y;$$

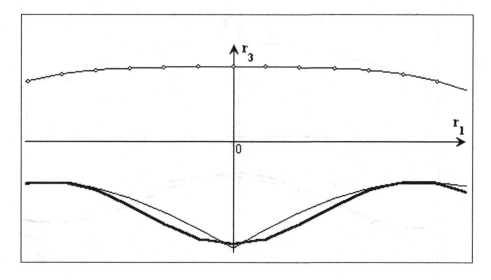

Figure 6.7. Example 6.3: $Y = (-4, 4) \times (-4, 4)$, $\delta = 0$, method (5.7.33), $\mathcal{M} = \mathcal{M}_4^{(t)}$

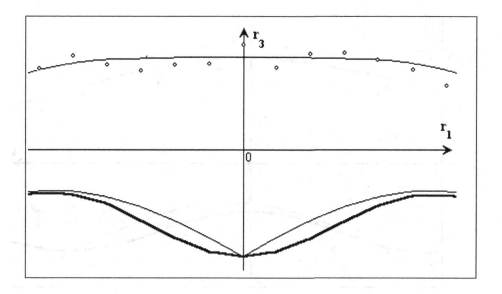

Figure 6.8. Example 6.3: $Y = (-4, 4) \times (-4, 4)$, $\delta = 0.5$, method (5.7.33), $\mathcal{M} = \mathcal{M}_4^{(t)}$

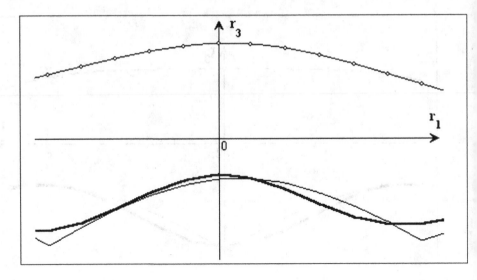

Figure 6.9.   Example 6.4: $Y = (-2, 6) \times (-4, 4)$, $\delta = 0$, method (5.7.33), $\mathcal{M} = \mathcal{M}_4^{(t)}$

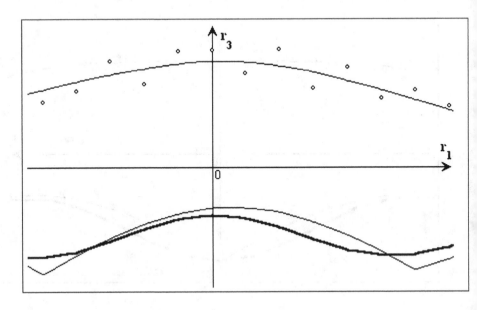

Figure 6.10.   Example 6.4: $Y = (-2, 6) \times (-4, 4)$, $\delta = 0.5$, method (5.7.33), $\mathcal{M} = \mathcal{M}_4^{(t)}$

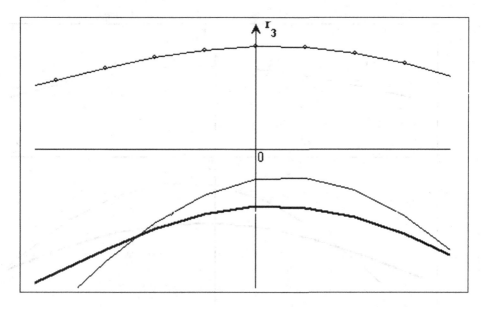

Figure 6.11.   Example 6.4: $Y = (-2, 6) \times (-4, 4)$, $\delta = 0$, method (5.7.33), $\mathcal{M} = \mathcal{M}_5^{(l)}$

the reference surface is $\overline{z}(r_1, r_2) = -6$, $(r_1, r_2) \in \mathbf{R}^2$.

Example 6.3. In the first experiment, the process (5.7.33) with $\mathcal{M} = \mathcal{M}_4^{(t)}$ and $\gamma = 0.25$ is applied to equation (1) in which $Y = (-4, 4) \times (-4, 4)$ and $\delta = 0$. The initial approximation is $z_0(r_1, r_2) = -6$. Let us denote $d_n = \|z_n - z^*\|_{L_2(Y)}$. We have $d_0 = 24.9012$, $d_{100} = 4.8799$, and $d_{500} = 3.0799$. Figure 6.7 presents the curves $r_3 = z^*(r_1, 0)$ (thin line) and $r_3 = z_{500}(r_1, 0)$ (bold line). At the top of Figure 6.7, the curve $r_3 = |g(r_1, 0)|$ is shown. A result obtained after 500 iterations in the case of $\delta = 0.5$ is given on Figure 6.8. Points at the top part of Figure 6.8 mark the perturbed values $|\widetilde{g}(r_1, 0)|$.

Example 6.4. The process (5.7.33) is applied to equation (1) with $Y = (-2, 6) \times (-4, 4)$ and $\delta = 0$. As above, $\mathcal{M} = \mathcal{M}_4^{(t)}$ and the stepsize $\gamma = 0.25$. The initial approximation is chosen as follows: $z_0(r_1, r_2) = -2$. We have $d_0 = 10.4129$, $d_{100} = 4.1037$, and $d_{500} = 3.1485$. On Figure 6.9 we present the curves $r_3 = z^*(r_1, 0)$ and $r_3 = z_{500}(r_1, 0)$. A result of 500 iterations with the same parameters and the error level $\delta = 0.5$ is shown on Figure 6.10.

Results for the same parameters with $\mathcal{M} = \mathcal{M}_5^{(l)}$ in the cases where $\delta = 0$ and $\delta = 0.5$ are presented on Figures 6.11 and 6.12.

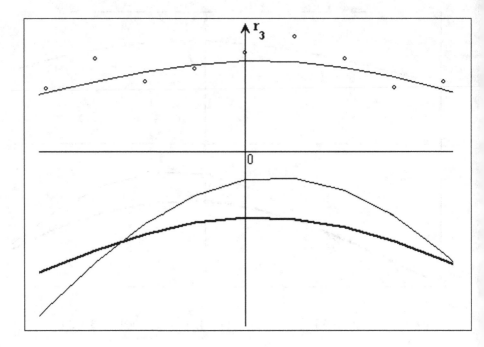

Figure 6.12. Example 6.4: $Y = (-2, 6) \times (-4, 4)$, $\delta = 0.5$, method (5.7.33), $\mathcal{M} = \mathcal{M}_5^{(l)}$

In the second series of tests (Examples 6.5–6.7), for solving equation (1) we use methods from Examples 4.1, 4.2, and 4.4. The methods are applied to equation (5.7.37) constructed by the original operator $\mathcal{F}(w) = F(\overline{z} + w)$ and two orthoprojectors $P_\mathcal{M}, Q \in L(L_2(Y))$. Let $P_\mathcal{M} = Q$ be the orthoprojector from $L_2(Y)$ onto $\mathcal{M} = \widetilde{\mathcal{M}}_N^{(t)}$, where

$$\widetilde{\mathcal{M}}_N^{(t)} = \left\{ w(r_1, r_2) = \sum_{m,n=1}^{N} c_{mn} \eta_m(r_1) \eta_n(r_2) : c_{mn} \in \mathbf{R} \right\};$$

$$\eta_s(r_i) = \sin \frac{\pi s(r_i - a_i)}{b_i - a_i}, \quad i = 1, 2, \quad s = 1, 2, \ldots.$$

We put $Y = (1, 2) \times (1, 2)$ and $\overline{z}(r_1, r_2) = -1.5$, $(r_1, r_2) \in Y$. The surface to be reconstructed is defined as follows: $z^*(r_1, r_2) = \overline{z}(r_1, r_2) + w^*(z_1, z_2)$,

$$w^*(z_1, z_2) = \begin{cases} A(r_1 - \overline{a}_1)(\overline{b}_1 - r_1)(r_2 - \overline{a}_2)(\overline{b}_2 - r_2), & (r_1, r_2) \in Y_1, \\ 0, & (r_1, r_2) \in Y \backslash Y_1; \end{cases}$$

$$Y_1 = (\overline{a}_1, \overline{b}_1) \times (\overline{a}_2, \overline{b}_2) \subset Y, \quad A > 0.$$

The initial approximation is $w_0(r_1, r_2) = 0$, $(r_1, r_2) \in Y$.

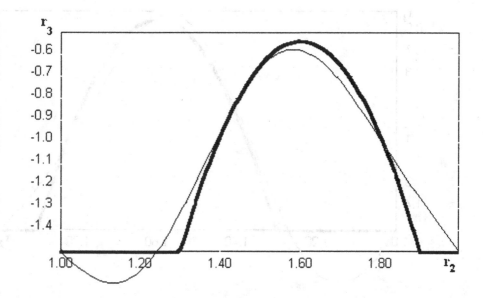

Figure 6.13. Example 6.5: $Y = (1, 2) \times (1, 2)$, $\delta = 0$, method (4.1.38)

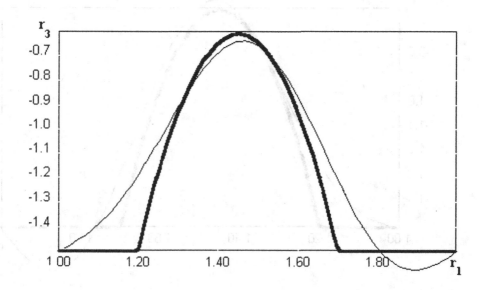

Figure 6.14. Example 6.5: $Y = (1, 2) \times (1, 2)$, $\delta = 0$, method (4.1.38)

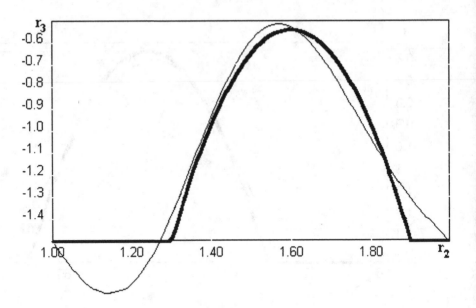

Figure 6.15. Example 6.5: $Y = (1, 2) \times (1, 2)$, $\delta = 0.05$, method (4.1.38)

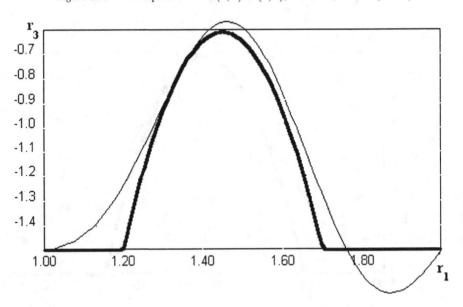

Figure 6.16. Example 6.5: $Y = (1, 2) \times (1, 2)$, $\delta = 0.05$, method (4.1.38)

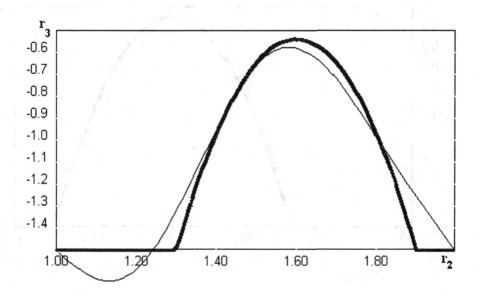

Figure 6.17. Example 6.6: $Y = (1, 2) \times (1, 2)$, $\delta = 0$, method (4.1.39)

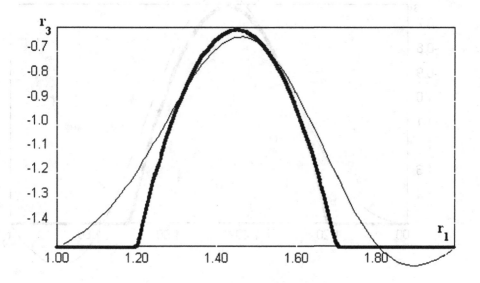

Figure 6.18. Example 6.6: $Y = (1, 2) \times (1, 2)$, $\delta = 0$, method (4.1.39)

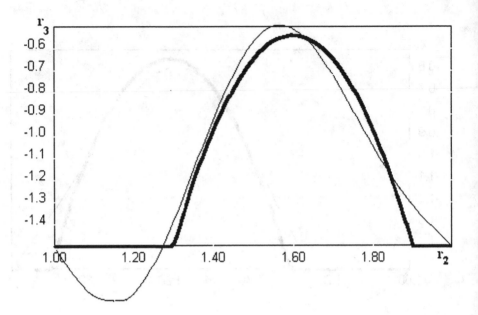

Figure 6.19. Example 6.6: $Y = (1, 2) \times (1, 2)$, $\delta = 0.05$, method (4.1.39)

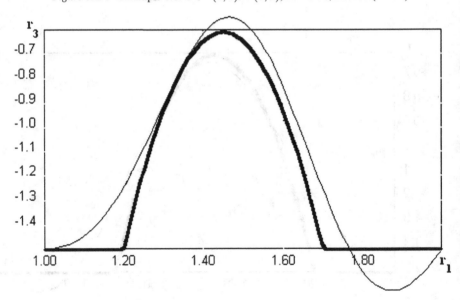

Figure 6.20. Example 6.6: $Y = (1, 2) \times (1, 2)$, $\delta = 0.05$, method (4.1.39)

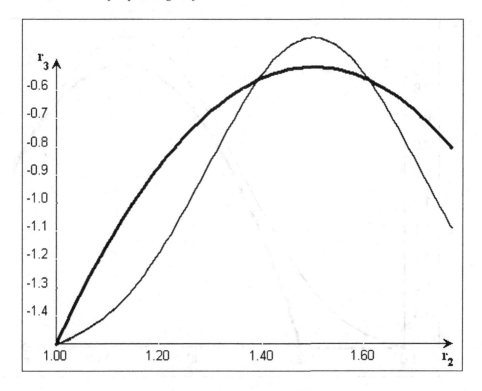

Figure 6.21. Example 6.7: $Y = (1, 2) \times (1, 2)$, $\delta = 0$, method (4.1.41)

Example 6.5. Let $Y_1 = (1.2, 1.7) \times (1.3, 1.9)$ and $\delta = 0$. The process (4.1.38) is applied to equation (1) with the solution $z^*(r_1, r_2)$ such that $\max_{(r_1, r_2) \in Y} z^*(r_1, r_2) = -0.5$. Recall that this process is based on Tikhonov's regularization of a linearized equation at each iteration. Setting $\mathcal{M} = \widetilde{\mathcal{M}}_3^{(t)}$ and $\alpha_n = 0.001 n^{-0.2}$, $n = 1, 2, \ldots$, we get $d_0 = 0.2922$ and $d_{10} = 0.0912$. On Figures 6.13 and 6.14 we present the curves $r_3 = z_{10}(1.5, r_2)$ (thin line), $r_3 = z^*(1.5, r_2)$ (bold line) and $r_3 = z_{10}(r_1, 1.5)$, $r_3 = z^*(r_1, 1.5)$. A result obtained after 10 iterations with $\delta = 0.05$ is shown on Figures 6.15 and 6.16.

Example 6.6. We apply to (1) the process (4.1.39) with $N = 3$. Here a linearized equation is regularized by the iterated Tikhonov's method. All input data are the same as in Example 6.5. On Figures 6.17 and 6.18 we present the curves $r_3 = z_{10}(1.5, r_2)$, $r_3 = z^*(1.5, r_2)$ and $r_3 = z_{10}(r_1, 1.5)$, $r_3 = z^*(r_1, 1.5)$ obtained after 10 iterations with $\delta = 0$. A result of 10 iterations in the case of $\delta = 0.05$ is given on Figures 6.19 and 6.20.

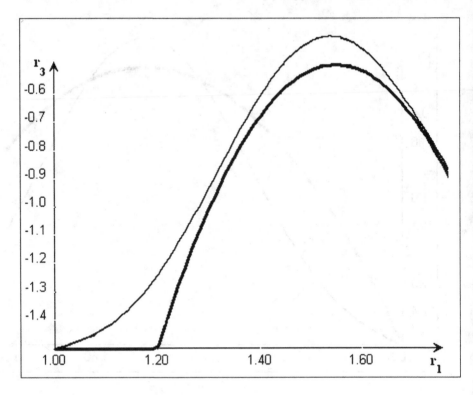

Figure 6.22.   Example 6.7: $Y = (1, 2) \times (1, 2)$, $\delta = 0$, method (4.1.41)

Example 6.7. We use the process (4.1.41) with $\mu_0 = 1.4$. At each iteration, a linearized equation is regularized by the explicit iterative method. Setting $Y_1 = (1.2, 1.9) \times (1, 2)$ and $\delta = 0$, we get $d_0 = 0.4462$, $d_{100} = 0.1376$, and $d_{200} = 0.1419$. On Figures 6.21 and 6.22 we present the curves $r_3 = z_{200}(1.5, r_2)$ (thin line), $r_3 = z^*(1.5, r_2)$ (bold line) and $r_3 = z_{200}(r_1, 1.5)$, $r_3 = z^*(r_1, 1.5)$. A result of 200 iterations in the case of $\delta = 0.05$ is shown on Figures 6.23 and 6.24.

6.3   Acoustic Inverse Medium Problem

This section is concerned with a numerical analysis of an inverse scattering problem. The problem is to determine a refractive index of an inhomogeneous medium, from measurements of far field patterns of scattered time–harmonic acoustic waves.

Assume that a 3D homogeneous medium contains a penetrable inhomogeneous inclusion localized within a bounded domain $R \subset \mathbf{R}^3$. Acoustic properties of the medium at a point $r \in \mathbf{R}^3$ are characterized by the sound speed

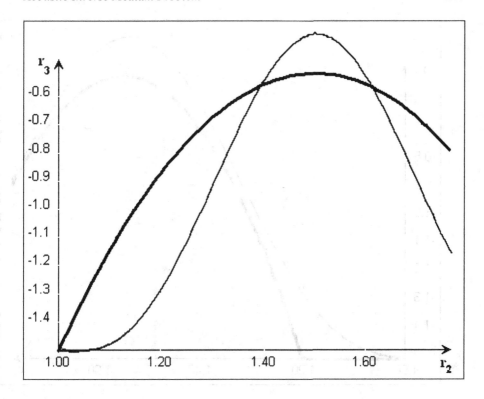

Figure 6.23. Example 6.7: $Y = (1, 2) \times (1, 2)$, $\delta = 0.05$, method (4.1.41)

$c(r)$. Suppose

$$c(r) = c_0, \quad r \in \mathbf{R}^3 \backslash R,$$

where $c_0 > 0$ is a given value. Then the acoustic medium can be completely characterized by the refractive index

$$n(r) = \frac{c_0^2}{c^2(r)}, \quad r \in R.$$

Denote $a(r) = 1 - n(r)$ and suppose that $a \in H^s(\mathbf{R}^3)$ with some $s > 3/2$. We obviously have $a(r) = 0$ for $r \in \mathbf{R}^3 \setminus R$, and $a \in H^s(R)$.

The scattering of an incident time–harmonic plane wave with the complex amplitude $u_{in}(r) = \exp(ik_0 d \cdot r)$ and the wavenumber $k_0 = \omega/c_0$ is governed by the equations

$$\Delta u(r) + k_0^2 n(r) u(r) = 0, \quad u(r) = u_{in}(r) + u_s(r), \quad r \in \mathbf{R}^3 \quad (1)$$

and the Sommerfeld radiation condition

$$\frac{\partial u_s(r)}{\partial \rho} - ik_0 u_s(r) = o\left(\frac{1}{\rho}\right), \quad \rho = |r| \to \infty. \quad (2)$$

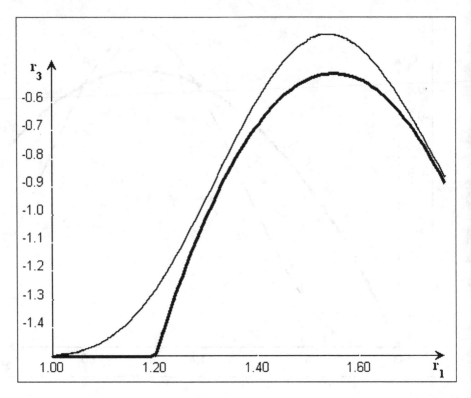

Figure 6.24. Example 6.7: $Y = (1,2) \times (1,2)$, $\delta = 0.05$, method (4.1.41)

By $u \cdot v$ we denote the Euclidean scalar product of vectors $u, v \in \mathbf{R}^3$; ω is a frequency and $d \in S^2 \equiv \{r \in \mathbf{R}^3 : |r| = 1\}$ a direction of moving of the incident wave; $u_s(r)$ is the complex amplitude of a scattered wave field. Throughout this section, the frequency ω is fixed while the direction d may vary within the sphere S^2. The Sommerfeld condition guarantees that the scattered wave $U_s(r,t) = \mathrm{Re}(u_s(r)\exp(-i\omega t))$ is outgoing.

For the scattered field we have the following asymptotic representation ([30, p.222]):

$$u_s(r) = \frac{e^{ik_0\rho}}{\rho}\left\{u_\infty(\hat{r},d) + \mathrm{O}\left(\frac{1}{\rho}\right)\right\}, \quad \hat{r} = \frac{r}{|r|}, \quad \rho = |r| \to \infty. \quad (3)$$

The factor $u_\infty(\hat{r},d) : S^2 \to \mathbf{C}$ in (3) is called the far field pattern or the scattering amplitude of $u_s(r)$; the unit vector $\hat{r}$ indicates the direction of observation.

In the direct medium scattering problem, the refractive index $n = n(r), r \in R$ and the direction $d \in S^2$ of incoming wave are given and the scattered wave field $u_s(r)$ in a bounded region of observation, or the scattering amplitudes

$u_\infty(\cdot, d)$ are to be found ([30]). It is well known ([30, Ch.8]) that this problem has a unique solution.

The inverse medium acoustic scattering problem is to determine the refractive index $n(r), r \in R$ or, equivalently, the function $a(r), r \in R$ given the scattering amplitudes $\{u_\infty(\hat{r}, d) : \hat{r} \in S^2, d \in S^2\}$. It can be shown [30, p.278] that the inverse acoustic problem in this setting has a unique solution. In other words, the refractive index $n(r)$ is uniquely determined by a knowledge of the far field patterns $u_\infty(\hat{r}, d), (\hat{r}, d) \in S^2 \times S^2$.

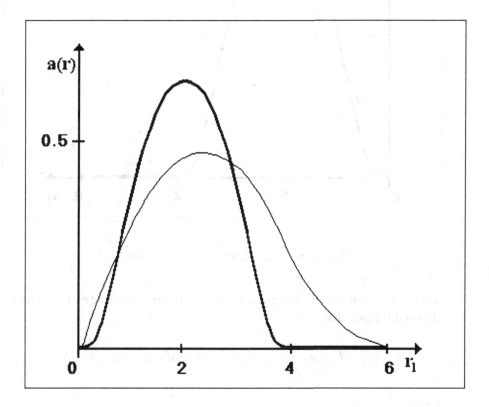

Figure 6.25. Example 6.8: $\delta = 0$, method (5.7.22)

Recall that the direct scattering problem (1)–(2) is equivalent to the Lippmann–Schwinger integral equation [30, p.216]

$$u(r) - k_0^2 \int_R \Gamma(r, r')a(r')u(r')dr' = u_{in}(r), \quad r \in R, \qquad (4)$$

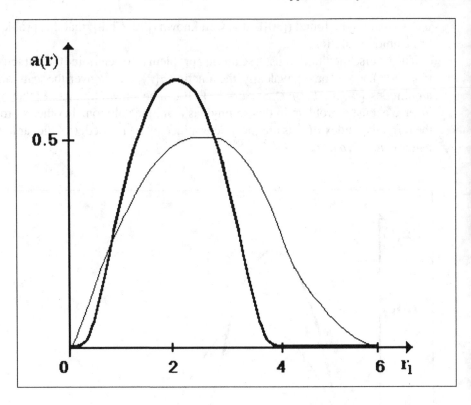

Figure 6.26.   Example 6.8: $\delta = 0.01$, method (5.7.22)

where $\Gamma(r, r')$ is the Green function (3.3.6). Let us denote $v(r) = a(r)u(r)$. By (4) it follows that

$$v(r) - k_0^2 a(r) \int_R \Gamma(r, r')v(r')dr' = a(r)u_{in}(r), \quad r \in R. \tag{5}$$

With the use of notation

$$(Vv)(r) = k_0^2 \int_R \Gamma(r, r')v(r')dr', \quad r \in R,$$

we can rewrite (5) as

$$(E - aV)v = au_{in}.$$

Here a denotes the operator of multiplication by $a(r)$. It is easy to see that if $v(r)$ solves equation (5), then the function $u(r) = u_{in}(r) + (Vv)(r)$ satisfies

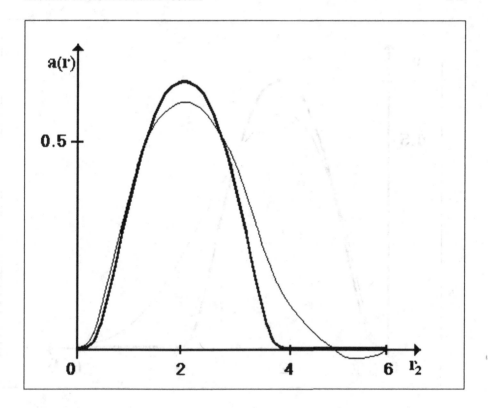

Figure 6.27. Example 6.8: $\delta = 0.01$, method (5.7.22)

(1) and (2). Since by (4),

$$u_s(r) = k_0^2 \int_R \Gamma(r, r')a(r')u(r')dr', \quad r \in \mathbf{R}^3,$$

for the scattering amplitude $u_\infty(\hat{r}, d)$ we have the representation [30, p.223]

$$u_\infty(\hat{r}, d) = -\frac{k_0^2}{4\pi}(Wv)(\hat{r}, d), \quad (\hat{r}, d) \in S^2 \times S^2, \tag{6}$$

where

$$(Wv)(\hat{r}, d) = \int_R e^{-ik_0\hat{r}\cdot r'} v(r')dr', \quad (\hat{r}, d) \in S^2 \times S^2.$$

Notice that $u_\infty(\hat{r}, d)$ is an analytic function of $(\hat{r}, d) \in S^2 \times S^2$.

Combining (5) and (6), we get the following nonlinear equation for the function $a = a(r), r \in R$:

$$-\frac{k_0^2}{4\pi}W(E - aV)^{-1}(au_{in})(\hat{r}, d) = u_\infty(\hat{r}, d), \quad (\hat{r}, d) \in S^2 \times S^2. \tag{7}$$

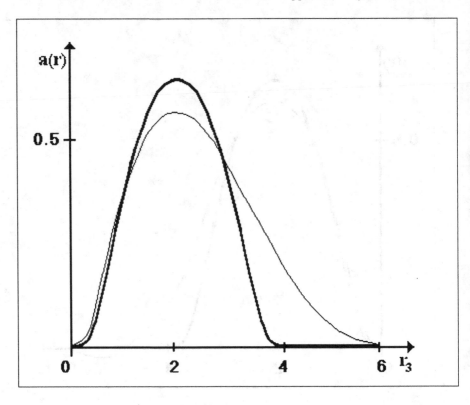

Figure 6.28. Example 6.8: $\delta = 0.01$, method (5.7.22)

It is well known [30, Ch.10; 62] that the problem (7) is irregular and ill–posed for each reasonable choice of the solution space and the space of input data.

Below we apply to (7) the iterative methods (5.7.22) and (5.7.24). Let us define the operator $F : H^s(R) \rightarrow L_2(S^2 \times S^2)$ as follows:

$$F(a)(\hat{r}, d) = \frac{k_0^2}{4\pi} W(E - aV)^{-1}(au_{in})(\hat{r}, d) + u_\infty(\hat{r}, d), \quad a \in H^s(R),$$

$$(\hat{r}, d) \in S^2 \times S^2.$$

The inverse problem now takes the form $F(a) = 0$.

Let $a^*(r)$ be a solution of (7) and $n^*(r) = 1 - a^*(r)$ the unknown refractive index. According to [62], the operator $F'(a^*) : H^s(R) \rightarrow L_2(S^2 \times S^2)$ is injective. Therefore Condition 5.8 is satisfied for each finite–dimensional subspace $\mathcal{M} \subset H^s(R)$. Assume that

$$R = (\alpha_1, \beta_1) \times (\alpha_2, \beta_2) \times (\alpha_3, \beta_3)$$

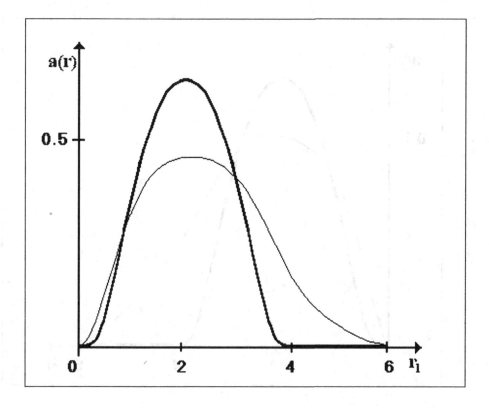

Figure 6.29. Example 6.9: $\delta = 0$, method (5.7.24)

and put $\mathcal{M} = \mathcal{M}_N$, where

$$\mathcal{M}_N = \left\{ a(r) = \sum_{l,m,n=1}^{N} c_{lmn}\eta_l(r_1)\eta_m(r_2)\eta_n(r_3) : c_{lmn} \in \mathbf{R} \right\};$$

$$\eta_s(r_i) = \sin\frac{\pi s(r_i - \alpha_i)}{\beta_i - \alpha_i}, \quad i = \overline{1,3}, \quad s = 1, 2, \ldots.$$

If the unknown solution a^* allows for a good approximation by the subspace $\mathcal{M}_N$, then in (5.7.22) and (5.7.24) we can set $\xi = 0$.

Suppose instead of the true amplitudes $u_\infty(\hat{r}, d)$ some approximations $\tilde{u}_\infty(\hat{r}, d)$, $(\hat{r}, d) \in S^2 \times S^2$ are available. Let us define the approximate operator

$$\tilde{F} : H^s(R) \to L_2(S^2 \times S^2),$$

$$\tilde{F}(a)(\hat{r}, d) = \frac{k_0^2}{4\pi}W(E - aV)^{-1}(au_{in})(\hat{r}, d) + \tilde{u}_\infty(\hat{r}, d).$$

(8)

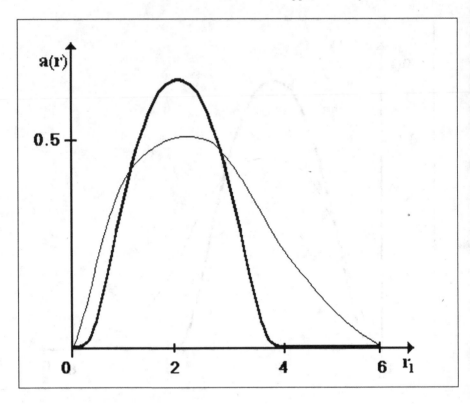

Figure 6.30.   Example 6.9: $\delta = 0.01$, method (5.7.24)

As in [62], it can easily be shown that if $\|(E - P_{\mathcal{M}})a^*\|_{H^s(R)}$ is sufficiently small, then the operators F and $\widetilde{F}$ are Fréchet differentiable with Lipschitz continuous derivatives on a neighborhood $\Omega_\varepsilon(P_{\mathcal{M}}a^*) \cap \mathcal{M}$, $\varepsilon > 0$.

Setting

$$a(r) = \sum_{l,m,n=1}^{N} c_{lmn}\eta_l(r_1)\eta_m(r_2)\eta_n(r_3),$$

we rewrite (5.7.3) as the finite–dimensional optimization problem

$$\min_{c \in \mathbf{R}^{N^3}} \widetilde{\psi}(c), \quad c = (c_{lmn})_{l,m,n=1}^{N},$$

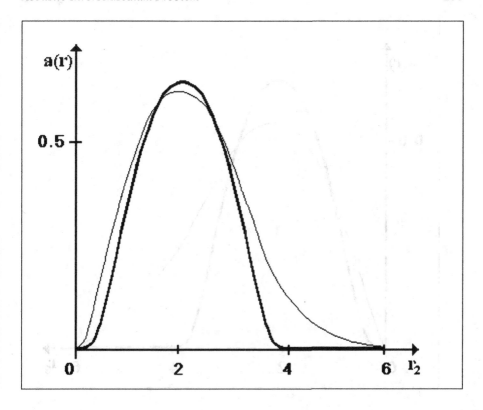

Figure 6.31. Example 6.9: $\delta = 0.01$, method (5.7.24)

where

$$\widetilde{\psi}(c) = \frac{1}{2} \left\| \widetilde{F} \left(\sum_{l,m,n=1}^{N} c_{lmn}\eta_l(r_1)\eta_m(r_2)\eta_n(r_3) \right) \right\|_{L_2(S^2 \times S^2)}^2 =$$

$$= \frac{1}{2} \int_{S^2} \int_{S^2} \left| \frac{k_0^2}{4\pi} W \left[E - \left(\sum_{l,m,n=1}^{N} c_{lmn}\eta_l\eta_m\eta_n \right) V \right]^{-1} \cdot \right. \tag{9}$$

$$\left. \cdot \left(\left(\sum_{l,m,n=1}^{N} c_{lmn}\eta_l\eta_m\eta_n \right) u_{in} \right) (\hat{r}, d) + \widetilde{u}_\infty(\hat{r}, d) \right|^2 ds_1 ds_2;$$

$$s_1 = s_1(\hat{r}), \quad s_2 = s_2(d); \quad \hat{r}, d \in S^2.$$

Partial derivatives of $\widetilde{\psi}(c)$ can be evaluated, e.g., by the finite–difference scheme

$$\frac{\partial \widetilde{\psi}(c)}{\partial c_{lmn}} \approx \frac{\widetilde{\psi}(c + he^{(lmn)}) - \widetilde{\psi}(c)}{h}, \quad e^{(lmn)} = (e_{ijk}^{(lmn)})_{i,j,k=1}^N; \tag{10}$$

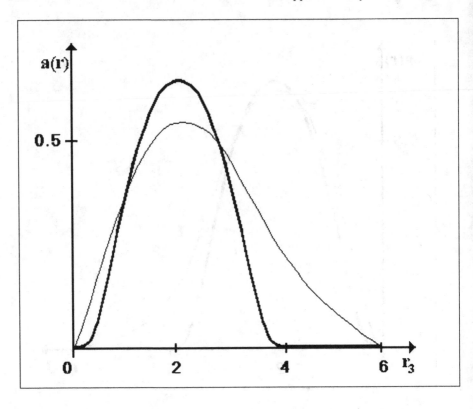

Figure 6.32. Example 6.9: $\delta = 0.01$, method (5.7.24)

$$e_{ijk}^{(lmn)} = \begin{cases} 0, & (i,j,k) \neq (l,m,n), \\ 1, & (i,j,k) = (l,m,n). \end{cases}$$

In our tests, the scheme (10) is used with $h = 10^{-3}$. An implementation of iterations (5.7.22) and (5.7.24) requires a multiple solution of the direct scattering problem (5). We solve equation (5) by the simple iteration process

$$v^0(r) \equiv 0, \quad v^{n+1}(r) = v^n(r)-$$

$$-\lambda \left\{ v^n(r) + k_0^2 a(r) \int_R \Gamma(r,r')v^n(r')dr' - a(r)u_{in}(r) \right\}, \quad r \in R. \quad (11)$$

In our computational experiments, the function

$$v(r) = \left[(E - aV)^{-1}(au_{in}) \right](r), \quad r \in R$$

is approximated by 8 iterations (11) with the stepsize $\lambda = 0.5$.

We consider a model inverse scattering problem with the solution $n^*(r) = 1 - a^*(r)$, where

$$a^*(r) = \begin{cases} 2 \cdot 10^{-6}(4r_1 - r_1^2)^3(4r_2 - r_2^2)^3(4r_3 - r_3^2)^3, & r \in R_0, \\ 0, & r \in R \setminus R_0. \end{cases}$$

Here $R_0 = (0, 4) \times (0, 4) \times (0, 4)$ is a domain actually containing the inhomogeneity and $R = (0, 6) \times (0, 6) \times (0, 6)$ a given domain where the inhomogeneity a priori lies. We set $k_0 = 1$. The direction of observation $\hat{r} = \hat{r}^{(i)}$ and direction of an incident wave $d = d^{(j)}$ range over the set of 14 directions

$$\Sigma = \left\{ (0, 0, \pm 1), (0, \pm 1, 0), (\pm 1, 0, 0), \left(\pm \frac{1}{\sqrt{3}}, \pm \frac{1}{\sqrt{3}}, \pm \frac{1}{\sqrt{3}} \right) \right\}$$

"uniformly" distributed over the sphere S^2. The integration over $S^2 \times S^2$ in (9) is performed by the rule

$$\int_{S^2} \int_{S^2} f(\hat{r}, d) ds_1 ds_2 \approx \left(\frac{4\pi}{m} \right)^2 \sum_{i,j=1}^{m} f(\hat{r}^{(i)}, d^{(j)})$$

with $m = 14$ and $\{\hat{r}^{(i)}\}_{i=1}^{m} = \{d^{(j)}\}_{j=1}^{m} = \Sigma$. For the integration over R in (6) and (11), we use the Gaussian cubature rule with 8 knots by each coordinate.

We put $\xi(r) = 0, r \in R$; the initial approximation is given by

$$a_0(r) = 0, \quad r \in R.$$

Perturbations of $\{u_\infty(\hat{r}^{(i)}, d^{(j)})\}_{i,j=1}^{m}$ are modelled by additive random terms so that in (8) and (9) we use the noisy data

$$\tilde{u}_\infty(\hat{r}^{(i)}, d^{(j)}) = u_\infty(\hat{r}^{(i)}, d^{(j)}) + \delta w^{(ij)}, \quad 1 \le i, j \le m; \quad \delta \ge 0,$$

where $w^{(ij)}$ are realizations of a random variable uniformly distributed on $[-1, 1]$.

Let us turn to results of numerical tests. We denote $d_n = \|\widetilde{F}(a_n)\|_{L_2(S^2 \times S^2)}$.

Example 6.8. In the first experiment, 160 iterations of the gradient process (5.7.22) are performed with the following values of parameters: $\gamma = 2$, $\mathcal{M} = \mathcal{M}_2$, $\delta = 0$. Figure 6.25 presents the true index $a^*(r_1, 2, 2)$ (bold line) and the reconstructed index $a_{160}(r_1, 2, 2)$, $r_1 \in [0, 6]$ (thin line). The initial and final discrepancies are $d_0 = 0.19$ and $d_{160} = 0.04$.

Figures 6.26–6.28 present results of 160 iterations of the same process with the maximal level of noise $\delta = 0.01$. The original and reconstructed functions $a^*(r_1, 2, 2)$, $a_{160}(r_1, 2, 2)$, $r_1 \in [0, 6]$ are shown on Figure 6.26; Figures 6.27 and 6.28 present the exact and reconstructed functions $a^*(2, r_2, 2)$,

$a_{160}(2, r_2, 2), r_2 \in [0, 6]$ and $a^*(2, 2, r_3)$, $a_{160}(2, 2, r_3), r_3 \in [0, 6]$. In this case we have $d_0 = 0.22$ and $d_{160} = 0.11$.

Example 6.9. In the second test we use the conjugate gradient method (5.7.24) with $\mathcal{M} = \mathcal{M}_2$. First, 40 iterations of the method are performed with $\delta = 0$. We obtain $d_{40} = 0.04$. On Figure 6.29 the exact function $a^*(r_1, 2, 2)$ and the reconstructed function $a_{40}(r_1, 2, 2), r_1 \in [0, 6]$ are shown. Then, 40 iterations of the same process are performed with $\delta = 0.01$. We obtain $d_0 = 0.21$ and $d_{40} = 0.11$. Figures 6.30–6.32 present the true solution and reconstructed functions $a^*(r_1, 2, 2)$, $a_{40}(r_1, 2, 2), r_1 \in [0, 6]$; $a^*(2, r_2, 2)$, $a_{40}(2, r_2, 2), r_2 \in [0, 6]$, and $a^*(2, 2, r_3)$, $a_{40}(2, 2, r_3), r_3 \in [0, 6]$ obtained in the case of noisy data.

Figures 6.25–6.32 demonstrate that methods (5.7.22) and (5.7.24) give qualitatively good approximations of the true refractive index.

6.4 Inverse Acoustic Obstacle Scattering. Far Field Observation

Let us turn to inverse scattering problems in the case where an unknown scatterer is impenetrable for acoustic waves. Imagine a homogeneous 3D acoustic medium characterized by a constant sound speed c_0 and let $D \subset \mathbf{R}^3$ be a bounded impenetrable domain. Suppose the boundary ∂D of the obstacle D is of class C^2. As in Section 6.3, assume that the incident acoustic field is a time–harmonic plane wave with the complex amplitude $u_{in}(r) = \exp(ik_0 d \cdot r)$, where $k_0 = \omega / c_0$ and $d \in S^2$. Then the scattering of the incident wave on D is governed by equations

$$\Delta u(r) + k_0^2 u(r) = 0, \quad u(r) = u_{in}(r) + u_s(r), \quad r \in \mathbf{R}^3 \backslash \overline{D} \qquad (1)$$

and the Sommerfeld radiation condition (3.2). Here $u_s(r)$ is the complex amplitude of a scattered wave. Throughout this section, the frequency ω of the incoming wave $u_{in}(r)$ is fixed, so is the direction d. The equations (1) and (3.2) should be accomplished by a boundary condition on ∂D in accordance with available information on physical properties of the scatterer D. In this section we consider the Dirichlet boundary condition

$$u(r) = 0, \quad r \in \partial D, \qquad (2)$$

which corresponds to so called sound–soft obstacles, and the impedance condition

$$\frac{\partial u(r)}{\partial \nu} + i\lambda(r) u(r) = 0, \quad r \in \partial D. \qquad (3)$$

Here $\lambda \in C(\partial D)$ is a real–valued function with $\lambda(r) > 0, r \in \partial D$ and $\nu = \nu(r)$ is the unit outward normal to ∂D at $r \in \partial D$. In the case where $\lambda \equiv 0$, condition (3) describes sound–hard obstacles.

Let us recall that the boundary value problems (1), (2), (3.2) and (1), (3), (3.2) have at most one solution [29, 30]. Existence of solutions to these and analogous problems is usually established by the method of boundary integral equations. According to this method, a solution is constructed in the form of a combination of acoustic double– and single–layer potentials [30, Chapter 3]. In particular, a solution to the problem (1), (2), (3.2) is sought in the form

$$u_s(r) = \int_{\partial D} \left\{ i\eta \Gamma(r, r') - \frac{\partial \Gamma(r, r')}{\partial \nu(r')} \right\} g(r') ds(r'), \quad r \in \mathbf{R}^3 \backslash \overline{D}; \quad (4)$$

$$\Gamma(r, r') = -\frac{\exp(ik_0|r - r'|)}{4\pi|r - r'|}$$

with a real coupling parameter $\eta \neq 0$ and an unknown complex–valued density $g \in C(\partial D)$.

Following [30], introduce the single– and double–layer operators

$$(Sg)(r) = -2 \int_{\partial D} \Gamma(r, r') g(r') ds(r'), \quad r \in \partial D; \quad (5)$$

$$(Kg)(r) = -2 \int_{\partial D} \frac{\partial \Gamma(r, r')}{\partial \nu(r')} g(r') ds(r'), \quad r \in \partial D. \quad (6)$$

Note that S and K are compact operators from $C(\partial D)$ into $C(\partial D)$. The combined potential (4) satisfies (1), (2), and (3.2), provided the density g solves the equation

$$[(E + K - i\eta S)g](r) = -2u_{in}(r), \quad r \in \partial D. \quad (7)$$

In its turn, the equation (7) has a unique solution $g = g^* \in C(\partial D)$ so that the problem (1), (2), (3.2) is also uniquely solvable. By (4) it follows that $u_s(r)$ satisfies (3.3) with the scattering amplitude

$$u_\infty(\hat{r}, d) = \frac{1}{4\pi} \int_{\partial D} \left\{ \frac{\partial e^{-ik_0 r' \cdot \hat{r}}}{\partial \nu(r')} - i\eta e^{-ik_0 r' \cdot \hat{r}} \right\} g^*(r') ds(r'), \quad (8)$$

$$\hat{r} = \frac{r}{|r|} \in S^2.$$

By S_0 we denote the single–layer operator (5) in the case of $k_0 = 0$:

$$(S_0 g)(r) = \frac{1}{2\pi} \int_{\partial D} \frac{g(r')}{|r - r'|} ds(r'), \quad r \in \partial D.$$

A solution to the impedance problem (1), (3), (3.2) we seek in the form

$$
u_s(r) = - \int_{\partial D} \left\{ \Gamma(r, r') g(r') + \right.
$$

$$
\left. + i\eta \frac{\partial \Gamma(r, r')}{\partial \nu(r')} (S_0^2 g)(r') \right\} ds(r'), \quad r \in \mathbf{R}^3 \backslash \overline{D}, \tag{9}
$$

where $S_0^2 g \equiv S_0 S_0 g$ and $\eta \neq 0$ is a real parameter. Let us introduce the normal derivative operators

$$
(K'g)(r) = -2 \int_{\partial D} \frac{\partial \Gamma(r, r')}{\partial \nu(r)} g(r') ds(r'), \quad r \in \partial D; \tag{10}
$$

$$
(Tg)(r) = -2 \frac{\partial}{\partial \nu(r)} \int_{\partial D} \frac{\partial \Gamma(r, r')}{\partial \nu(r')} g(r') ds(r'), \quad r \in \partial D. \tag{11}
$$

Then the combined potential (9) gives a solution to the problem (1), (3), (3.2) if the density g satisfies the equation [29, 117]

$$
[(E - K' - i\eta T S_0^2 - i\lambda S + \eta\lambda(E + K)S_0^2)g](r) =
$$

$$
= 2 \frac{\partial u_{in}(r)}{\partial \nu} + 2i\lambda(r) u_{in}(r), \quad r \in \partial D. \tag{12}
$$

It can be shown that equation (12) has a unique solution $g = g^* \in C(\partial D)$ [29]. From (9), for the scattering amplitude of $u_s(r)$ we get

$$
u_\infty(\hat{r}, d) = \frac{1}{4\pi} \int_{\partial D} \left\{ e^{-ik_0 r' \cdot \hat{r}} g^*(r') + i\eta \frac{\partial e^{-ik_0 r' \cdot \hat{r}}}{\partial \nu(r')} (S_0^2 g^*)(r') \right\} ds(r'). \tag{13}
$$

Let us remark that the scattering amplitudes (8) and (13) are analytic functions of $\hat{r} \in S^2$.

By a direct obstacle scattering problem is usually meant a problem of evaluating the scattered field $u_s(r), r \in Y$ or the scattering amplitudes $u_\infty(\hat{r}, d), \hat{r} \in S^2$ for a given scatterer D, direction $d \in S^2$, and region of observation $Y \subset \mathbf{R}^3 \backslash \overline{D}$. In the inverse obstacle scattering problem, which will be considered below, a sound–soft or an impedance scatterer D is to be reconstructed by information on the scattering amplitudes $\{u_\infty(\hat{r}, d) : \hat{r} \in S^2\}$ with a fixed direction $d \in S^2$. It is known [30, p.108] that a sound–soft scatterer of a small diameter can be uniquely determined by this information. To be more precise, if a ball of radius $a < \pi/k_0$ contains two sound–soft scatterers D_1 and D_2 and the scattering amplitudes $u_\infty(\hat{r}, d)$ related to these obstacles coincide for all directions $\hat{r} \in S^2$, then $D_1 = D_2$. As to sound–hard and impedance obstacles, we can only assert that the equality $D_1 = D_2$ without any restrictions on diameters of D_1 and D_2 follows by the coincidence of corresponding amplitudes $u_\infty(\hat{r}, d)$ for all $\hat{r} \in S^2$ and all $d \in S^2$ [30, p. 112]. Nevertheless, below we shall treat the impedance case in assumption that the observable data are $\{u_\infty(\hat{r}, d) : \hat{r} \in S^2\}$ with a fixed d.

Suppose an unknown obstacle D is star–shaped with respect to the origin of $\mathbf{R}^3$. Using the spherical coordinates, we can represent the boundary ∂D as

$$r_1 = \rho(\theta, \varphi) \sin\theta \cos\varphi, \quad r_2 = \rho(\theta, \varphi) \sin\theta \sin\varphi,$$
$$r_3 = \rho(\theta, \varphi) \cos\theta; \qquad (14)$$
$$\rho(\theta, \varphi) \geq 0, \quad (\theta, \varphi) \in \Omega \equiv [0, \pi] \times [0, 2\pi]$$

or, equivalently, in the form

$$r = \rho(\hat{r})\hat{r}, \quad \hat{r} \in S^2 \quad (r = (r_1, r_2, r_3) \in \partial D),$$

where

$$\hat{r} = (\sin\theta \cos\varphi, \quad \sin\theta \sin\varphi, \quad \cos\theta).$$

We consider only sound–soft and impedance obstacles D defined by functions $\rho \in H^s(\Omega), s > 3$ with $\rho(\theta, \varphi) > 0 \;\; \forall(\theta, \varphi) \in \Omega$. Then by Sobolev's embedding theorem, $\rho \in C^2(\Omega)$. Now denote by F_∞ the mapping that takes each function $\rho \in D(F_\infty)$,

$$D(F_\infty) = \{\rho \in H^s(\Omega) : \rho(\theta, \varphi) > 0 \;\; \forall(\theta, \varphi) \in \Omega\}$$

to the far field pattern $u_\infty(\cdot, d) \in L_2(S^2)$ for the problem (1), (2), (3.2) or the problem (1), (3), (3.2) with D given by (14). Then the inverse obstacle scattering problem takes the form

$$F(\rho) \equiv F_\infty(\rho) - u_\infty = 0, \quad \rho \in D(F) \equiv D(F_\infty), \qquad (15)$$

where $u_\infty = u_\infty(\cdot, d) \in L_2(S^2)$. Suppose instead of true scattering amplitudes $u_\infty(\hat{r}, d)$, $\hat{r} \in S^2$ some their approximations $\widetilde{u}_\infty(\hat{r}, d)$ are given. Define the approximate operator $\widetilde{F} : D(\widetilde{F}) \rightarrow L_2(S^2)$ by

$$\widetilde{F}(\rho) \equiv F_\infty(\rho) - \widetilde{u}_\infty, \quad \rho \in D(\widetilde{F}) \equiv D(F).$$

Since $H^s(\Omega)$, $s > 3$ is continuously embedded into $C^2(\Omega)$, the domain of definition $D(F)$ is open in $H^s(\Omega)$. Let $\rho^* \in D(F)$ be a solution to (15). Then the unknown scatterer D is given by

$$\partial D = \{r \in \mathbf{R}^3 : r = \rho^*(\hat{r})\hat{r}, \quad \hat{r} \in S^2\}.$$

According to [30, 98], the mapping $F : H^s(\Omega) \to L_2(S^2)$ is Fréchet differentiable on a sufficiently small neighborhood $\Omega_\varepsilon(\rho^*) \subset H^s(\Omega)$. Using the scheme of [30], the reader will easily prove that the derivative $F'(\rho)$ is Lipschitz continuous, at least for $\rho \in \Omega_\varepsilon(\rho^*)$ with small $\varepsilon > 0$. It should be pointed out that equation (15) is ill–posed and irregular with respect to any reasonable choice of spaces of solutions ρ and input data u_∞ ([30]).

For solving (15) we use the iterative processes (5.7.22) and (5.7.24). Let us remark that the operator $F(\rho)$ defined by the Dirichlet boundary condition (2) has an injective derivative $F'(\rho^*)$ [30]. Therefore Condition 5.8 is satisfied for each finite–dimensional subspace $\mathcal{M} \subset H^s(\Omega)$. The same is true for the impedance condition (3), provided the function $\lambda(r)$ takes sufficiently large values on ∂D [98]. Let us put $\mathcal{M} = \mathcal{M}_N$, where $\mathcal{M}_N$ is the subspace of spherical functions (1.5). In the assumption that $\|(E - P_\mathcal{M})\rho^*\|_{H^s(\Omega)}$ is sufficiently small, that is, the projection $P_\mathcal{M}\rho^*$ provides a good approximation of ρ^*, we can put

$$\xi(r) = 0, \quad r \in \Omega.$$

The optimization problem (5.7.3) now takes the form

$$\min_{c \in \mathbf{R}^{N^2}} \widetilde{\psi}(c); \quad c = (c_{mn}), \quad m = \overline{0, N}, \quad n = \overline{-m, m},$$

where

$$\widetilde{\psi}(c) = \frac{1}{2} \left\| F_\infty \left(\sum_{m=0}^{N} \sum_{n=-m}^{m} c_{mn} Y_m^n(\theta, \varphi) \right) - \widetilde{u}_\infty^0 \right\|_{L_2(S^2)}^2. \tag{16}$$

When implementing the methods (5.7.22) and (5.7.24), it is necessary to find $\widetilde{\psi}(c)$ and hence $F_\infty(\rho)$ several times per iteration. To this end, having a function ρ we first solve equation (7) or (12) with

$$\partial D = \{r \in \mathbf{R}^3 : r = \rho(\hat{r})\hat{r}, \hat{r} \in S^2\}$$

for the density $g^*(r), r \in \partial D$ and then obtain $F_\infty(\rho) = u_\infty(\cdot, d)$ by formula (8) or (13). The equations (7) and (12) are solved by the simple iteration processes

$$g^0(r) = 0,$$

$$g^{n+1}(r) = g^n(r) - \mu \Big([(E + K - i\eta S)g^n](r) + 2u_{in}(r) \Big); \tag{17}$$

$$g^{n+1}(r) = g^n(r) -$$
$$- \mu \Big(\big[(E - K' - i\eta T S_0^2 - i\lambda S + \eta\lambda(E + K)S_0^2)g^n \big](r) -$$
$$- 2\frac{\partial u_{in}(r)}{\partial \nu} - 2i\lambda(r)u_{in}(r) \Big); \quad r \in \partial D. \tag{18}$$

Convergence of the iterations (17) and (18) is guaranteed by an appropriate choice of the stepsize $\mu \in \mathbf{C}$ with small $|\mu|$. In all experiments we put $\mu = 0.2 + 0.2i$ and perform 10 iterations (17) or (18).

The direction of observation $\hat{r} = \hat{r}^{(i)}$ ranges over the same set

$$\Sigma = \left\{ (0,0,\pm 1), (0,\pm 1,0), (\pm 1,0,0), \left(\pm \frac{1}{\sqrt{3}}, \pm \frac{1}{\sqrt{3}}, \pm \frac{1}{\sqrt{3}} \right) \right\}$$

as in Section 6.3. We evaluate the norm in (16) using the quadrature rule

$$\int_{S^2} f(\hat{r}) ds(\hat{r}) \approx \frac{4\pi}{m} \sum_{i=1}^{m} f(\hat{r}^{(i)})$$

with $m = 14$ and $\{\hat{r}^{(i)}\}_{i=1}^{m} = \Sigma$. To compute the surface integrals over ∂D in (17) and (18) (see (5), (6), (10), and (11)), we first reduce them to integrals over the rectangle Ω and then apply the Simpson rule with 15 knots by θ and 21 knots by φ. Errors of input data $\{u_\infty(\hat{r}^{(i)}, d)\}_{i=1}^{m}$ are modelled by additive random perturbations:

$$\tilde{u}_\infty(\hat{r}^{(i)}, d) = u_\infty(\hat{r}^{(i)}, d) + \delta \omega^{(i)}, \quad 1 \le i \le m; \quad \delta \ge 0.$$

Here $\omega^{(i)}$ are realizations of a random variable uniformly distributed on the segment $[-1, 1]$. Partial derivatives of $\tilde{\psi}(c)$ are approximated by the finite-difference scheme (3.10) with $h = 10^{-3}$.

We put

$$d = \left(\frac{1}{\sqrt{3}}, \frac{1}{\sqrt{3}}, \frac{1}{\sqrt{3}} \right), \quad k_0 = 0.25, \quad \eta = 1,$$

and $\mathcal{M} = \mathcal{M}_2$. A sound–soft or impedance scatterer D to be recovered is given by

$$\rho^*(\theta, \varphi) = \sum_{m=0}^{2} \sum_{n=-m}^{m} c_{mn}^* Y_m^n(\theta, \varphi) + c_{3,-3}^* Y_3^{-3}(\theta, \varphi).$$

We see that the approximation error $\|(E - P_\mathcal{M})\rho^*\|_{H^s(\Omega)}$ is small, provided so is the coefficient $c_{3,-3}^*$. Below we set

$$c_{0,0}^* = 1, \quad c_{1,-1}^* = 0.2, \quad c_{1,0}^* = \cdots = c_{2,2}^* = 0, \quad c_{3,-3}^* = 0.3.$$

As an initial approximation to ∂D we choose a sphere of radius 0.5 centered at the origin, that is, $\rho_0(\theta, \varphi) = 0.5$, $(\theta, \varphi) \in \Omega$.

Example 6.10. First, we test the processes (5.7.22) and (5.7.24) in application to equation (15) for a sound–soft obstacle. We put $\delta = 0$, $\gamma = 0.02$ and perform 30 iterations of the gradient process (5.7.22). Denote $d_n =$

$\left\| \widetilde{F}(\rho_n) \right\|^2_{L_2(S^2)}$. We obtain $d_0 = 1.8776$, $d_{10} = 0.1486$, $d_{20} = 0.1211$, and $d_{30} = 0.1055$. Figures 6.33 and 6.34 present the curves $\rho = \rho^*(\theta, 0)$ (bold line), $\rho = \rho_0(\theta, 0)$ (small circle, thin line), $\rho = \rho_{30}(\theta, 0)$ (thin line) and $\rho = \rho^*(\theta, \pi/2)$ (bold line), $\rho = \rho_0(\theta, \pi/2)$ (small circle, thin line), $\rho = \rho_{30}(\theta, \pi/2)$ (thin line) respectively. Results obtained in the case of noisy input data with the error level $\delta = 0.04$ are shown on Figures 6.35 and 6.36. Here we get $d_{30} = 0.1584$

The conjugate gradient process (5.7.24) with $\delta = 0$ gives $d_5 = 0.1584$ and $d_{10} = 0.1423$. On Figures 6.37 and 6.38 we show sections $\varphi = 0$ and $\varphi = \pi/2$ of true and resulting surfaces obtained after 10 iterations. Figures 6.39 and 6.40 present a result of the same iteration in the case of $\delta = 0.04$.

Example 6.11. Suppose the unknown obstacle is an impedance scatterer with $\lambda(r) = 0.5, r \in \partial D$. The gradient process (5.7.22) with $\delta = 0$ and $\gamma = 0.1$ gives $d_0 = 1.2446$, $d_5 = 0.0457$, $d_{10} = 0.0342$, $d_{15} = 0.0278$, and $d_{20} = 0.0242$. On Figures 6.41 and 6.42 we present the result obtained after 20 iterations. In the case of noisy input data with $\delta = 0.04$ we get $d_5 = 0.1137$, $d_{10} = 0.1104$, $d_{15} = 0.1088$, and $d_{20} = 0.1080$ (see Figures 6.43 and 6.44 for sections $\varphi = 0$ and $\varphi = \pi/2$ of true and resulting surfaces).

Using the conjugate gradient method (5.7.24) with $\delta = 0$, we obtain the discrepancies $d_5 = 0.0497$ and $d_{10} = 0.0292$. The results of 10 iterations are presented on Figures 6.45 and 6.46. In the case of noisy data with $\delta = 0.04$ we have $d_5 = 0.1150$ and $d_{10} = 0.0986$ (see Figures 6.47 and 6.48).

6.5   Inverse Acoustic Obstacle Scattering. Near Field Observation

In this section we slightly modify the setting of an inverse acoustic scattering problem given in the previous section. Let $D \subset \mathbf{R}^3$ be an impenetrable sound–soft scatterer with a boundary ∂D of class C^2. Suppose $Y \subset \mathbf{R}^3, Y \cap \overline{D} = \emptyset$, and $\mathrm{int} Y \neq \emptyset$. As observational data, we now take values of the scattered wave field $u_s(r)$ for points $r \in Y$. So the problem is to reconstruct the obstacle D having the near–field data $\{u_s(r) : r \in Y\}$. As above, the scattering process is governed by equations (4.1), (4.2), and (3.2); the incident acoustic field is $u_{in}(r) = \exp(ik_0 d \cdot r)$. Since the scattered field $u_s(r)$ outside D is uniquely determined by both the far field patterns $u_\infty(\hat{r}, d), \hat{r} \in S^2$ and the restriction $u_s|_Y(r) \equiv u_s(r), r \in Y$, the presented inverse problem has a unique solution, at least in the class of scatterers embedded into a ball of radius $a < \pi/k_0$.

Suppose a scatterer is bounded and star–shaped with respect to the origin; then we have

$$\partial D = \{r \in \mathbf{R}^3 : r = \rho(\hat{r})\hat{r}, \quad \hat{r} \in S^2\}.$$

Assume that $\rho \in H^s(\Omega)$, where $s > 3$ and $\Omega = [0, \pi] \times [0, 2\pi]$. By F_Y we denote the mapping that takes each function $\rho \in D(F_Y)$,

$$D(F_Y) = \{\rho \in H^s(\Omega) : \rho(\theta, \varphi) > 0 \quad \forall (\theta, \varphi) \in \Omega\},$$

to the restriction $u_s|_Y$. Note that the function $u_s = u_s(r)$ is analytic on $\mathbf{R}^3 \backslash \overline{D}$. Let us consider F_Y as an operator from $H^s(\Omega)$ into $L_2(Y)$. Denote

$$F(\rho) = F_Y(\rho) - u_s|_Y, \quad \rho \in D(F);$$

$$D(F) = D(F_Y),$$

where $\{u_s|_Y(r) : r \in Y\}$ are input data of the problem. Then the inverse problem takes the form

$$F(\rho) = 0, \quad \rho \in D(F). \tag{1}$$

Below we apply to equation (1) the process (5.7.22). Necessary smoothness properties of the operator F can be established by the scheme of [30, Section 5.3]. Moreover, using the explicit representation of $F'(\rho^*)$ obtained in [37], it is not difficult to prove that the operator $F'(\rho^*) : H^s(\Omega) \to L_2(Y)$ is injective. Therefore Condition 5.8 is valid with an arbitrary finite–dimensional subspace $\mathcal{M} \subset H^s(\Omega)$.

To evaluate $F_Y(\rho)(r), r \in Y$, we first solve equation (4.7) and then obtain $F_Y(\rho)(r) = u_s(r), r \in Y$ according to (4.4). For finding a solution of (4.7) we use 15 iterations of the process (4.17) with the stepsize $\mu = 0.2 + 0.2i$. In this section we restrict ourselves with the case of exact input data $u_s|_Y$. Let us set $\xi(r) = 0, r \in \Omega$ and rewrite the problem (5.7.3) as

$$\min_{c \in \mathbf{R}^{N^2}} \psi(c); \quad c = (c_{mn}), \quad m = \overline{0, N}, \quad n = \overline{-m, m},$$

where

$$\psi(c) = \frac{1}{2} \left\| F_Y \left(\sum_{m=0}^{N} \sum_{n=-m}^{m} c_{mn} Y_m^n(\theta, \varphi) \right) - u_s|_Y \right\|_{L_2(Y)}^2. \tag{2}$$

We put

$$Y = [5, 6] \times [5, 6] \times [5, 6], \quad d = \left(\frac{1}{\sqrt{3}}, \frac{1}{\sqrt{3}}, \frac{1}{\sqrt{3}} \right), \quad k_0 = 0.25.$$

The norm in (2) is evaluated by the Simpson rule with 5 knots by each coordinate. All integrations over ∂D are performed by reduction to corresponding integrals over the rectangle Ω with subsequent application of the Simpson rule with 9 knots by θ and 17 knots by φ. Derivatives of $\psi(c)$ are computed by formulae

(3.10) with $h = 10^{-3}$. The subspace $\mathcal{M} = \mathcal{M}_2$ is defined according to (1.5) with $N = 2$.

A true sound–soft scatterer is given by the formula

$$\rho^*(\theta, \varphi) = 1 + 0.1 \sin \theta \cos \varphi + 0.4 \left(\frac{5}{2} \cos^3 \theta - \frac{3}{2} \cos \theta \right) \sin \theta;$$

as an initial approximation to ∂D we choose the sphere

$$\rho_0(\theta, \varphi) = 2, \quad (\theta, \varphi) \in \Omega.$$

Since the spherical harmonic

$$Y_3^0(\theta, \varphi) = \frac{5}{2} \cos^3 \theta - \frac{3}{2} \cos \theta$$

is orthogonal to $\mathcal{M}_2$ in the metric of $L_2(S^2)$ [30, p.25], the function $Y_3^0(\theta, \varphi) \sin \theta$ and the subspace $\mathcal{M}_2$ are orthogonal in the sense of $L_2(\Omega)$. Taking into account that $1 + 0.1 \sin \theta \cos \varphi \in \mathcal{M}_2$, we obtain the following lower estimate for $d = \|\rho - \rho^*\|_{L_2(\Omega)}$, $\rho \in \mathcal{M}_2$:

$$\|\rho - \rho^*\|_{L_2(\Omega)} \geq \operatorname{dist}(\rho^*, \mathcal{M}_2) =$$
$$= 0.4 \|Y_3^0(\cdot, \cdot\cdot) \sin(\cdot)\|_{L_2(\Omega)} = 0.4228 \quad \forall \rho \in \mathcal{M}_2. \tag{3}$$

Example 6.12. Using the process (5.7.22), we get approximations $\rho_n(\theta, \varphi) \in \mathcal{M}_2$ to $\rho^*(\theta, \varphi)$ such that $d_0 = 4.4742$, $d_{25} = 0.8617$, $d_{50} = 0.6938$, $d_{75} = 0.6664$, and $d_{100} = 0.6530$. Let us remark that the last value doesn't differ significantly from the lower bound (3). On Figure 6.49 we show sections $r_3 = 1 - j/9$, $j = 1, \ldots, 9$ of the true obstacle $r = r^*(\theta, \varphi)$ and its approximation $r = \rho_{100}(\theta, \varphi)$.

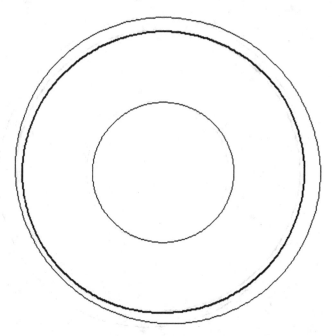

Figure 6.33. Example 6.10: $\delta = 0$, method (5.7.22), $\varphi = 0$

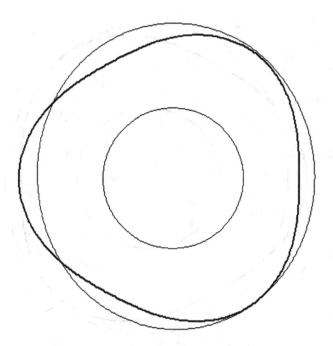

Figure 6.34. Example 6.10: $\delta = 0$, method (5.7.22), $\varphi = \pi/2$

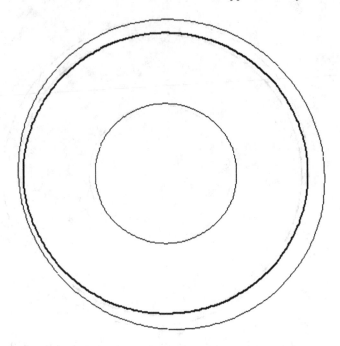

Figure 6.35.   Example 6.10: $\delta = 0.04$, method (5.7.22), $\varphi = 0$

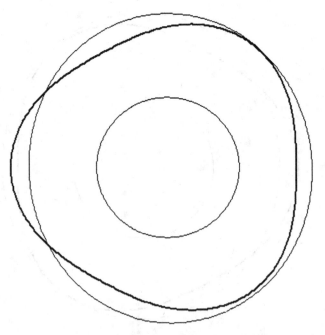

Figure 6.36.   Example 6.10: $\delta = 0.04$, method (5.7.22), $\varphi = \pi/2$

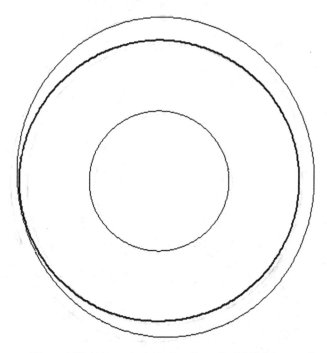

Figure 6.37. Example 6.10: $\delta = 0$, method (5.7.24), $\varphi = 0$

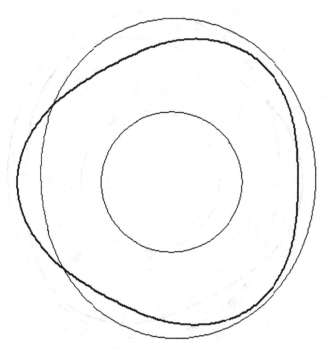

Figure 6.38. Example 6.10: $\delta = 0$, method (5.7.24), $\varphi = \pi/2$

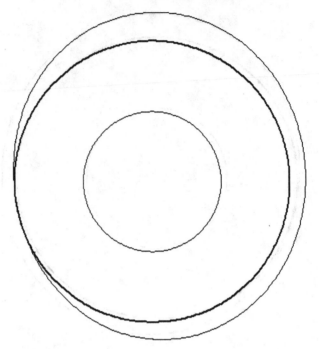

Figure 6.39.   Example 6.10: $\delta = 0.04$, method (5.7.24), $\varphi = 0$

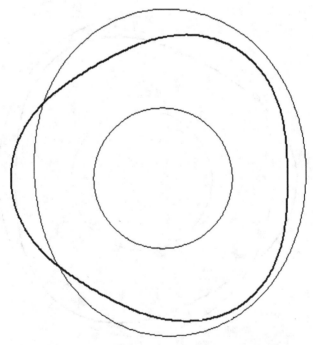

Figure 6.40.   Example 6.10: $\delta = 0.04$, method (5.7.24), $\varphi = \pi/2$

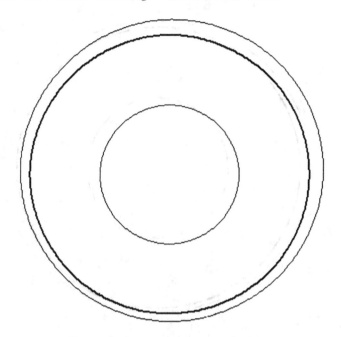

Figure 6.41. Example 6.11: $\delta = 0$, method (5.7.22), $\varphi = 0$

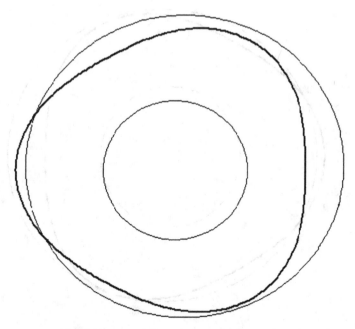

Figure 6.42. Example 6.11: $\delta = 0$, method (5.7.22), $\varphi = \pi/2$

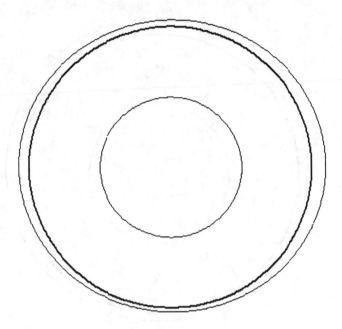

Figure 6.43. Example 6.11: $\delta = 0.04$, method (5.7.22), $\varphi = 0$

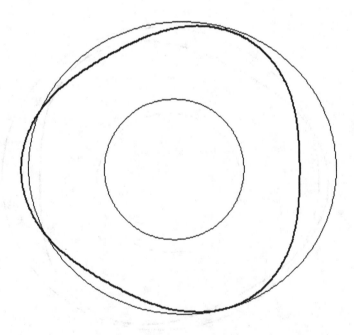

Figure 6.44. Example 6.11: $\delta = 0.04$, method (5.7.22), $\varphi = \pi/2$

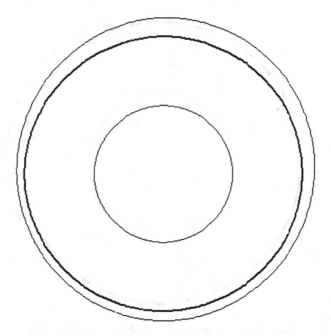

Figure 6.45. Example 6.11: $\delta = 0$, method (5.7.24), $\varphi = 0$

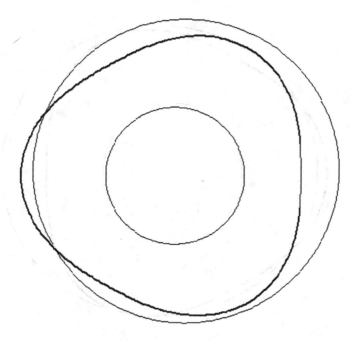

Figure 6.46. Example 6.11: $\delta = 0$, method (5.7.24), $\varphi = \pi/2$

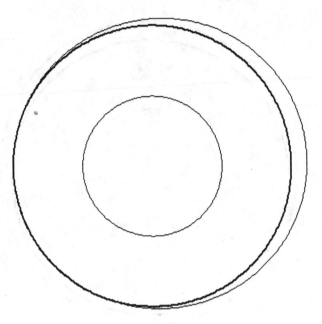

Figure 6.47. Example 6.11: $\delta = 0.04$, method (5.7.24), $\varphi = 0$

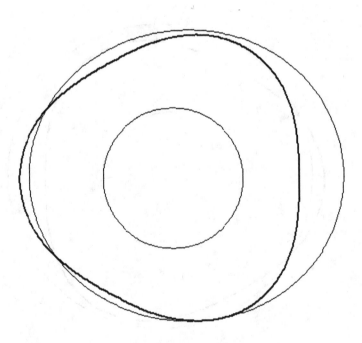

Figure 6.48. Example 6.11: $\delta = 0.04$, method (5.7.24), $\varphi = \pi/2$

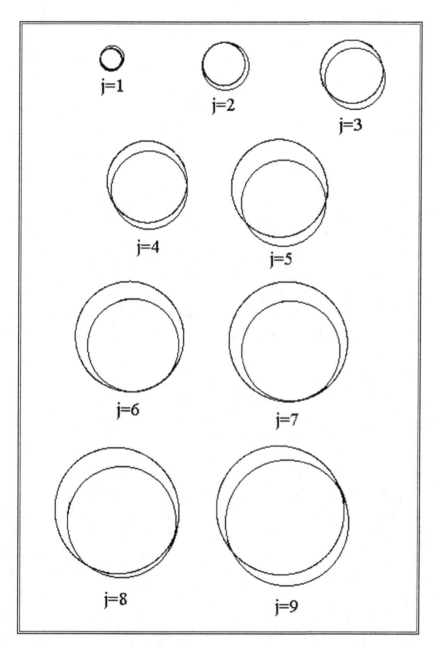

Figure 6.49. Example 6.12: $r_3 = 1 - j/9, j = 1, \dots, 9$

Chapter 7

NOTES

Chapter 1.

Section 1.1. Some standard books on basic functional analysis as it is needed for our purposes are, for example, [38, 58, 123].

Section 1.2. There are several excellent textbooks and monographs on mathematical theory of ill–posed problems, e.g., [42, 49, 68, 137]. More details can be found in [17, 45, 78, 106, 133, 139, 146].

Chapter 2.

Section 2.1. The class of methods (2.1.4) has been proposed by Bakushinsky first in the case where $\xi = 0$ [6]. More general case where ξ is an arbitrary element of the solution space was subsequently studied by Vainikko and Veretennikov [144]. Converse results similar to those of Theorems 2.4 and 2.5 have been originally established by Neubauer [112]; we refer to [42] for a detailed presentation. In the case of logarithmic sourcewise representation (2.1.48), direct and converse theorems on the rate of convergence of the methods (2.1.4) and (2.1.11) have been first obtained by Hohage [61]. Our presentation of Theorems 2.4–2.9 follows the papers [90, 91] by Kokurin and Yusupova. Finite–dimensional versions of regularization methods (2.1.35) with different rules of the regularization parameter choice were constructed and studied by Vainikko, Pereverzev, Solodkii et al [104, 130, 143]. In past several decades, in theory of linear ill–posed problems considerable attention has been paid to regularization methods on the basis of classical gradient and conjugate gradient type processes. An overview of results in this direction and further references can be found in [44, 45, 110].

Section 2.2. The class of methods (2.2.2) for linear ill–posed equations in a Banach space has been introduced by Bakushinsky first in the particular case where $\xi = 0$ [7]. In [7], Bakushinsky has established convergence of methods (2.2.2) under the condition (2.2.5) with $p = 1$. Theorems 2.10–2.13

concerning methods (2.2.2) under more general power sourcewise representation condition (2.2.5) have been obtained by Kokurin [83, 84]. Theorem 2.14 has been established by Kokurin and Karabanova [87]. For several special generating functions $\Theta(\lambda, \alpha)$, Plato and Hämarik [114] proposed and studied regularization algorithms of the form (2.2.14) with a posteriori choice of the regularization parameter.

Section 2.3. The most of results in this section have been obtained by Bakushinsky and Kokurin [18]. Theorem 2.19 has been established by Kokurin and Kljuchev [88].

Chapter 3.

Section 3.1. For a systematic presentation of the classical Newton–Kantorovich and Gauss–Newton methods we refer to [34, 35, 77, 113].

Section 3.2. The parametric linearization scheme (3.2.1) for smooth irregular equations was introduced and studied by Bakushinsky [11, 13].

Section 3.3. This section is based on the results of [11, 13].

Chapter 4.

Section 4.1. The class of iterative methods (4.1.3) has been proposed by Bakushinsky [8–10, 12]. In [13, 15], Bakushinsky established convergence results closely related to Theorems 4.1 and 4.2; lately these results were revised and improved by Kokurin and Yusupova [89]. Presentation of Section 4.1 is based on the improved convergence theorems. Using the sourcewise representation (4.1.4) with an arbitrary $p > 0$ and structural conditions of type (5.6.4), Engl, Scherzer, Deuflhard, Kaltenbacher, and Tautenhahn obtained analogous convergence results for the additive error model where

$$ F(x) = G(x) - y, \quad \widetilde{F}(x) = G(x) - y_\delta, \quad \|y_\delta - y\|_{X_2} \leq \delta $$

[36, 73, 134]. Our convergence theorems don't exploit structural conditions on operators' nonlinearities but we assume that $p \geq 1/2$. In [36, 60], under similar structural conditions on operators and a logarithmic sourcewise representation condition, Deuflhard, Engl, Scherzer, and Hohage have established logarithmic rate of convergence estimates for the iteration (4.1.3).

Section 4.2. Necessary conditions for the rate of convergence estimates (4.2.1) and (4.2.23) of iterative methods (4.1.3) and Tikhonov's method (4.2.22) have been established by Bakushinsky, Kokurin, and Yusupova [19]. For a theoretical foundation of Tikhonov's method we refer to [17, 48, 49, 106, 137, 138]; some recent results are presented in [136]. In this book we don't touch approaches to irregular equations exploiting monotonicity or other special properties of operators. An overview of results concerning methods for solving ill–posed monotone equations and variational inequalities can be found in [17, 135].

Section 4.3. The class of iterative processes (4.3.1) and regularization methods (4.3.34) in a Banach space have been first proposed and studied by Bakushinsky and Kokurin [20].

Section 4.4. Results of this section have been obtained in [21].

Section 4.5. The presentation of Section 4.5 follows recent papers by Bakushinsky and Kokurin [22, 86]. The particular case of (4.5.4) with $\Theta(\lambda, \alpha) = (\lambda + \alpha)^{-1}$ and several closely related schemes of continuous approximation in a Hilbert space were studied in [2, 75, 121].

Chapter 5.

Section 5.1. Gradient type methods (5.1.9) have been first suggested to use for smooth irregular equations in [14, 16] (see also [17]). The standard gradient method (Landweber's iteration), conjugate gradient and quasi–Newton type methods in irregular situation were intensively studied by Engl, Hanke, Neubauer, Scherzer, Deuflhard, Kaltenbacher et al [36, 42, 74, 126, 127]. It should be pointed out that additional structural conditions on operators' nonlinearities, e.g., (5.6.4) play a key role in the convergence theory developed by these authors; the smoothness conditions alone are not sufficient for convergence of the methods when applied to irregular equations. In [70, 76, 128], several finite–dimensional versions of gradient iterations are proposed; the related convergence analysis also is based on conditions like (5.6.4).

Sections 5.2, 5.3. The presentation of these sections follows the paper [79].

Section 5.4. The scheme (5.4.4) of constructing stable iterative processes on the basis of parametric approximations has been proposed in [23].

Section 5.5. This section presents results of the paper [85] by Kokurin.

Section 5.6. The results of this section have never been published. An alternative approach to iterative implementation of Tikhonov's method in the case of additive errors has been recently developed by Ramlau [119]. As opposite to the iteration (5.6.3), the method of [119] involves a two–level iterative process.

Section 5.7. The approach to constructing stable iterative methods presented in this section has been suggested by Kokurin and Kozlov (see [92]).

Chapter 6.

Applied inverse problems considered in this chapter are often used in computational experiments. This is particularly true for the inverse acoustic problems. In many instances, the emphasis in investigations of these problems is on numerical approximation of corresponding direct problems and on algorithmic implementation of solution methods while less efforts are devoted to rigorous analysis of convergence of these methods to solutions of the inverse problems. Let us call the reader's attention to the following widespread phenomenon that arises when solving various irregular (ill–posed) problems in infinite–dimensional spaces. Suitable discretization of a problem, that is, transition to a finite–dimensional analog of the original model, and subsequent application of standard iterative processes say, the classical gradient method or Newton's method, often give quite satisfactory approximations to a solution although the original problem is ill–posed. In this connection we refer to [71], where the phenomenon of regularization by a transition to an appropriate subspace is exploited in application to multidimensional inverse problems for hyperbolic equations. The results of Section 5.7 can be considered as an attempt to give a theoretical justification of this approach to numerical solution of general irregular operator equations, not necessarily smooth. Numerical experiments of Chapter 6 partially support the theory developed in Section 5.7. It should be pointed out however that inaccurate discretization of the original equation can lead to arbitrarily large deviations of approximations from a solution [137].

References

[1] Adams, R.A. *Sobolev spaces*, Academic Press, New York, 1975.

[2] Airapetyan, R.G., Ramm, A.G., and Smirnova, A.B. Continuous methods for solving nonlinear ill–posed problems, Operator theory and its applications, American Mathematical Society, Providence R.I., Fields Inst. Commun., 25 (2000), 111–137.

[3] Anger, G. *Inverse problems in differential equations*, Plenum, New York, 1990.

[4] Anikonov, Yu.E. *Multidimensional inverse and ill-posed problems for differential equations*, VSP, Utrecht, 1995.

[5] Babin, A.V. and Vishik, M.I. *Attractors of evolution equations*, North–Holland, Amsterdam, 1992.

[6] Bakushinsky, A.B. A general method of constructing regularizing algorithms for a linear ill–posed equation in Hilbert space, Comput. Math. Math. Phys., 7 (1967), 279–288.

[7] Bakushinsky, A.B. The problem of constructing linear regularizing algorithms in Banach spaces, Comput. Math. Math. Phys., 13 (1973), 261–270.

[8] Bakushinsky, A.B. On a convergence problem of the iterative–regularized Gauss–Newton method, Comput. Math. Math. Phys., 32 (1992), 1353–1359.

[9] Bakushinsky, A.B. Iterative methods for solving nonlinear operator equations without regularity. A new approach, Russian Acad. Sci. Dokl. Math., 47 (1993), 451–454.

[10] Bakushinsky, A.B. Iterative methods without saturation for solving degenerate nonlinear operator equations, Dokl. Akad. Nauk, 344 (1995), 7–8.

[11] Bakushinsky, A.B. On the problem of linear approximation of solutions of nonlinear operator equations, Comput. Math. Math. Phys., 36 (1996), 1169–1174.

[12] Bakushinsky, A.B. Iterative methods for solving nonlinear operator equations without the regularity property, Fundam. Prikl. Mat., 3 (1997), 685–692.

[13] Bakushinsky, A.B. Universal linear approximations of solutions to nonlinear operator equations and their application, J. Inverse Ill–Posed Probl., 5 (1998), 507–521.

[14] Bakushinsky, A.B. Iterative methods of gradient type for nonregular operator equations, Comput. Math. Math. Phys., 38 (1998), 1884–1887.

[15] Bakushinsky, A.B. On the rate of convergence of iterative processes for nonlinear operator equations, Comput. Math. Math. Phys., 38 (1998), 538–542.

[16] Bakushinsky, A.B. Gradient-type iterative methods with projection onto a fixed subspace for solving irregular operator equations, Comput. Math. Math. Phys., 40 (2000), 1387–1390.

[17] Bakushinsky, A. and Goncharsky, A. *Ill–posed problems: theory and applications*, Kluwer, Dordrecht, 1994.

[18] Bakushinsky, A.B. and Kokurin, M.Yu. Conditions of sourcewise representation and rates of convergence of methods for solving ill-posed operator equations. Part I, Advanced Computing. Numerical Methods and Programming, 1, Sec. 1, (2000), 62–82.

[19] Bakushinsky, A.B., Kokurin, M.Yu., and Yusupova, N.A. Necessary conditions for the convergence of iterative methods for solving nonlinear operator equations without the regularity property, Comput. Math. Math. Phys., 40 (2000), 945–954.

[20] Bakushinsky, A.B. and Kokurin, M.Yu. Iterative methods for solving nonlinear irregular operator equations in Banach space, Numer. Funct. Anal. Optim., 21 (2000), 355–378.

[21] Bakushinsky, A.B. and Kokurin, M.Yu. Sourcewise representation of solutions to nonlinear operator equations in a Banach space and convergence rate estimates for a regularized Newton's method, J. Inverse Ill–Posed Probl., 9 (2001), 349–374.

[22] Bakushinsky, A.B. and Kokurin, M.Yu. Continuous methods for stable approximation of solutions to nonlinear equations in the Banach space based on the regularized Newton–Kantarovich scheme, Siberian Journal of Numerical Mathematics, 7 (2004), 1–12.

[23] Bakushinsky, A.B., Kokurin, M.Yu., and Yusupova, N.A. Iterative Newton–type methods with projecting for solution of nonlinear ill–posed operator equations, Siberian Journal of Numerical Mathematics, 5 (2002), 101–111.

[24] Balakrishnan, A.V. Fractional powers of closed operators and the semigroups generated by them, Pacif. J. Math., 10 (1960), 419–437.

[25] Batrakov, D. and Zhuck, N. Solution of general inverse scattering problem using the distorted Born approximation and iterative technique, Inverse Problems, 10 (1994), 39–54.

[26] Bukhgeim, A.L. *Introduction to the theory of inverse problems*, VSP, Utrecht, 2000.

[27] Chadan, K. and Sabatier, P. *Inverse problems in quantum theory*, Springer, Berlin, 1977.

[28] Clement, Ph., Heijmans, H.J.A.M., Angenent, S., van Duijn, C.J., and de Pagter, B. *One–parameter semigroups*, North–Holland, Amsterdam, 1987.

[29] Colton, D. and Kress, R. *Integral equation methods in scattering theory*, J. Wiley & Sons, New York, 1983.

[30] Colton, D. and Kress, R. *Inverse acoustic and electromagnetic scattering theory–2 ed*, Springer, Berlin, 1998.

[31] Dalec'kii, Ju.L. and Krein, M.G. *Stability of solutions of differential equations in Banach space*, American Mathematical Society, Providence, R.I., 1974.

[32] Dem'yanov, V.F. and Rubinov, A.M. *Approximate methods in optimization problems*, Elsevier, New York, 1970.

[33] Denisov, A.M. *Elements of the theory of inverse problems*, VSP, Utrecht, 1999.

[34] Dennis, J.E. and Schnabel, R.B. *Numerical methods for unconstrained optimization and nonlinear equations*, Prentice–Hall, New Jersey, 1983.

[35] Deuflhard, P. *Newton methods for nonlinear problems*, Springer, Berlin, 2004.

[36] Deuflhard, P., Engl, H.W., and Scherzer, O. A convergence analysis of iterative methods for the solution of nonlinear ill–posed problems under affinely invariant conditions, Inverse Problems, 14 (1998), 1081–1106.

[37] Djellouli, R. and Farhat, C. On the characterization of the Frechet derivative with respect to a Lipschitz domain of the acoustic scattered field, J. Math. Anal. Appl., 238 (1999), 259–276.

[38] Dunford, N. and Schwartz, J.T. *Linear Operators. Part I: General Theory*, Interscience Publishers, London, 1958.

[39] Dunford, N. and Schwartz, J.T. *Linear operators. Part II: Spectral theory. Self adjoint operators in Hilbert space*, J. Wiley & Sons, New York, 1963.

[40] Dunford, N. and Schwartz, J.T. *Linear operators. Part III: Spectral operators*, Interscience Publishers, New York, 1971.

[41] Edwards, R.E. *Fourier series. Vol. 2. A modern introduction*, Springer, Berlin, 1982.

[42] Engl, H.W., Hanke, M., and Neubauer, A. *Regularization of inverse problems*, Kluwer, Dordrecht, 1996.

[43] Eremin, Yu.A. On the problem of the existence of an invisible scatterer in diffraction theory, Differentsial'nye Uravneniya, 24 (1988), 684–687.

[44] Gilyazov, S.F. Regularization of ill–posed constrained minimization problems by iteration methods, Numer. Funct. Anal. Optim., 19 (1998), 309–327 .

[45] Gilyazov, S.F. and Gol'dman, N.L. *Regularization of ill–posed problems by iteration methods*, Kluwer, Dordrecht, 2000.

[46] Goncharsky, A.V. and Romanov, S.Yu. On a three–dimensional diagnostic problem in the wave approximation, Comput. Math. Math. Phys., 40 (2000), 1308–1311.

[47] Griwank, A. On solving nonlinear equations with simple singularities or nearly singular solutions, SIAM Review, 27 (1985), 537–563.

[48] Groetch, C.W. *The theory of Tikhonov regularization. Fredholm equations of the first kind*, Pitman, Boston, 1984.

[49] Groetsch, C.W. *Inverse problems in mathematical sciences*, Vieweg, Braunschweig, 1993.

[50] Gutman, S. and Klibanov, M. Iterative methods for multi–dimensional inverse scattering problems at fixed frequency, Inverse Problems, 10 (1994), 573–599.

[51] Hale, J.K. *Asymptotic behavior of dissipative systems*, American Mathematical Society, Providence, R.I., 1988.

[52] Hammerlin, G. and Hoffman, K.U. *Improperly posed problems and their numerical treatment*, Birkhauser, Basel, 1983.

[53] Hanke, M. Regularizing properties of a truncated Newton CG algorithm for nonlinear inverse problems, Numer. Funct. Anal. Optim., 18 (1997), 971–993.

[54] Hanke, M., Neubauer, A., and Scherzer, O. A convergence analysis of the Landweber iteration for nonlinear ill–posed problems, Numer. Math., 72 (1995), 21–37.

[55] Henry, D. *Geometric theory of semilinear parabolic equations*, Springer, Berlin, 1981.

[56] Hettlich, F. The Landweber iteration applied to inverse conductive scattering problems, Inverse Problems, 14 (1998), 931–947.

[57] Hettlich, F. and Rundell, W. Iterative methods for the reconstruction of an inverse potential problem, Inverse Problems, 12 (1996), 251–266.

[58] Hille, E. and Phillips, R.S. *Functional analysis and semi–groups*, American Mathematical Society, Providence, R.I., 1957.

[59] Hofmann, B. *Regularization for applied inverse and ill–posed problems*, Teubner, Leipzig, 1986.

[60] Hohage, T. Logarithmic convergence rates of the iteratively regularized Gauss–Newton method for an inverse potential and inverse scattering problem, Inverse Problems, 13 (1997), 1279–1299.

[61] Hohage, T. Regularization of exponentially ill–posed problems, Numer. Funct. Anal. Optim., 21 (2000), 439–464.

[62] Hohage, T. On the numerical solution of a three–dimensional inverse medium scattering problem, Inverse Problems, 17 (2001), 1743–1763.

[63] Hohage, T. and Schormann, C.A. Newton–type method for a transmission problem in inverse scattering, Inverse Problems, 14 (1998), 1207–1227.

[64] Hoy, A. and Schwetlick, H. Some superlinearly convergent methods for solving singular nonlinear equations. Lectures in applied mathematics. V.26. (E.L.Allower et al Eds.), American Mathematical Society, Providence, R.I., 1990, 285–300.

[65] Ioffe, A.D. and Tihomirov, V.M. *Theory of extremal problems*, North–Holland, Amsterdam, 1979.

[66] Isakov, V. *Inverse source problems*, American Mathematical Society, Providence, R.I., 1990.

[67] Isakov, V. *Inverse problems for partial differential equations*, Springer, Berlin, 1998.

[68] Ivanov, V.K., Vasin, V.V., and Tanana, V.P. *Theory of linear ill–posed problems and its applications*, VSP, Utrecht, 2002.

[69] Izmailov, A.F. and Tret'yakov, A.A. *2–regular solutions of nonlinear problems*, Fiziko–Matematicheskaya Literatura, Moscow, 1999.

[70] Jin, Q. and Amato, U. A discrete scheme of Landweber iteration for solving nonlinear ill–posed problems, J. Math. Anal. Appl., 253 (2001), 187–203.

[71] Kabanikhin, S.I. *Projection difference methods for the determination of coefficients of hyperbolic equations*, Nauka, Novosibirsk, 1988.

[72] Kabanikhin, S.I. and Lorenzi, A. *Identification problems of wave phenomena–theory and numerics*, VSP, Utrecht, 2000.

[73] Kaltenbacher, B. Some Newton–type methods for the regularization of nonlinear ill–posed problems, Inverse Problems, 13 (1997), 729–753.

[74] Kaltenbacher, B. On Broyden's method for the regularization of nonlinear ill–posed problems, Numer. Funct. Anal. Optim., 19 (1998), 807–833.

[75] Kaltenbacher, B., Neubauer, A., and Ramm, A.G. Convergence rates of the continuous regularized Gauss–Newton method, J. Inverse Ill–Posed Probl., 10 (2002), 261–280.

[76] Kaltenbacher, B. and Schicho, J. A multi–grid method with a priori and a posteriori level choice for the regularization of nonlinear ill–posed problems, Numer. Math., 93 (2002), 77–107.

[77] Kantorovich, L.V. and Akilov, G.P. *Functional analysis in normed spaces*, The Macmillan Co., New York, 1964.

[78] Kaplan, A. and Tichatschke, R. *Stable methods for ill–posed variational problems*, Akademie Verlag, Berlin, 1994.

[79] Karabanova, O.V., Kozlov, A.I., and Kokurin, M.Yu. Stable finite–dimensional iterative processes for solving nonlinear ill–posed operator equations, Comput. Math. Math. Phys., 42 (2002), 1073–1085.

[80] Klibanov, M. and Malinsky, J. Newton–Kantorovich method for three–dimensional potential inverse scattering problem and stability of the hyperbolic Cauchy problem with time–dependent data, Inverse Problems, 7 (1991), 577–596.

[81] Klibanov, M.V. and Timonov, A. *Carleman estimates for coefficient inverse problems and numerical applications*, VSP, Utrecht, 2004.

[82] Kokurin, M.Yu. *Operator regularization and investigation of nonlinear monotone problems*, MarSU, Yoshkar–Ola, 1998.

[83] Kokurin, M.Yu. Source representability and estimates for the rate of convergence of methods for the regularization of linear equations in a Banach space. I, Russian Math. (Iz. VUZ), 45 (2001), no. 8, 49–57.

[84] Kokurin, M.Yu. Source representability and estimates for the rate of convergence of methods for the regularization of linear equations in a Banach space. II, Russian Math. (Iz. VUZ), 46 (2002), no. 3, 19–27.

[85] Kokurin, M.Yu. Approximation of solutions to irregular equations and attractors of non-linear dynamical systems in Hilbert spaces, Advanced Computing. Numerical Methods and Programming, 4, Sec. 1, (2003), 207–215.

[86] Kokurin, M.Yu. Continuous Methods of Stable Approximation of Solutions to Nonlinear Equations in Hilbert Space Based on the Regularized Gauss–Newton Scheme, Comput. Math. Math. Phys., 44 (2004), 6–15.

[87] Kokurin, M.Yu. and Karabanova, O.V. Regularized projection methods for solving linear operator equations of the first kind in a Banach space, Izv. Vyssh. Uchebn. Zaved. Mat., (2003), no. 7, 35–44.

[88] Kokurin, M.Yu. and Klyuchev, V.V. Necessary conditions for a given convergence rate of iterative methods for solution of linear ill–posed operator equations in a Banach space, Siberian Journal of Numerical Mathematics, 5 (2002), 295–310.

[89] Kokurin, M.Yu. and Yusupova, N.A. On nondegenerate estimates for the rate of convergence of iterative methods for solving some ill–posed nonlinear operator equations, Comput. Math. Math. Phys., 40 (2000), 793–798.

[90] Kokurin, M.Yu. and Yusupova, N.A. On necessary conditions for the qualified convergence of methods for solving linear ill–posed problems, Russian Math. (Iz. VUZ), 45 (2001), no. 2, 36–44.

[91] Kokurin, M.Yu. and Yusupova, N.A. On necessary and sufficient conditions for the slow convergence of methods for solving linear ill–posed problems, Russian Math. (Iz. VUZ), 46 (2002), no. 2, 78–81.

[92] Kozlov, A.I. A class of stable iterative methods for solving nonlinear ill–posed operator equations, Advanced Computing. Numerical Methods and Programming, 3, Sec. 1, (2002), 180–186.

[93] Krasnoselskii, M.A. *The operator of translation along the trajectories of differential equations*, American Mathematical Society, Providence, R.I., 1968.

[94] Krasnoselskii, M.A., Vainikko, G.M., Zabreiko, P.P., Rutitskii, Ya.B., and Stetsenko, V.Ya. *Approximate solution of operator equations*, Wolters–Noordhoff Publishing, Groningen, 1972.

[95] Krasnoselskii, M.A. and Zabreko, P.P. *Geometrical methods of nonlinear analysis*, Springer, Berlin, 1984.

[96] Krasnoselskii, M.A., Zabreiko, P.P., Pustylnik, E.I., and Sobolevskii, P.E. *Integral operators in spaces of summable functions*, Noordhoff International Publishing, Leyden, 1976.

[97] Krein, S.G. *Linear differential equations in Banach space*, American Mathematical Society, Providence, R.I., 1971.

[98] Kress, R. Newton's method for inverse obstacle scattering meets the method of least squares, Inverse Problems, 19 (2003), S91–S104.

[99] Ladyzhenskaya, O.A. Finding minimal global attractors for the Navier–Stokes equations and other partial differential equations, Uspekhi Mat. Nauk, 42 (1987), no. 6, 25–60.

[100] Ladyzhenskaya, O.A. *Attractors for semigroups and evolution equations*, Cambridge University Press, Cambridge, 1991.

[101] Lavrentiev, M.M., Avdeev, A.V., Lavrentiev, M.M., Jr., and Priimenko, V.I. *Inverse problems of mathematical physics*, VSP, Utrecht, 2003.

[102] Lavrent'ev, M.M., Romanov, V.G., and Shishatskij, S.P. *Ill–posed problems of mathematical physics and analysis*, American Mathematical Society, Providence, R.I., 1986.

[103] Lebedev, N.N. *Special functions and their applications*, Prentice Hall, Englewood Cliffs, N.J., 1965.

[104] Mathé, P. and Pereverzev, S. Discretization strategy for linear ill–posed problems in variable Hilbert scales, Inverse Problems, 19 (2003), 1263–1277.

[105] Melnikova, I.V. and Filinkov, A. *Abstract Cauchy problems: three approaches*, Chapman & Hall/CRC, Boca Raton, 2001.

[106] Morozov, V.A. *Regularization methods for ill-posed problems*, CRC Press, Boca Raton, 1993.

[107] Nashed, M.Z. *Ill–posed problems: theory and practice*, Reidel, Dordrecht, 1985.

[108] Natterer, F. *The mathematics of computerized tomography*, J. Wiley & Sons, Stuttgart, 1986.

[109] Natterer, F. An error bound for the Born approximation, Inverse problems, 20 (2004), 447–452.

[110] Nemirovskii, A.S. Regularizing properties of the conjugate gradient method in ill–posed problems, Zh. Vychisl. Mat. i Mat. Fiz., 26 (1986), 332–347.

[111] Neubauer, A. Tikhonov regularization for nonlinear ill–posed problems: optimal convergence rates and finite–dimensional approximation, Inverse Problems, 5 (1989), 541–557.

[112] Neubauer, A. On converse and saturation results for Tikhonov regularization of linear ill–posed problems, SIAM J. Numer. Anal., 34 (1997), 517–527.

[113] Ortega, J.M. and Rheinboldt, W.C. *Iterative solution of nonlinear equations in several variables*, Academic Press, New York, 1970.

[114] Plato, R. and Hämarik, U. On pseudo–optimal parameter choices and stopping rules for regularization methods in Banach spaces, Numer. Funct. Anal. Optim., 17 (1996), 181–195.

[115] Polak, E. *Computational methods in optimization. A unified approach*, Academic Press, New York, 1971.

[116] Potthast, R. *Point sources and multipoles in inverse scattering theory*, Chapman & Hall / CRC, Boca Raton, 2001.

[117] Potthast, R., Sylvester, J., and Kusiak, S. A 'range test' for determining scatterers with unknown physical properties, Inverse Problems, 19 (2003), 533–547.

[118] Ramlau, R. Modified Landweber method for inverse problem, Numer. Funct. Anal. Optim., 20 (1999), 79–98.

[119] Ramlau, R. TIGRA–an iterative algorithm for regularizing nonlinear ill–posed problems, Inverse Problems, 19 (2003), 433–465.

[120] Ramm, A.G. *Multidimensional inverse scattering problems*, Longman, New York, 1992.

[121] Ramm, A.G., Smirnova, A.B., and Favini, A. Continuous modified Newton's–type method for nonlinear operator equations, Annali di Matematica, 182 (2003), 37–52.

[122] Remis, R.F. and Berg, P.M. On the equivalence of the Newton–Kantorovich and distorted Born methods, Inverse Problems, 16 (2000), L1–L4.

[123] Riesz, F. and Sz.–Nagy, B. *Functional analysis*, Frederick Ungar Publishing Co., New York, 1955.

[124] Romanov, V.G. *Investigation methods for inverse problems*, VSP, Utrecht, 2002.

[125] Romanov, V.G. and Kabanikhin, S.I. *Inverse problems for Maxwell's equations*, VSP, Utrecht, 1994.

[126] Scherzer, O. A convergence analysis of a method of steepest descent and a two–step algorithm for nonlinear ill–posed problems, Numer. Funct. Anal. Optim., 17 (1996), 197–214.

[127] Scherzer, O. A modified Landweber iteration for solving parameter estimation problem, Appl. Math. Optim., 38 (1998), 45–68.

[128] Scherzer, O. An iterative multi level algorithm for solving nonlinear ill–posed problems, Numer. Math., 80 (1998), 579–600.

[129] Schröder, E. Über unendlich viele Algorithmen zur Auflösung der Gleichungen, Math. Ann., 2 (1870), 317–365.

[130] Solodkij, S.G. On a modification of a projection scheme for solving ill–posed problems, Russ. Math., 42 (1998), no. 11, 79–86.

[131] Suetin, P.K. *Classical orthogonal polynomials. 2nd enl. ed*, Nauka, Moscow, 1979.

[132] Szegö, G. *Orthogonal polynomials*, American Mathematical Society, Providence, R.I., 1975.

[133] Tanana, V.P. *Methods for solution of nonlinear operator equations*, VSP, Utrecht, 1997.

[134] Tautenhahn, U. On a general regularization scheme for nonlinear ill–posed problems: II. Regularization in Hilbert scales, Inverse Problems, 14 (1998), 1607–1616.

[135] Tautenhahn, U. On the method of Lavrentiev regularization for nonlinear ill–posed problems, Inverse Problems, 18 (2002), 191–207.

[136] Tautenhahn, U. and Jin, Q. Tikhonov regularization and a posteriori rules for solving nonlinear ill–posed problems, Inverse Problems, 19 (2003), 1–21.

[137] Tikhonov, A.N. and Arsenin, V.Ya. *Solution of ill–posed problems*, J. Wiley & Sons, New York, 1977.

[138] Tikhonov, A.N., Goncharsky, A.V., Stepanov, V.V., and Yagola, A.G. *Numerical methods for the solution of ill–posed problems*, Kluwer, Dordrecht, 1995.

[139] Tikhonov, A.N., Leonov, A.S., and Yagola, A.G. *Nonlinear ill–posed problems. V.1 and V.2*, Chapman & Hall, London, 1998.

[140] Titchmarsh, E.C. *The theory of functions*, Oxford University Press, London, 1975.

[141] Triebel, H. *Interpolation theory, function spaces, differential operators*, North–Holland, Amsterdam, 1978.

[142] Vainberg, M.M. *Variational method and method of monotone operators in the theory of nonlinear equations*, J. Wiley & Sons, New York, 1973.

[143] Vainikko, G. On the discretization and regularization of ill–posed problems with non-compact operators, Numer. Funct. Anal. Optim., 13 (1992), 381–396.

[144] Vainikko, G.M. and Veretennikov, A.Yu. *Iterative procedures in ill–posed problems*, Nauka, Moscow, 1986.

[145] Vasil'ev, F.P. *Methods for solving extremal problems*, Nauka, Moscow, 1981.

[146] Vasin, V.V. and Ageev, A.L. *Ill–posed problems with a priori information*, VSP, Utrecht, 1995.

[147] Wang, T. and Oristaglio, M.L. An inverse algorithm for velocity reconstruction, Inverse Problems, 14 (1998), 1345–1351.

[148] Warga, J. *Optimal control of differential and functional equations*, Academic Press, New York, 1972.

[149] Zhuk, V.V. and Kuzyutin, V.F. *Approximation of functions and numerical integration*, SPbGU, St. Petersburg, 1995.

Index